EINSTEIN FOR THE 21ST CENTURY

EINSTEIN FOR THE 21ST CENTURY

HIS LEGACY IN SCIENCE, ART, AND MODERN CULTURE

Peter L. Galison, Gerald Holton,
and Silvan S. Schweber,
Editors

PRINCETON UNIVERSITY PRESS | PRINCETON AND OXFORD

Published by Princeton University Press, 41 William Street, Princeton,
New Jersey 08540

In the United Kingdom: Princeton University Press, 6 Oxford Street,
Woodstock, Oxfordshire OX20 1TR

First paperback printing, 2018
Paper ISBN 978-0-691-17790-8

The Library of Congress has cataloged the cloth edition as follows:

Einstein for the twenty-first century: His legacy in science, art, and modern
culture / Peter L. Galison, Gerald Holton, and Silvan S. Schweber, editors.
p. cm.
Includes bibliographical references and index.
ISBN 978-0-691-13520-5 (hardcover : acid-free paper)   1. Einstein, Albert,
1879–1955—Influence.   I. Galison, Peter Louis.   II. Holton, Gerald James.
III. Schweber, S. S. (Silvan S.)   IV. Title: Einstein for the 21st century.
QC16.E5E446 2008
530.092—dc22   2007034853

British Library Cataloging-in-Publication Data is available

This book has been composed in Aldus and Trajan

Printed on acid-free paper. ∞

press.princeton.edu

CONTENTS

PART

2

ART AND WORLD

INTRODUCTION

Peter L. Galison, Gerald Holton, and Silvan S. Schweber

AMONG THE MANY THOUSANDS of Albert Einstein's letters, very few show him to be really angry. However, a rather fierce one was directed at the publisher of one of his books who had had the temerity to give that volume the title *Mein Weltbild*. That choice, Einstein wrote, was "tasteless and misleading," for the book (later available in English as *Ideas and Opinions*) was only a collection of some of his separate essays on a variety of topics, ranging from science to politics and pacifism. What Einstein had called elsewhere "the proud term *Weltbild*" was the essential determinant and goal of one's intellectual life and character, a deeply held commitment into which, as he put it, "one could place the very center of gravity of one's emotional life"—thus providing a coherent conception that would explain "every natural process, including life itself."

The book before you, which had its origin in lectures presented at the Berlin Einstein Symposium of 2005, is a unique attempt along two lines: the "inputs" to and "outputs" from Einstein's worldview. The first line of inquiry garners, from the results of decades of scholarly research across a wide spectrum of fields, the variety of main elements that, in the cauldron of Einstein's inner sanctum, formed his unified view of the world. These inputs included, by his own account, his study of the work of *scientists* from Newton, Faraday, Maxwell, Lorentz to Boltzmann, Planck, and other contemporaries; of *philosophers*—he referred to Spinoza, Hume, Kant, Schopenhauer, Mach, among others; of *literature and music*—his love for Goethe, Heine, Bach, Mozart; and of *social and political affairs* that touched his humanity.

The other kind of presentations in these chapters deals with outputs, the effects streaming from Einstein's work and persona, from his *Weltbild*, into the active life of his contemporaries and ours. More than any scientist since Newton and Darwin, Einstein has continued to inspire others to draw on his legacies, again across a wide variety of fields: *scientists* of course, from Minkowski, von Laue, Schrödinger, Born, to some of today's most distinguished researchers; *philosophers and theologians; writers and visual artists;* and the whole range of people, from humble individuals to statesmen, touched by Einstein's beliefs and action in *social and political affairs*. One may well say that Einstein's continuing influences on our culture are of a quality and quantity not less than those on which he himself drew.

PART I: SOLITUDE AND WORLD

Our guiding question for this volume is this: How is it that Einstein, a physicist of the early 20th century, remains such a figure of fascination for so many fields of work, from the sciences to the humanities? Gerald Holton's essay opens our volume with some reflections on how Einstein saw himself at that world historical moment just after the Nazis came to power: "A man, a good European, a Jew." And as Holton swiftly adds, a scientist—these four overlapping identities return in many of the chapters that follow. As a man, Einstein saw himself as singularly isolated, never quite able, at least in his later, world-famous life, to form the human bonds he saw linking others. Perhaps that separateness, that life apart, speaks to others' fears and romantic wishes. As a good European, he both absorbed deep currents of central European culture and furiously resisted being enlisted to nationalistic strivings. He maintained that internationalism even under the enormous pressure of two world wars and a cold one. Later in life he found a cultural significance in Judaism, always resisting both too literal a God or too exclusivist a national striving within Zionism. To the rest of us, Einstein became a magnet by his constant battle to navigate alone while absorbing so much from—and giving back so much to—the broader world.

Lorraine Daston takes up Holton's first announced identity: Einstein's search for a "paradise beyond the personal." This life-long quest for relief from the "merely personal" was, for Einstein, surely in part a psychological liberation. It was also a feature of his science and his philosophy. Schopenhauer, one of Einstein's favorite philosophers, emphasized over and over how important it was for true knowledge to be had through a kind of will to willlessness, a removal of the individual's wants. Einstein's own physics bore the mark of this desire to find an objective world not dependent on us, neither our particular perspective (frame of reference), nor more broadly still, our consciousness or measurement decisions. These struggles over objectivity in quantum mechanics, and in the broader world, are with us still.

By contrast, Gutfreund, Elkana, and Ezrahi reveal Einstein's striking subjectivity, as in his relationship with what he called his "tribe," in his need to speak out on the social and intellectual issues of the day, and in the influence his work and that of other 20th-century scientists have had on democratic institutions. Thus Hanoch Gutfreund traces, in terms of roots and consequences, Einstein's "discovery" of being a Jew outside doctrinal orthodoxies. As in most of his other attitudes, Einstein again surprises us by his idiosyncratic responses, both in his analysis of the causes of rampant anti-Semitism and in his growing attachment to a special version of Zionism. On that last point, Isaiah Berlin once wrote with admiration about the fact that "Einstein lent the *prestige mondial* of his great name, and in fact gave his heart, to the movement which created the state of Israel."

Yehuda Elkana, drawing also on the treasure trove of Einstein's correspondence and nonscientific publications, daringly entitled his essay simply *Einstein and God*. He presents, with commentary, a coherent series of Einstein's own public remarks that document his life-long struggle to sense the transcendental, beckoning to him from beyond mere natural phenomena. As in some other respects, on this task, too, Einstein and Newton agreed. When Niels Bohr once famously asked Einstein to stop referring so often in the same breath to science and to God, Bohr simply revealed one of the many deep-seated thematic differences between those two great scientists.

But on at least one other topic, the major scientists largely agreed. Yaron Ezrahi takes up that topic in his challenging essay on *Einstein's Unintended Legacy*: the changing relevance of modern science for the culture and epistemological framework of democratic politics. His analysis begins with reference to the "Newtonian moment" characterizing earlier centuries, when the authority of a largely independent, consensual polity could be said to rest in good part on the concept that authority, both in society as well as in the understanding of nature itself, rested "not on hypotheses but on evidence"—on the widely accessible evidence considered to operate within the frame of commonsense realism.

Hence one could argue, as politicians of the time did, for a mutually reinforcing bond between the laws and worldview of their science and those of their democratic society. But by and by, and especially with the more esoteric developments of 20th-century science and technology, this old relationship has been unraveling. Ezrahi notes that, as a result, the clear distinction between fact and fiction, which science had appeared to guarantee to the lay public in the past, has been blurred. "The gap between commonsense realism and the [new] scientific notions of reality was bound to grow." Now we must contemplate the disruptive effects: on democratic institutions, on the one hand, where some politicians easily scoff at and dismiss what they call the "reality-based community," and on the other hand, where the scientific community finds itself more and more "radically disempowered."

Philosopher Susan Neiman, whose work aims to join philosophy and politics, is relieved to find in Einstein nothing of the *Luftmensch*, the distracted, disengaged spirit that so often clutters our memory of him. Over these last years, through the work of many scholars, we have increasingly come to see Einstein as very engaged politically—from his protests against militarism and nationalism in World War I, through his anti-Nazi advocacy during the later interwar period, and throughout the Cold War. Einstein was proud to know Paul Robeson, supported individual resistance to authoritarianism, advocated a Zionism that would treat both Jews and Arabs fairly, and spent much of his post-Hiroshima life supporting control of nuclear weapons. For Neiman, Einstein's political engagement was not accidental; it was tied to his deepest philosophical commitments. Einstein—and his commentators—have long emphasized that he shared with Kant a critical stance toward space and time, recognition that naïve realism was not enough. But Neiman insists that Einstein's passion to right the world, to count justice as constitutive even when it clashed with the real in *Realpolitik*, was also a fundamental component of Kantianism. Justice comes from us, not merely from experience. Would that the political problems Einstein grappled with were but historical artifacts. Needless to say, they remain very much with us.

Of all these threats, as Silvan Schweber stresses, Einstein saw an existential threat for humanity. Schweber tracks Einstein's role in the famous letter to Roosevelt warning that the Germans might be after a nuclear bomb, and he shows how strongly Einstein reacted to the threat of annihilation that lay behind the H-bomb and its proliferation. Most strikingly we see here the powerful cross-current between Einstein's physical and political reasoning. In both cases, he was after a widening, ever more embracing set of principles: special relativity, general relativity, unified field theories. In world politics, Einstein too sought principles that would run deeper and carry humanity further toward the elimination of war. Those principles were for a transnational government and a world court; nothing short, he thought, could keep war, and in particular war with nuclear weapons, from destroying humankind.

PART II: ART AND WORLD

Einstein continues to live in debates that swirled in his time and in ours around politics, philosophy, and physics. That is no less true in the arts. Over the last quarter century, art historians have grappled with Einstein and delivered to us two complementary messages: First, they have offered a cautionary note warning against a naïve expectation that Einstein's work propelled the high modernism of cubism. Second, they have shown with specificity how Einstein's work fits into the wider artistic and architectural sphere in an astonishing number of instances, not only between and following the world wars, but into the early 21st century. In this section our art historians, Linda Henderson and

Caroline Jones, are joined by artist Matthew Ritchie, musician Leon Botstein, and author E. L. Doctorow to reflect on the ways in which Einstein does (and doesn't) bear on contemporary problems of art and artistic creation.

In major books and articles, Linda Henderson has issued clarion calls against attributing every artist's mention of space-time or a fourth dimension to Einstein. As she argues here, there was a long and vigorous tradition of popular writings about perception of space and higher geometrical dimensions that long preceded Einstein and were not at all tied inexorably to relativity. For example, Edwin Abbot's *Flatland* and Henri Poincaré's *Science and Hypothesis* were hugely popular among artists. In her chapter, Henderson takes the story far forward, drawing together the myriad of ways in which artists have been and are engaging with Einstein's ideas. Some use Einstein to back a move toward the organic form, others to bolster their mechanistic aesthetics; some are fascinated by the publicity and difficulty of Einstein's theories, others by their scent of the up-to-the-moment science. But the story is not over—as she, Caroline Jones, and Matthew Ritchie make very clear indeed.

According to Ritchie, the artist and the scientist take similar positions. Both are trying to bring the construction of the world to the center of their emotional lives. To artists of the early modern period, Einstein offered a point of reference to cover what they were doing, even when they had been doing those same things before Einstein. On this he agrees completely with Linda Henderson. For the postmodernists, Einstein and other physicists appear in bent reflection, scientific diagrams altered to provide not so much guidance as an aesthetic of resistance. For the present generation of artists who invoke Einstein explicitly or implicitly (among whom Ritchie counts himself), at issue is something else: not so much artists *referring* to Einstein as artists *instantiating* Einsteinian paths. Ritchie sees his own work (aptly enough, e.g., *The Fine Constant* or *The Hierarchy Problem*) as playing with timelines, as if all could be put into "a single, continuously evolving object." Not only the cosmological scale of Einstein's work appeals to Ritchie; more generally, he, and other artists with whom he identifies, are searching for ways to make "story," "time," and "place" central.

Like Henderson, Jones is very dubious about the way in which Einstein has been used to explain far too much in the history of art. Her essay complements Henderson's: Where Henderson demands precision in what exactly is and isn't borrowed, alluded to, or transmogrified from Einstein, Jones steps back to contrast an artistic-philosophic perspective on time before and after Einstein. First she explores the ways in which time and motion preoccupied the artist in the latter part of the 19th century (using Monet as her touchstone), and then she turns, intriguingly, to shared concerns of contemporary artists and scientists—among them, as more contemporary *points de repère*,

Matthew Ritchie and Philip Glass. Schematizing her view, one might say that the pre-Einsteinian (Monetian) artistic stance toward time was one that aimed to be both systematic and fundamentally experiential: time caught between subjective flux and atomized, mathematical moments. The post-Einsteinian stance is necessarily different, with its (Einsteinian) stress on the variability of narrative frame of reference.

Ritchie's interest in Einstein's work is in the physics, or an artistic analogue of the physics, not Einstein's (sparsely expressed) views about the visual arts. By contrast, when Leon Botstein looks at Einstein, he has to confront Einstein's very public engagement with music, especially his violin playing. Botstein argues that Einstein's direct, technical accomplishments in performance were those of an accomplished amateur; in fact, his musical enthusiasm, and even his socializing through music were altogether standard among *fin-de-siècle* cultured Germans, especially German Jews. But Botstein mines deeper into Einstein's musical choices. There Einstein was quite conservative, sticking to Mozart, for example, against modernists whom he largely ignored. Indeed, in Einstein's lifelong preference for Mozart, Botstein sees a search for a purity of genre set against more Romantic, eclectic, or contemporary musical tendencies. Here, as in Einstein's physics, was a modernism found through a purifying form of conservatism. Most strikingly (Botstein contends), Einstein's insistence that science was something impersonal, objective, outside us is mirrored in long-standing debates about tone, pitch, and musical meaning that have long preoccupied musicians and musical theorists. These Einsteinian issues of a reality beyond us, an impersonal musical aesthetic, Botstein concludes, are still with us.

So too is the impersonal aesthetic present for E. L. Doctorow as he sets literary and scientific creation side by side. Doctorow is fascinated by Einstein's remark that the scientific creator, and Einstein in particular, saw what the individual produced as the "impersonal product of his generation." The creative act produces just such a sentiment, a sense that, in the moment of significant writing, one is outside oneself, as if "taking dictation." That externalism makes the creative act not so much one of supreme ego-enhancement but of its opposite. As Doctorow comments, "You are [in the creative act] less than the person you usually are." But that reduction in self-promotion was not for Einstein a matter of despair; on the contrary, Einstein's sense of reduced self was part and parcel of a thoroughgoing engagement with the world. There was, in the contact with the real, with the external, with the "beyond the personal," a deep-lying pleasure that binds work across the science-literature divide, and continues to do so today.

Part III: Science and World

In 1949, in famous and oft quoted remarks when responding to the essays that had been contributed to the volume devoted to him in Paul Schilpp's *Library of Living Philosophers*, Einstein stated:

> The reciprocal relationship of epistemology and science is of noteworthy kind. They are dependent upon each other. Epistemology without contact with science becomes an empty scheme. Science without epistemology is—insofar as it thinkable at all—primitive and muddled. However, no sooner has the epistemologist, who is seeking a clear system, fought his way through to such a system, than he is inclined to interpret the thought content of science in the sense of his system and to reject whatever does not fit into his system. The scientist, however, cannot afford to carry his striving for epistemological systematic this far. He accepts gratefully the epistemological conceptual analysis; but the external conditions, which are set for him by the facts of experience, do not permit him to let himself be too much restricted in the construction of his conceptual world by the adherence to an epistemological system. He therefore must appear to the systematic epistemologist as a type of unscrupulous opportunist: he appears as *realist . . .* , as *idealist . . .* , as *positivist. . . .* He may even appear as *Platonist* or *Pythagorean. . . .*

Peter Galison's chapter on Einstein's friendship with physicist, epistemologist, and political militant Friedrich Adler puts us at a cross-section of many of our themes. Einstein and Adler attended the Zurich Polytechnic (ETH) together, they remained friends in the years following when they and their new families shared an apartment in Zurich. Their discussions ranged widely, but included intense discussions about foundational questions of physics that bore directly on the research Einstein was to produce in 1905—questions that appear to have orbited around Mach's physics and philosophy. When a job opened up at the University of Zurich, both applied for it. Then in 1916, Adler assassinated the Prime Minister of Austria, Count Stürgkh. Einstein leapt to the defense of his terrorist friend. In the worst days of World War I, the two found themselves in intense correspondence ranging through fundamental questions of physics, politics, and the philosophy of physics. Intense political instability and fundamental questions of physics crossed for these two friends—not abstractly but in their midst. All the while Adler sat in prison, first accused of and then condemned for the assassination, and wrote a tract that sought to restrict Einstein's claims to a truly relativistic concept of time.

In his illuminating presentation, Michael Friedman illustrates Einstein's linking of physics and philosophy with an analysis of how the question "What is the geometry of space?" was answered by Riemann, by Helmholtz, by Poincaré, and by Einstein, and how their differing answers were analyzed and criticized by "epistemological systematicists," in particular, by the logical positivists, Carnap, Schlick, and Reichenbach. His essay elucidates the road taken by Einstein in arriving at general relativity and clarifies the role that philosophical issues played in this. His paper is an example of how valuable a similar analysis would be of some of the theoretical advances of late 20th-century physics.

As we will see in the next chapters of Part III, the papers by Dudley Herschbach, Jürgen Renn, Jürg Fröhlich, Douglas Stone, David Gross, and Lisa Randall each in their own way address the relevance of Einstein's life and works for the 21st century. Dudley Herschbach focuses on the young Einstein—growing up in Munich, his stay with the Wintelers to attend the Aarau Cantonal High School, his experiences at the ETH in Zurich and at the Patent Office in Bern—to try to ascertain the enabling conditions that nurtured his wondrous creativity, culminating with his five "miraculous" articles during his *annus mirabilis*. Herschbach then makes use of the insights thus obtained to suggest needed reforms in the graduate education of present-day scientists.

In a parallel undertaking, Jürgen Renn explores what can be learned from a critical analysis of the transformation of the systems of knowledge induced by the special and the general theories of relativity. Renn sees Einstein not as the isolated initiator of much of 20th-century physics but rather as "the one who completed classical physics in a way that uprooted its foundations." In formulating the special theory of relativity Einstein left much of the technical aspects of classical physics in place but dramatically altered its physical interpretation. Both from his analysis of the formulation of special relativity, and in his more detailed examination of how Einstein arrived at the general theory, Renn concludes that Einstein's innovations can only be understood if the long-term development of scientific knowledge is taken into account, only if "Einstein's revolution is understood as the result of a successful integration of shared knowledge resources," and only if one recognizes that the development of knowledge "does not only consist of enriching a given architecture but also comprises processes of reflection by which this architecture can be transformed." Renn then draws two conclusions from his review of the Einsteinian revolution: "that the world of knowledge, of which science is but the tip of the iceberg, is a great interconnected system that is subject to continuous transformation," and that "reflection, that is thinking about thinking, is the crucial mechanism enabling structural changes in such systems of knowledge." He concludes his stimulating essay with some reflections on the transformative potentialities of the Internet.

Jürg Fröhlich's point of departure is an analysis of the four (dimensional) constants in terms of which all physical quantities can be expressed: Boltzmann's constant, Planck's constant, the speed of light, and the Planck length (which is related to Newton's gravitational constant). Each of these constants is associated with a revolution in the physics of the 20th century to which Einstein contributed in a fundamental manner. We tend to think of Einstein's creation of the theories of relativity as making him the equal of Newton. But his contributions to the development of quantum theory and to statistical mechanics were as consequential, and according to Einstein himself, more revolutionary. Fröhlich concentrates on this facet of his work: Einstein's demonstration of the universality of $\hbar$, Planck's constant. Fröhlich stresses the role played by pure mathematics in each of the revolutions, and conversely, the fruitfulness for mathematics of the advances ushered by the revolutions in physics.

When in 1913 Bohr formulated his quantum theory of the hydrogen atom, he only considered circular orbits. Soon thereafter Arnold Sommerfeld and Paul Epstein generalized Bohr's quantization conditions to apply to elliptical orbits, but their quantization rules depended on particular choices of coordinate systems. Douglas Stone analyzes the important, but relatively little-known, 1917 paper by Einstein entitled *On the Quantization Condition of Sommerfeld and Epstein*, which overcame the dependence of the quantization rules on particular coordinate systems and formulated them in a coordinate-invariant manner—clearly the legacy of the mathematics Einstein had learned for his work on general relativity. The technique that Einstein invented, which was clear proof of his impressive mathematical capabilities, is still used in contemporary research. Einstein's paper also contained a brilliant insight into the limitations of the old quantum theory when the mechanical system in question is, in modern terminology, "chaotic." Einstein's paper was seminal: it influenced both Louis de Broglie and Erwin Schrödinger and paved the way toward the formulation of wave mechanics.

Einstein once summarized his "proper life's work" as attempting to answer three key questions: How does the representation of a light ray depend on the state of motion of the coordinate system to which it is referred? What is the basis for the equality of the inertial and gravitational mass of bodies? Can the gravitational field and the electromagnetic field be theoretically grasped in a unified manner?

Already as a young student at the ETH, Einstein had written to his friend Marcel Grossmann of the "glorious feeling to recognize the unity of a complex of phenomena, which appear . . . as quite distinct things." Shortly after developing his general theory of relativity, Einstein started thinking about formulating a unified field theory of electromagnetism and gravitation, the two basic natural forces then known. For the next forty years, he tried to give an answer to this problem, but was never able to do so in a manner that was convincing to both himself and the physics community.

In a readily accessible paper, David Gross reviews Einstein's struggles with the problems of unifying electromagnetism and gravitation at the classical level, and gives possible reasons for his failure. He then expounds the virtues of string theory, the most hopeful approach to date to the problem of unifying the four known forces of nature. In the process he gives a comprehensive and comprehensible account of the history of physicists trying "to arrive at those fundamental laws of nature from which the cosmos can be built up by pure deduction" during the 20th century.

David Gross' essay provides the background for Lisa Randall's introduction to contemporary research in cosmology and string theory. One of Einstein's legacies was a scientific cosmology. Astrophysical observational data of the past three decades—such as the fine structure of the 3K cosmic radiation, the properties of the radiation from distant quasars, the rotational motions of galaxies—have provided a dramatic new picture of our universe. There was a Big Bang, then a period of rapid inflation, and the universe is expanding and seemingly will continue to expand. Dark matter and dark energy exist in the universe and constitute the greater part of the energy content of the observable cosmos. The impressive instrumental and technological advances in obtaining data from space probes have resulted in a unification of particle physics and cosmology, with the early universe becoming the laboratory for particle physics at ultra-high energies. The equally impressive theoretical advances in string theory and the insights they have yielded into the relation between general relativity and quantum theory have similarly been important factors in addressing and answering cosmological questions. String theory stipulates space-time to be ten-dimensional (nine components of space and one of time), but the world around us appears to be four-dimensional. Where are the extra six dimensions? Since general relativity implies that gravity operates in all the dimensions, do these extra dimensions have observable consequences? Randall introduces us to the ways physicists have dealt with these problems, and in particular to the fascinating and startling world of the brane, a region of space-time that extends through only a slice of the bulk of space. There is no better introduction to the role of imagination and of "free creation" in building theories that Einstein so emphasized than Randall's exposition.

A hundred years after Einstein's *annus mirabilis*, the most startling miracle of all is that this icon of the early 20th century remains with us, in so many powerful ways, as a figure of the early 21st.

EINSTEIN FOR THE 21ST CENTURY

PART

I

SOLITUDE AND WORLD

1

WHO WAS EINSTEIN?
WHY IS HE STILL SO ALIVE?

Gerald Holton

THE SPEAKERS FOR OUR SYMPOSIUM have come from six nations to consider the work and influence of a man who, only a century ago, started to overturn the scientific *Weltbild* of the time. Of all twentieth-century scientists, only he could possibly be the subject of such a wide-ranging meeting as ours, with nineteen contributions to follow, from art historians and chemists, political scientists and philosophers, musicologists and physicists, historians of science, and more. Not since Isaac Newton's *Principia* can one imagine an analogous symposium to mark a physical scientist's legacy in such a wide spectrum of fields.

This fact demands some commentary: Who was Einstein? And why, fifty years after his death, is he still so alive, with the year 2005 having been declared an International Year of Physics by the United Nations, by UNESCO, even by the U.S. House of Representatives, all of them citing his publications in 1905 as the main reason for their decision?

My task as introducer of this volume requires me to begin with a precaution: When we contemplate the work and influence of a person with such gigantic and manifold characteristics as Einstein, a remark of Werner Heisenberg is appropriate here: "The space in which a person developed as an intellectual/spiritual being [*geistiges Wesen*] has more dimensions than the space which he occupied physically." Einstein himself, in a letter of 1914, gave us an even better metaphor. He wrote in high spirits: "I succeeded in proving . . . that the hypothesis of the equivalence of acceleration and the gravitational field is absolutely correct. Now the harmony of the mutual relationships in the theory is such that I no longer have the slightest doubt about its correctness." But then he added at once: "Nature shows us of the lion only the tail. But there is no doubt in my mind that the lion belongs with it, even if he cannot reveal himself to the eye all at once because of his huge dimensions." It seems to me that the Einstein phenomenon itself is, as it were, a grand, multidimensional lion, and that in this conference, acting together, we shall try to coax the lion from his lair. Although we cannot hope to comprehend the whole n-dimensional being once and for all, each of us, from our individual perspectives, may try to map one or another of the lion's dimensions and influence, from his days to ours.

Who was Einstein? We need not be discouraged by the obvious gap between him and those who study and write about his work. After all, Einstein too did not fully and properly describe himself. He tried to do so in his Royal Albert Hall speech in London, in October 1933, where he said only: "I am a man, a good European, a Jew." One must honor this heartfelt self-description, but Einstein's most obvious omission was his role as *scientist*. Being a scientist was also the reason he had accepted the offer to come to Berlin in 1914, to *the* place where, at that time, physical science was the best in the world, *the* place to be for a physicist. Even during the war years and in the hard decade that followed, despite all the hardships, Berlin could boast of a constellation of extraordinary physical scientists, and the exciting atmosphere in their colloquia and publications.

How much did these facts contribute to Einstein's unique ability and daring to develop, between 1915 and late 1917, his General Relativity Theory in Berlin? Could he have done so if he had accepted a grand offer from a city in another country? My own answer is: No other man than Einstein could have produced General Relativity, and in no other city than in Berlin, with its critical mass of close colleagues at the Academy and the University—Max Planck, Walther Nernst, Max von Laue, Fritz Haber, among many—all setting for themselves and one another the highest standards and expectations. Moreover, during that "Great War," which let loose the hounds of hell for the rest of the century, during that war when much of humanity devoted itself to senseless destruction, Einstein, even though wracked by severe challenges to body and spirit, completed his work on General Relativity by near superhuman effort and so revealed the outlines of the grand construction of the universe. That must count as one of the most moral acts of its day.

The science that Einstein left us appeared in some 300 publications. But those are not sitting on some dusty shelf as research material for historians. No. Although there was often a substantial delay before Einstein's ideas could be tested or used, they are alive today among active scientists around the globe, in a great variety of new work that testifies to his genetic role, in the explicit and implicit citations of new publications as well as in the rise of new technologies. Thus the so-called ether drift, which Einstein dismissed in 1905 in one sentence, has now been experimentally determined to be absent, with extraordinary accuracy. The gravitational lensing effect, which he published in 1936, turned out later to work also for galaxies and much else besides. The Bose-Einstein condensate, predicted in 1925, helped to explain superfluidity in 1928, and only a few years ago even trapped light. Einstein's 1924 prediction that matter waves would show interference effects was fulfilled three years later. In many parts of the world, a good portion of the electricity used daily comes from $E = mc^2$. The Equivalence Principle of General Relativity, initially only a courageous speculation, was confirmed most elegantly in the experiments by Robert V. Pound and his students more than five decades after Einstein had first intuited it. Gravitational waves, predicted in an article in 1918, have now become near certainty, as demonstrated by remarkable experimental techniques unknown in Einstein's days. Again and again, the headlines shout, Einstein was right.

To be sure, his image is also alive in banal advertisements, on T-shirts, in the fantasies of people who know nothing about physics. I will have to say more about this puzzling phenomenon. But a significant factor in Einstein's ubiquity among lay persons is surely that today's *scientists* find it safe and necessary to build many of their theories and experiments on what he achieved so many years ago.

In the last few decades, scholars have had glimpses of how Einstein's mind worked when he was doing science. I was privileged to come upon many such glimpses. In the late 1950s, the Estate of Einstein asked me to help put together the vast collection of his correspondence and manuscripts, then kept at the Institute for Advanced Study in Princeton, and to convert it into an archive accessible to scholars. Soon after having immersed myself at length in the glorious materials, two things struck me most. One was Einstein's utter self-confidence, despite many setbacks. Max Planck called it *freudige Sicherheit* ["joyful certainty"]. Of course, Einstein knew well that eventually good experiments would decide. For example, the very first response in the *Annalen der Physik* to Einstein's Relativity paper of 1905 came in early 1906. It claimed Einstein's theory to be empirically a failure, revealed by the foremost experimental physicist in the field, Walter Kaufmann of Göttingen. Einstein paid no attention to it for two years. Somehow he knew Kaufmann's to be a bad experiment, which eventually turned out to be the case. In the meantime Einstein left it to Planck to defend Relativity in the absence of experimental confirmation. Planck, who said he valued "simplicity and intuitiveness," was pushed to the wall in a scientific meeting in 1906, and finally had to fall back on declaring why he believed in Einstein's paper: *Mir ist das eigentlich sympatischer* [To me it is really more sympathetic]. On his part, Einstein was apt to let himself be guided by what he called his *Fingerspitzengefühl*, rather than by the inductive method taught in schoolbooks—or, as he put it in a letter to Max Born: "I try to capture [the 'objectively existing world'] in a wildly speculative way." Indeed he was one of the rare scientists who had, again and again, an insight of what was still around the corner—a talent that Hans Christian Oersted had so memorably named "an anticipatory consonance with Nature."

The other point that struck me early in reading his documents in his *Nachlass* was that Einstein very often let himself be guided, through thick and thin, by a few thematic presuppositions, which he called "non-Kantian categories," and above all by these seven: unity (Kant's own first category), simplicity, generalization, logical parsimony, deterministic causality, completeness, and the continuum.

So, Einstein's self-description should have been, at least: *scientist* and man.

Man

I shall later add under the heading of *Man*, but one main point must suffice for the moment: It is the disparity between, on the one hand, the humanitarian, kindly person with those eyes of a saint, always generous and vulnerable to pity, from his Berlin years on, constantly using his fame on behalf of equality, liberty, moral conviction—and on the other hand, the puzzling and chilling picture Einstein often gave of himself. The louder the acclaim from the world outside, the more did he feel lonely, isolated, unable to have truly close relationships like those he had had joyfully in the early decade of life with Mileva Marić, and with a few close friends, with Michele Besso, Marcel Grossmann, Paul Ehrenfest, and Max von Laue.

Moreover, while seemingly always approachable, those who knew him well noted that Einstein would sometimes suddenly seem to leave our world for a time, withdrawing into his own, the other one—perhaps the kind of transformation which Goethe called "a loving self-drowning into Nature."

Einstein's complexity is hinted at by these and other apparent internal opposites that we shall encounter again. In studying the lives and works of others, such as Kepler, Bohr, and Fermi, I came in each case also upon puzzling diametricals, whether in their science or in their personal characteristics. But I suspect that such perceptions, made by us earth-bound people, may often be only optical illusions. What appear to us down here as contrary parts may well be, up there, elements that combine and help to *produce* that extraordinary scientist's particular brilliance. It is analogous to what one may call the Rainbow Illusion, because that bright, intangible display, appearing to us in its very different colors, is not really up in the sky. It exists only on our own retinas.

Perhaps just because Einstein could live with, and bridge, what seem to us puzzling contradictions in his life and character, this man was able to find unities among the contradictions and dualities in the *physics* of his time, such as removing the antitheses between the electromagnetic and mechanistic worldviews, between space and time, between inertial mass and energy—all resolved by Relativity Theory; attacking also the antithesis between the wave theory of light and photoelectric emission; and even overcoming the epistemological differences between empiricism and rationalism, as well as emotionally, within himself, the contrary pulls of realism and romanticism.

So: scientist, man, and now good European.

A Good European

The historian Fritz Stern once wrote: "Einstein and Germany: they illuminate each other." Especially because we are meeting in Berlin, we must not overlook Einstein's apparent complexity on the topic of his nationality. Certainly he was a German, born to a family that on both sides could trace its origins in southern Germany to at least the seventeenth century. He was educated to his mid-teens in Munich, and later, starting from age thirty-five, was the holder of very distinguished academic positions in Berlin for nearly twenty years. As I noted, during the worst part of World War I, Einstein refused a very attractive offer from abroad, saying he would not want to separate himself from his excellent colleagues in Berlin. And in the immediate aftermath of that war, he worked energetically against the isolation of German scholars.

But it is equally well established that, when it came to declaring a choice, more often than not he rejected that nationality label, starting with his early flight from Munich to Italy and Switzerland, renouncing his German citizenship at the time. Especially from 1914, most memorably in 1933, after having again discarded his German citizenship, he declared himself to be a European at a time when Europe meant barely more than a geographical entity. Long before the pioneering vision of Konrad Adenauer, Jean Monnet, Robert Schuman, and Paul Henry Spaak, Europe existed as a political and economic entity chiefly in the imagination of the likes of G. F. Nicolai, whose ill-fated manifesto of October 1914 called for the creation of "an organic unity of Europe," or in Count Coudenhove-Kalergi and his supranational Pan-Europa movement of the 1920s. That movement counted among its members Sigmund Freud, Thomas Mann, Rainer Maria Rilke, Miguel de Unamuno—and Einstein, who even wrote an article on Pan-Europa and spoke out in defense of what he called "European civilization." Again, as in his science, Einstein prophesied Europe's eventual unification, just when that was thought by almost all others to be blatantly utopian and impossible.

Today, with all its difficulties, there is of course an EU, and given the current hegemonic ambitions on other continents, there will soon have to be a stronger EU. But Einstein, the most famous and self-declared internationalist of his time, looked even beyond an organic unity of Europe, and lent his fame to that cause. He lived to see with pleasure the beginnings of one of the most promising developments in Western history—the rise of a previously unimaginable set of internationalizing institutions. With all their flaws and faults, with their wrong starts and mistakes, today there is a United Nations and a UNESCO, a World Health Organization, and similar ones for Food, Trade, and Banking, an International Criminal Court, international protocols on the Environment, on Arms Inspection, and on and on. Equally significantly, in science itself, where Einstein contributed to chemistry, cosmology, mathematics, and inventive engineering as well as to physics, there is forming now a kind of Bose-Einstein condensate. To cite only one parochial example of a worldwide trend toward interdisciplinary research: A new building with the remarkable name "Laboratory for Integrated Science and Engineering" is now going up next to our old physics building at my university. That lab can unite in common research faculty and students from about a dozen fields. So, both in international relations and in science, something is trying to be born, just as the man from Ulm had intuited, had hoped for, had worked for, long ago.

Still, there is here again another dimension to Einstein. While in his years in Germany he was, in sociopolitical terms, only ambivalently a German, preferring to be a good European and even a world citizen, he presents another, different view if we look at another characteristic that he did not mention in his brief self-description: Einstein was quite recognizably a German *Kulturträger*.

THE PREPARATION OF A *KULTURTRÄGER*

To be sure, Einstein's reputation as an obstinate, antiauthoritarian nonconformist and defiant rebel—even as a vagabond and gypsy, as he repeatedly described himself—is solidly grounded in many of his actions, and is lively in the popular imagination. But we find equal evidence for viewing Einstein as a cultural traditionalist, even of the kind that the sociologist Karl Mannheim had identified as a free-floating intellectual [*freischwebende Intelligenz*], one without a well-defined anchor in society. More than that, there is evidence that even Einstein's science itself had roots in the standard *Kultur* of his youth and his early years, in the European and especially in the German literary and philosophical cultural tradition.

I need not linger long over Einstein's German-based *scientific* education as part of his cultural preparation. In his very early youth, he read with enthusiasm popular science books by Ludwig Büchner, Aaron Bernstein, and Alexander von Humboldt. We know from his so-called love letters to Mileva, written while he was hatching those key papers during the very period we are celebrating, that this bookworm—as even his mother called him—who always preferred self-cultivation, was carefully studying books and articles in physics. Here is the list of works he said he studied, in the order in which he gave them in those early letters to Mileva: Paul Drude, Hermann von Helmholtz, Heinrich Hertz, Ludwig Boltzmann ("absolutely magnificent"), Ernst Mach, Wilhelm Wien, Gustav Kirchhoff, Wilhelm Ostwald, Planck, Philipp Lenard, Hendrik Lorentz, Woldemar Voigt, and as we know from other sources, importantly, August Föppl. It is a splendid, almost entirely German diet for any young physicist in Wilhelmian Germany.

Bildung, the process by which the German of his day acquired the products and attitudes of *Kultur*, far exceeded learning from science books. A biographer who was also a member of the family (Rudolf Kaiser) revealed that in the home of Einstein's childhood, evenings would typically include listening to his mother playing classics on the piano or getting her help with his violin lessons from age six, perhaps introducing him already then to his life-long favorites, Bach and Mozart. His father would assemble the family around the lamplight to read aloud from writers such as Friedrich Schiller and Heinrich Heine. Through a regular guest of the family, Max Talmey, the precocious youngster was introduced to the philosophy of Immanuel Kant, starting with the *Critique of Pure Reason*, at the tender age of thirteen. He reread it at sixteen, and while a student at the Technical Institute in Zürich, young Einstein enrolled in an optional lecture course on Immanuel Kant. In 1918 he wrote to Max Born that he was reading Kant's *Prolegomena*, saying he was "beginning to comprehend the enormous suggestive power that emanated from that fellow [*von diesem Kerl*]." Later, Einstein wrote a lengthy review on a book analyzing Kant's philosophy, and he referred repeatedly to Kant's ideas in his conversations and correspondence.

All this of course did not at all make Einstein a follower of transcendental idealism; but it is the background to Einstein's own twist. He explicitly freed Kantian Categories from their unalterable a priori, letting his own version be chosen freely, and thus making them a central tool in his epistemology.

One could further elaborate the *Bildung* of this German *Kulturträger*. Here it must suffice to refer to some rough indicators, such as the curricula at his Munich *Gymnasium* and in his school in Aarau, designed for the preparation of young aspirants to the *Bildungsbürgertum*. We have also the extensive reading list of books Einstein discussed at length with two friends in Bern at their private *Akademie Olympia*, meeting sometimes several times a week, during Einstein's most creative period in science: there we find, for example,

Baruch Spinoza, David Hume, Georg Riemann, Henri Poincaré, Mach, Kirch-hoff, Helmholtz, as well as literary classics from Sophocles on. Throughout his life Einstein was a man of the book, to a much higher degree than other scientists whom I have studied. The remarkably diverse collection of volumes in his library grew constantly. Even if we look only at the German-language books published before 1910 that survived in Einstein's Princeton household, the list includes much of the canon of the time: Boltzmann, Büchner, Friedrich Hebbel, the collected works of Heine in two editions, Helmholtz, von Humboldt, many books of Kant, Gotthold Lessing, Mach, Friedrich Nietzsche, and Arthur Schopenhauer. But what looms largest are the collected works of Johann von Goethe in a thirty-six volume edition and another of twelve volumes, plus two volumes on his Optics, one on the exchange of letters between Goethe and Friedrich von Schiller, and a separate volume of *Faust*. So it is no surprise that young Einstein, while a student at the Zürich Polytechnic and still regarding himself as a kind of bohemian at war with the "philistines," had also taken a second optional course while preparing to become a high school science teacher. The title of the course was nothing less than *Goethe, Werke und Weltanschauung*.

Goethe was of course the prime exemplar in the cultural context of the time from which, I maintain, some major German scientists seemed to draw courage for their originality. Some of Goethe's writings were to be thought about, written about, and quoted to each other, usually without attribution, like a secret Masonic handshake of mutual recognition and cultural legitimation. Physicists of those days, such as Wien, Boltzmann, Arnold Sommerfeld, Born, and Erwin Schrödinger, interspersed their lectures and books with quotations from Goethe. In Helmholtz's volume of popular lectures, mostly on science—on which, by the way, Einstein wrote a review—the first and last essays are on Goethe. And when Einstein wrote to Sigmund Freud, he listed three "moral and spiritual leaders"—Jesus, Kant, and of course Goethe, the iconic center of a movement of German idealism and neo-Romanticism.

Within that aesthetic-philosophical movement—to which Einstein, the obsessed questor for *Verallgemeinerung*, was exposed all his life—the insistent message was, as David Cassidy put it, the search for

> some sort of transcendent higher unity, for the existence of permanent ideas or forces that supersede or underlie the transient, ephemeral world of natural phenomena, practical applications, and the daily struggle of human existence. The scholar, the artist, the poet, the theoretical physicist, all strove to grasp that higher reality, a reality that because of its permanence and transcendence must reveal ultimate "truth," and hence serve as a unifying basis for comprehending . . . the broader world of existence in its many manifestations.

And the historian of science, Anne Harrington, agreed when she wrote: "Goethe's resulting aesthetic-teleological vision of living nature would subsequently function as one of the later generation's recurrent answers to the question of what it 'meant' to be a holistic scientist in the grand German style."

To these messages, emanating from the surrounding culture, Einstein resonated in his central preoccupation: above all, the search for a general *Verallgemeinerung*, for unity. As he put it in an essay on one of his heroes, Johannes Kepler, the aim of the search for unity in science was to find "the mysterious harmony of the world into which we are born."

When considering Einstein as a traditional carrier of culture, one must also not omit mentioning his love for Spinoza and for Schopenhauer. He often read, referred to, and quoted both of these authors. They furnished emotional and intellectual support for his desperate hold on the thema of scientific determinism, even while the large majority of his contemporaries in physics went triumphantly over to the counter thema, indeterminism. Today almost all scientists regard Einstein to have been wrong on that point, misled by that particular thematic loyalty.[1]

So, Einstein as scientist, as man, as European, as culture carrier.

A JEW

And, last but not least, as on his own list, Einstein the Jew—but a Jew whose theological views few rabbis would approve. On that dimension in Einstein, with typical fluctuations between extremes, I need say little here. He reported on the first page of his *Autobiographical Notes* that as a child, through reading "the stories of the Bible," he "came to a deep religiosity," which he called his first "religious paradise of youth," "a first attempt to free myself from the chains of the 'merely personal.'" He then abandoned it on finding his second paradise, science. When about fifteen, he had himself formally withdrawn from the roles of the synagogue. For some seventeen years more, he was essentially without denomination, or *konfessionslos*, as he wrote on his 1901 application for Swiss citizenship. But his friend and biographer, Philipp Frank, told me that when Einstein was appointed Professor at the German University in Prague in 1911, he had to fill out a questionnaire before delivering his oath of loyalty to the reigning monarch, Kaiser Franz Joseph, mandatory for all civil servants. For his religious affiliation, Einstein began to write *konfessionslos*. A friend stopped him, saying, "You must have some religion, otherwise you can't take the oath." So Einstein wrote, reluctantly, "Mosaisch," and he later joked, "It was Franz Joseph who made me into a Jew."

But what really reestablished Einstein's life-long identification with what he called "our tribe" was this city, Berlin. As Einstein put it, "there I discovered for the first time that I *was* a Jew." From 1914 on, soon after his arrival in Berlin and more and more intensely in the wake of his fame of 1919, he became all too aware of the increasingly brutal anti-Semitism and the vicious attacks on him from extreme segments of the public, as well as from scientists, including one Nobel Prize scientist, Philipp Lenard, who openly called Einstein's theories "a Jewish fraud."

Much of the rest of the world was also and indeed still is infected by some variant of that persistent virus. But of course, Germany was a special case. In 1919, General Paul von Hindenburg had to testify why Germany had lost the war. He had a ready answer, one that resonated for decades: Germany's brave soldier had been betrayed, "stabbed in the back," by communists, Jews, and women. In 1920, the German ambassador in London felt it necessary to write confidentially to his Ministry of Exterior that "Professor Einstein is just at this time for Germany a cultural factor of first rank. . . . We should not drive such a man out of Germany with whom we can carry on real *Kultur* propaganda." One year later, with Nazi gangs already roaming the streets, Einstein told Philipp Frank that he was unlikely to remain in Germany for another ten years. As usual, his prediction was close to the mark. By May 1933, his books were among those being burned and his property confiscated. In 1922, he left Berlin for some months after his friend, Germany's Foreign Minister Walther Rathenau, had been assassinated in June of that year, in one of about 300 of such murders since 1918 by right-wing Nationalist fanatics. Einstein's name was reported to be on the list of the next intended victims. He cancelled his lectures, some of which had been interrupted by rowdies, and left, feeling it wise to accept invitations from abroad that required a lengthy absence from Germany. As he put it frankly in a letter at the time, he felt he had to "escape the increasing danger."

After his journey to Japan, during which he heard about the award of his Nobel Prize, he went in February 1923 to Jerusalem. There he gave what was effectively the first major lecture at Hebrew University, whose establishment had become one of Einstein's passions. After his return to Berlin in 1923, Einstein was told of a continuing danger to his life, and he felt he had to flee again, this time to Leyden in the Netherlands.

From that time on, with the stubbornness he often declared to be one of his most evident characteristics, Einstein became more and more openly and emotionally explicit about and involved in a Jewish identity. He wrote finally to Abba Eban in 1952: "My relationship to the Jewish people has become my strongest human bond." From his Berlin days on, Einstein was attracted to Zionism for two major reasons. One was that he thought a Jewish commonwealth in Palestine, then under British rule, would be a solution for the suffering and the precarious predicament of young Jews who were being denied access to education and employment in much of Europe, but who might flourish at the planned Hebrew University in Jerusalem. The other reason was to give the widely dispersed fellow religionists a spiritual home. As Isaiah Berlin put it in an essay, *Einstein and Israel*, Einstein's "Zionism was grounded in the belief that basic human needs create a right to their satisfaction: men have an inalienable right to freedom from hunger, nakedness, insecurity, injustice, and from homelessness too." One may imagine that if Einstein somehow returned today, he would be most preoccupied with two topics—surely with the brilliance of today's physics, but also with the tragedy of the Middle East.

Concluding this brief sketch of Einstein's complex Jewish identity, one must note that, just as he devised his own physics in resonance with German science and German culture, he also invented his mature religiosity in resonance with earlier European philosophy and theology, above all with that of Spinoza, whom Einstein read and quoted and reread for half a century and whose conception of the Deity as Reason embodied in Nature he fully accepted. In several essays starting in 1930, Einstein described his own so-called Cosmic Religion, in which he tried, as usual, to bring together seeming incommensurables, this time the awe of religiosity and the passion for science. Einstein fashioned thereby what effectively was his Third Paradise, the unification of his early, first, and second ones. In the pursuit of his elevated thoughts in science, he had found himself caught up in an emotionally deep religious experience, a glimpse of the vision of a divine Nature in its unity and harmony. He once defined his search for a Cosmic Religion in highly charged words: "The individual feels the vanity of human desires and aims, and the nobility and marvelous order which are revealed in Nature and in the world of thought. He feels the individual destiny as an imprisonment, and seeks to experience the totality of existence as a unity full of significance."

Those few lines seem to me to describe well one main theme running through Einstein's life and thoughts, from beginning to end.

Epilogue

A *Kulturträger*, properly speaking, not only imbues and represents the culture of his time and place, but also stimulates others, widens their horizon, imagination, and vocabulary. That process certainly was the fate of Einstein's ideas and publications as others used them, often to his surprise and puzzlement. One can trace Einstein's influence on the shaping of the imagination of his contemporaries and their followers in a great variety of fields, from theologians such as Paul Tillich, philosophers including Henri Bergson, Alfred North Whitehead, and Ernst Cassirer, to logical empiricists in Europe and pragmatists in the United States. All adopted or struggled with some of Einstein's ideas as they understood them. So did cultural anthropologists and psychologists, such as Claude Lévy-Strauss and Jean Piaget. Einstein had special trouble with some art historians who wanted him to be the father of cubism. And most of us have seen an Einstein impersonated on the stage, in plays and in an opera. Poets and novelists by the dozen tried their hands at celebrating or incorporating some Einsteiniana.

A special case is Thomas Mann, whose skillful transference of ideas from physics, biology, and medicine ran through many of his books, and particularly his *Zauberberg*. Mann recorded as an emotional highpoint of his life his meeting with Einstein in 1939, when he received from the physicist-philosopher the Einstein Medal. In his response, Mann said, "I am able at least to divine that in physics, of which Albert Einstein is the world renowned representative, there happen things more fantastic than all that fiction can invent, and more important, more transformative for mankind and its world picture than everything literature can bring about."

This accolade hints at one answer to the question, why is Einstein still so alive today, in the imagination of people high and low, in all segments of the globe—a fact that Einstein, who experienced that phenomenon constantly, was himself completely unable to explain, dismissing it as a case of mass hysteria. But a good part of the explanation may be this: The lives and works of some scientists project to the wider populace a charismatic view of science. Building on Max Weber's original discussion of "charismatic authority," this concept has been the subject of scholarly study extending the concept to scientists, for example in Joseph Ben-David's *Scientific Growth* and earlier in Robert K. Merton's book, *Science, Technology, and Society in Seventeenth-Century England*. The social scientist Bernard H. Gustin elaborated on this perception, writing that science at the highest level is charismatic because scientists devoted to such tasks are "thought to come into contact with what is essential in the universe."

This observation catches precisely why so many people the world over, despite or perhaps because they know little about Einstein's science, still seek after him and feel somehow elated, uplifted, when contemplating his iconic image. They extend to their conception of Einstein what the historian Jacob Burckhardt identified long ago as the essential "power of veneration within us." Happily, mankind has a need to admire and seeks out objects that satisfy that need.

I began by positing the question, Who is Einstein for us, and what accounts for his being still so alive. I have hinted at the rich complexity of that great lion. He will of course never be fully fathomed, but we can now dare to set our sights, one by one, on his many extraordinary roles in history. Let our symposium begin.

CHAPTER ONE

2

A SHORT HISTORY OF EINSTEIN'S PARADISE BEYOND THE PERSONAL

Lorraine Daston

INTRODUCTION: "A PERSON OF MY TYPE"

In an autobiographical sketch written late in life—an "obituary," as he wryly called it—Albert Einstein relates how the religious vision of transcendence of his youth had in adulthood been transformed into a scientific "paradise" beyond the "merely personal." This remarkable passage is worth quoting in full:

> It is quite clear to me that the religious paradise of youth, which was thus lost, was a first attempt to free myself from the chains of the "merely-personal," from an existence which is dominated by wishes, hopes and primitive feelings. Out yonder there was this huge world, which exists independently of us human beings and which stands before us like a great, eternal riddle, at least partially accessible to our inspection and thinking. The contemplation of this world beckoned like a liberation, and I soon noticed that many a man whom I had learned to esteem and to admire had found inner freedom and security in devoted occupation with it. The mental grasp of this extra-personal world within the frame of the given possibilities swam as highest aim half consciously and half unconsciously before my mind's eye. Similarly motivated men of the present and of the past, as well as the insights which they had achieved, were the friends which could not be lost. The road to this paradise was not as comfortable and as alluring as the road to the religious paradise; but it has proved itself as trustworthy, and I have never regretted having chosen it.[1]

In the next breath, Einstein admitted that this *confession de foi* no doubt oversimplified the actual course of his intellectual and emotional development, but he insisted on its essential truth—that it captured not only his own

15

essence, but that of a whole subspecies of humanity, those people "of my type": "In a man of my type the turning-point of the development lies in the fact that gradually the major interest disengages itself to a far-reaching degree from the momentary and the merely personal and turns towards the striving for a mental grasp of things."[2] Einstein did not invent this paradise beyond the merely personal; as he himself notes, he had models and companions who pointed the way to him. He is but one of a "type," someone who flees the personal but who does so in good company. My aim here is to explore the history of this "type" and its object of yearning, Einstein's paradise beyond the personal.

This will be a short history, and not only because of length restrictions. Although Einstein himself has been likened to the ancient models of the seer, the sage, and the saint, the "type" he identifies and identifies with is a relatively recent category, still new in the last quarter of the 19th century when Einstein received his education and formation. Any number of other scientists who came of age around the turn of the 20th century could and did express similar longings to extinguish all but the thinking self, in the name of a larger—indeed cosmic—community of like-minded researchers. They too often reached for religious analogies to describe and dignify their quest for a paradise beyond the personal. But the kind of transcendence they sought was not merely Judaism or Christianity in another guise, God tricked out as the "great, eternal riddle." Nor was the self to which they aspired, despite some suggestive resonances, a revamped version of Socratic or Stoic or humanist models of the enlightened philosopher, indifferent to the slings and arrows of outrageous fate. It was a genuinely new ideal of how to be and know in the world, as novel as the words "scientist" and *Wissenschaftler* that came into currency at about the same time to identify a persona distinct from that of the scholar, the philosopher, or the *Gelehrter*.[3]

This scientific persona was distinguished by two paradoxical traits. First, it was a persona that eschewed the merely personal, a self that aimed at self-effacement. All that stamped an individual with personality, all that marked an individual *as* an individual—not only quirks, but any quality that varied from person to person, including unusual skills and abilities—was to be suppressed for these purposes. The core self that remained was not a Cartesian ego freed from its body; scientists were far too concerned with the material world to deny their own materiality. But it was a self whose primary activity was thinking, because thinking was what it had in common with other selves.

Second, it was a persona that practiced solitude in the name of community. The solitude was in relation to the actual flesh-and-blood human beings of daily encounters. The community was constituted of those fellow seekers, scattered over time and space, who had also distilled themselves (or been distilled by history) into thinking selves dedicated to fathoming nature. These were Einstein's "friends who could not be lost." Some might indeed be friends in the usual sense, contemporaries with whom one talked, laughed, argued, and grieved. But others were long dead, like Newton and Kepler, or yet to be born. This was a ghostly community, a utopia—in the literal sense of being nowhere as well as in the figurative sense of being idealized. Here is how Einstein understood Newton: "To think of him is to think of his work. For such a man can be understood only by thinking of him as a scene on which the struggle for eternal truth took place."[4] But as spiritualists know, ghosts still experience and provoke passion, and this is also the case for the inhabitants of the paradise beyond the personal.

This persona is familiar to us as both an epistemological stance and a sociological convention. Philosopher Thomas Nagel has coined the vivid phrase "view from nowhere" for the epistemological stance, otherwise known as objectivity: "A view or form of thought is more objective than another if it relies less on the specifics of the individual's makeup and position in the world, or on the character of the particular type of creature he is. . . . We may think of reality as a set of concentric spheres, progressively revealed as we detach gradually from the contingencies of the self."[5] Note that the precondition for revelations about reality, for knowledge, is to shed "the contingencies of the self." The sociological convention goes by the blander name of "scientific community"—a term perhaps first used by the American philosopher and scientist Charles Sanders Peirce in the late 19th century,[6] although it does not seem to have become common currency until after World War II, when it was taken up by the chemist and philosopher Michael Polanyi.[7] This is a far-flung community of conviction rather than proximity, with none of the cozy associations of *Gemeinschaft*. Face-to-face interactions are the exception rather than the rule; Newton is in principle as close a neighbor as the colleague down the hall, perhaps closer.

So familiar are both epistemological stance and sociological convention that we may fail to recognize them for the distinctly odd cultural innovations they are, especially in tandem; we may read Einstein's musings over the paradise beyond the personal as the sort of thing that might have been written by Newton, if not Seneca. There is something about the diction of this passage, with its appeals to the mysterious and sublime riddle of the universe and its mythological associations of a fateful choice of paths (Hercules at the crossroads, choosing between vice and virtue), that invites a timeless reading. In the teeth of this temptation, I shall attempt in what follows to render the familiar strange, first by contrasting it with earlier learned personae and then by describing the circumstances under which it first emerged in the mid–19th century, all in very brief compass. I hope to provide a historical context for

Einstein's utterances and thereby to make them fresh and startling, to rescue them from the realm of the edifying and the platitudinous—and therefore to make it possible to disagree vehemently with them, rather than to nod piously.

To avoid misunderstandings, a few caveats about what I will *not* do are in order here. I have no intention of trying to ascertain whether Einstein's actual conduct matched his ideal. There is good evidence that Einstein was in fact sensitive to and appreciative of the purely personal qualities of his friends and colleagues. One thinks for example of his warm reminiscences of the French physicist Paul Langevin, "universally beloved for his devotion to every good cause, for his understanding kindness toward all creatures,"[8] or his hurt disappointment upon learning that Newton had most ungraciously tried to deny credit to Robert Hooke, as Galileo had failed to acknowledge the work of Kepler—egregious but common intrusions of vanity into the paradise beyond the personal.[9] These divergences between ideal and fact are surely of interest for the biographer, but not necessarily for the historian of a scientific persona. I take Einstein's paradise beyond the personal to have been a genuine ideal, i.e., one sincerely held and with the power to shape aspirations and behavior, even if not to determine them. As is the case with moral norms, there is a vast difference between a society that, for example, professes an ideal of generosity, however honored in the breach, and one that does not recognize the claims of such a value. What I am after here is the history of such an articulated and acknowledged ideal, and the persona called into existence in an attempt—however partial and even futile—to realize it.

The Philosopher, the Sage, and the Scholar

Just how novel was the persona who strives toward the paradise beyond the personal? The description of learned personae stretches back to antiquity; the best-known work in this genre, the model for all other such collective portraits of celebrated thinkers well into the modern period, was Diogenes Laertius' compendium of the *Lives and Opinions of Famous Philosophers*, composed in Greek probably sometime in the 3rd century B.C.E.[10] The key characteristic of all of these works, from Diogenes Laertius down to 19th-century collections such as George Craik's *The Pursuit of Knowledge under Difficulties* (1858), is precisely the interweaving of philosophical or scientific doctrines with biographical anecdote. Diogenes Laertius not only tells us about Socratic teachings on the Good, he also retails anecdotes about how Socrates learned to play the lyre in old age and danced regularly to stay in shape, how he was drenched with water by his shrewish wife Xanthippe, how he walked out of a play by Euripides because it made light of virtue.[11] Modern authors in this genre related how Newton was a tolerably competent Master of the Mint,[12] how Goethe tucked into his meals with gusto,[13] how the French physicist André-Marie Ampère was scorned by a Parisian hostess for coming to dinner with chemical-stained hands.[14]

The point of these stories was that the life and works of the learned were of a piece; to evaluate the one, knowledge of the other was considered essential. Each of these stories was told to make a point about an exemplary life of the mind—exemplary both in the sense of providing a model worthy of emulation, an exemplum, but also in the sense of illuminating teachings by a specific example of their application in daily life. This tradition of regarding life and works as an indissoluble whole has proved extraordinarily tenacious. Historians of science do not need to be reminded that anecdotes, most of them apocryphal, about Newton's apple, Kant's daily walk, or Einstein's lackluster Gymnasium grades persist not only in popular lore, but also among scientists. Although life and works of philosophers and scientists were officially rent asunder in the 19th century, they are still constantly if covertly conjoined. And sometimes not so covertly: witness the highly publicized squabbles over the bearing of Freud's character on the truth of his theories of the mind among psychoanalysts and historians.[15] There is nothing trivial or self-evident or even venerable about the ideal of an intellectual paradise beyond the personal; rather, it contradicts a millennia-old conception of the *vita contemplativa* as a *vita*, a lived life full of personality and incident. Einstein was not giving voice to some ancient ethos of the intellectual.

Nor was his vision of the scientist simply an updated version of older learned personae. The history of such personae is a long and rich one. Here I can do no more than sketch three of its variants to show how sharply Einstein's "person of my type" diverged from them: the philosopher, the sage, and the scholar.

The philosopher was originally a "lover of wisdom"—not only a "lover of truth," as post-17th-century philosophers have come to redefine themselves. This means that ancient philosophy was necessarily wedded to a way of life, in addition to or even to the exclusion of an inquiry into the constitution and workings of the world. The prototypical philosopher was and remains Socrates. He is notoriously a man of masks, a master of irony, and a most elusive character.[16] Nonetheless, Plato's portrait of him, put into the mouth of Alcibiades in the *Symposium*, is vivid and sharply etched, and all the more useful for our purposes in being most probably a mythic creation rather than a factual report; that is, it captures the essence of a persona, not the specifics of a person.

Alcibiades describes Socrates as strange, as not caring about the things that drive other men. Socrates is neither an ascetic nor other-worldly; he falls in love with pretty young men like Alcibiades, and he has served Athens bravely as a soldier. But his priorities lie elsewhere, and in cases of conflict, he is ruthless about sacrificing lust, ambition, and wealth to the quest for wisdom. He pursues Alcibiades but then refuses to be seduced by him—as the smitten Alcibiades realizes too late—in order to initiate the young man into another kind of eros, the love of wisdom. The priestess Diotima, Socrates' own informant about love, makes Eros resemble Socrates himself, barefoot and dirty. The eros of philosophy draws students to master like moths to a flame; it is as all-devouring as any other kind of love and just as indifferent to convention and comfort. Alcibiades describes Socrates standing motionless as a statue all day and all night as he ruminated upon a philosophical problem, astonishing even the hardened soldiers by his fortitude.[17] This is not the mortification of flesh for its own sake, no more than Socrates' rags and bare feet are the tokens of a vow of poverty. Socrates' absorption and his strangeness are the symptoms of being in love, and the relationship between master and disciples in search of wisdom is also erotic. The eros may be between souls rather than bodies, but it is just as consuming and particularized as more humdrum romantic love. Socrates is indeed so particularized as to be unclassifiable, a personality *sui generis*, corresponding to no known type, as Alcibiades says.[18]

Even from this skeletal summary, it should be clear that Socrates, the emblematic philosopher, and the philosophical community that ideally crystallized around him contrast starkly with the inhabitants of Einstein's paradise beyond the personal. Far from obliterating the merely personal, the philosopher accentuates it; it is at once his primary qualification for his vocation, a beautiful soul, and the magnet that draws disciples to him. Eros demands personality, megawatt personality. The relationship among Einstein's "friends who cannot be lost" is that of kindred spirits, but narrow-band kindred spirits: they are united not by affection or affinity, but rather by shared thoughts and interests. It would be ludicrous to apply the word "seduction" to their ties, even if the seduction of souls is what is meant. But if Alcibiades' testimony in the *Symposium* is to be credited, the Athenians who accused Socrates of seducing youths were speaking only half-metaphorically.

It would be reasonable at this point to object that the proper case for comparison would be not the Socratic philosopher, but the Stoic sage. The Stoics, after all, professed a physics as well as an ethics; they exhorted their followers to extinguish their unruly passions in the contemplation of reason manifest in the order of nature. One can indeed detect a faint echo of Seneca's advice in the *Natural Questions* (ca. 63 C.E.) in Einstein's flight from "wishes, hopes, and primitive emotions" into the world independent of all human desire. For Seneca and other Stoics, the aim of the contemplation of nature was to liberate the mind from fear, especially the fear of death. All mortals must die, Seneca reminds his readers; why therefore fear death by earthquake or drowning more than by some other more banal cause? "The earth is split and burst by a great power of I know not what calamity and carries me off into the immense depths. So what? Is death easier on a level surface? What do I have to complain about if nature does not want me to lie in an ordinary death, if she places upon me a part of herself?"[19] The habit of imagining the earth as seen from a remote heavenly vantage point teaches the vanity of riches and ambition: "The mind says to itself: 'Is this that pinpoint which is divided by fire and sword among so many nations?'"[20] Whereas Socrates the philosopher sought an unattainable wisdom, Seneca the sage taught a way of life through the exercise of reason.

The word "exercise" must be underscored, for the ideals of the Stoic sage were to be realized by a strict regimen of what historian of ancient philosophy Pierre Hadot has called "spiritual exercises," repeated routines of the mind designed to strengthen and shape the soul as athletic exercises strengthen and shape the body: "Generally, they consist, above all, of self-control and meditation. Self-control is fundamentally being attentive to oneself: an unrelaxing vigilance for the Stoics; the renunciation of unnecessary desires for the Epicureans."[21] The meditations of Marcus Aurelius train the imagination to dwell upon scenes of human insignificance (the forgotten rulers of past epochs, the processes of decay already at work even among the living), much as Seneca took a cosmic perspective to shrink the entire planet earth to a pinpoint.

Despite the deliberate attempt at estrangement from human affairs in Stoic spiritual exercises, they are anything but negations of the self. On the contrary; they are cultivations of the self, gymnastics of the soul. They do not reject the merely personal; they reform it. The Stoic sage was intensely, perhaps even obsessively, concerned with the self, since only by changing the self could fear be conquered and fortune overcome. The contemplation—not, it should be noted, the active investigation—of nature was a means toward the reformation of the self, not the other way around. Moreover, the relationship between master and disciple was just that, with all of the usual associations of authority and dogmatism, as well as a cult of personality surrounding the master.[22] Einstein's paradise beyond the personal once again recedes from view, far from view.

It seems closer to the precepts and practices of the Republic of Letters of Renaissance and Enlightenment scholars. Here we encounter a virtual community sustained by reading and letter writing, defined in explicit opposition to the real communities of family, religion, nationality, and even historical period in which the scholars in fact found themselves. These contacts across barriers of confession and creed were deliberately narrowed to avoid conflicts. The distances, both literal and figurative, that defined relationships within the Republic of Letters were viewed by many Enlightenment scholars as a safeguard of impartial judgment. Proximity in time and space was perceived to be in inverse relation to that impartiality, and for this reason the good opinion of posterity and of foreigners was particularly prized. Hence intimate ties to foreigners took on an ambivalent cast. As the French mathematician and encyclopedist Jean d'Alembert remarked in 1759 apropos of French anglomania, "the closer one becomes to foreigners, the more they lose that character of posterity for which the distance of space is at least necessary, in default of the distance of time."[23] On the side of posterity, this distance also had its advantages. In the company of books, one could ignore the less amiable aspects of an author's character. In his 1837 essay on Francis Bacon, British historian Thomas Babington Macaulay pointed out the advantages of communion with dead authors in the library over conversations with live colleagues: "With the dead there is no rivalry. In the dead there is no change. Plato is never sullen. Cervantes is never petulant. Demosthenes never comes unseasonably. Dante never stays too long."[24]

This distanced relationship to colleagues dispersed in time and space had its equivalent on the home front, especially among the northern European scholars who married and established households in increasing numbers after the 15th century. The incompatibility of marriage and scholarship is familiar from ancient and medieval lore of learned lives, such as Diogenes Laertius' portrait of the unhappy marriage of Socrates and Xanthippe and the letters of the ill-starred medieval lovers, Abelard and Heloise. As Heloise wrote reproachfully to the brilliant logician Abelard when he proposed to forsake holy orders to marry her and thereby legitimate their infant son Astrolabe: "Who can concentrate on thoughts of Scripture or philosophy and be able to endure babies crying, nurses soothing them with lullabies, and all the noisy coming and going of men and women about the house?"[25] For Abelard and countless other scholars thereafter, the main issue was not chastity but concentration: how to reconcile the distractions and burdens of a populous household with expenditures of time, money, and above all attention demanded of the scholar? Historian Gadi Algazi has studied the habitus of "learned forgetfulness" that married Renaissance scholars cultivated to work amidst a large and loud family[26]: a pattern of selective attention that neglected familial, religious, and civic duties for peculiar objects of study—Greek verbs, insects, comets—studied with peculiar intensity. By the late 17th century, this habitus was familiar enough to be ridiculed by moralists like Jean de La Bruyère in his *Characters*:

> Another man loves insects. . . . You've chosen the wrong time to visit him; you find him sunk in deep despair; he is in the blackest and bitterest of moods, and his whole family are the victims of it; for he has suffered an irreparable loss. Go up and look at what he's showing you on his finger, a lifeless object that has just breathed its last: it is a caterpillar, and what a caterpillar![27]

These habits of distance, cultivated through learned correspondence and studious and studied absentmindedness, established the preconditions for a community that screened out many aspects of the personal in the interests of impartiality and concentration. The fairest judges of scholarly merit were those farthest away in space and time; the duties and cares of daily life were best kept at arm's length, with the study door barred to family and visitors. Life and works were beginning to come apart. But membership in such a community did not demand detachment from oneself, only from one's immediate surroundings. And although the Republic of Letters endorsed disinterested judgment, it did not go so far as to claim that collective inquiry was epistemologically superior to individual efforts, a surer way of achieving truth. On the contrary: intellectual progress, especially in the sciences, was tightly linked to the achievements of towering individuals like Newton or Linnaeus or Euler. Neither scholar nor philosopher nor sage diagnosed the very fact of *having* a self as an obstacle to knowledge; instead they sought the *right kind* of self, cared for in the right way. It is only in the middle decades of the 19th century that Einstein's paradise beyond the personal came into its own.

Independent of You or I

What kind of self could block the way to truth? There were venerable answers to this question: a self in the grip of passion, one bamboozled by religious authority or seduced by a glittering system, one blinded by ignorance and prejudice. But by about 1860, scientists had come up with a new answer: the self *tout court* was at fault—or rather the private, individualized self that in the course of the first half of the 19th century become synonymous with subjectivity. If the psychologists, physiologists, statisticians, criminologists, and other 19th-century scientists like Francis Galton (or Alphonse Bertillon or Wilhelm Wundt or a host of others) investigated individual differences so assiduously, surely one reason was that the individualized self had become historically vivid and significant to them in ways that it had not been to their predecessors.

Armed with cameras, collimators, chronometers, and calipers, mid-19th-century scientists studied the speed of nervous transmission, color sensations, attention spans, and even logic and mathematics as psychophysiological phenomena. Their own researches in sensory physiology and experimental psychology documented the variability of mental representations and intuitions. These variations invaded science itself: in astronomy and geodesy, observers were forced to acknowledge the existence of personal equations that resisted every attempt to eliminate them by training and technology. The "personality" of an astronomer's observations was discovered to be as indelibly individual as a signature.[28] Logic fared little better at the hands of the psychophysiologists.[29] Not only were scientists acutely aware, thanks to the revelations of sensory physiology, that their impressions of the external world were highly mediated by the human nervous system; they also confronted the fact, established by numerous psychological experiments in newly built laboratories, that even elementary sensations, such as color, differed significantly from person to person.[30] The self had become a fortress, locked away from nature and other minds alike.

The recommended countermeasures to scientific solipsism emphasized renunciation: giving up one's own sensations and ideas in favor of formal structures accessible to all thinking beings—the realm of objective thought. The sine qua non of the objective was communicability. According to the logician Gottlob Frege, anything that was picturable, subject to the laws of association, and above all private was ipso facto "psychological" and could not be modified by the adjective "objective."[31] Nor could it be scientific: "Thus, I can also acknowledge thoughts as independent of me. Other men can grasp as much as I: I can acknowledge a science in which many can be engaged in research. We are not owners of thoughts as we are owners of our ideas [*Vorstellungen*]."[32] The French mathematical physicist Henri Poincaré boiled this down to the lapidary "pas de discours, pas d'objectivité."[33] "Discours" referred to relations—mathematical, logical, in any case formal—that were invariant over individual psyches. Everything else about the self, anything that individuated the individual, was to be jettisoned in the name of science.

Variability had a historical as well as a personal dimension for late 19th-century scientists, who were acutely aware of the *prestissimo* pace of scientific change. Theories succeeded one another at an ever-accelerating pace; facts pointed to contradictory conclusions. The history of science would not stay written, for at any moment a theory solemnly pronounced dead might be revived. The expectations for scientific progress voiced in the early 19th century had not been disappointed; rather, they had been fulfilled with a vengeance. Poincaré himself had been overwhelmed by the rush of events in his review of electrodynamics in 1902. Only a few years before, "the edifice of electrodynamics had seemed, at least in its large outlines, definitively constructed; everything presented itself under the most satisfactory aspect." But this tranquility had been disturbed by recent experiments, and Poincaré refused to hazard even a guess at the outcome, lest events take yet another turn "between the day when I give this book to the publisher and that when it appears at the bookstore."[34] The Austrian physicist Ernst Mach contended that the principal value of the history of science lay precisely in its revelation that no theory endures, that "science is incomplete, mutable."[35]

Confronted with the variability of individuals and the mutability of history, scientists like Peirce, Poincaré, and others envisioned science as an intrinsically collective undertaking, one that must stretch into the indefinite future. Peirce set forth a new brand of realism, "independent, not indeed of thought in general, but of all that is arbitrary and individual in thought; is quite independent of how you, or I, or any number of men think."[36] No single human mind, no single human lifetime would suffice to reach the truth; the ideal knower was a "communion of minds" by no means "limited to men in this earthly life or to the human race, but . . . including some probably whose senses are very different from ours, so that in that consent no predication of a sensible quality can enter"[37]—or indeed any quality that would differentiate one mind from another. Max Planck caught something of the same intergalactic vision in a 1906 lecture in which he contended that "the system of theoretical physics demands validity not merely for the inhabitants of this earth, but also for the inhabitants of other planets."[38] This truly cosmic community demanded the sacrifice of the "merely personal," which both Peirce and Poincaré regarded as a moral act, the submersion of the self in a greater whole:

> And science renders us another service; it is a collective work and cannot be otherwise; it is like a monument the construction of which requires centuries and to which each brings his stone; and that stone sometimes costs him his life. It thus gives us the sentiment of a necessary cooperation, of the solidarity of our efforts and those of our contemporaries, and even of that of our predecessors and successors.[39]

With Planck and Poincaré, we have reached Einstein's contemporaries and colleagues, who shared essential aspects of his vision of a scientific paradise beyond the personal. This was a vision forged in the historical context of late 19th-century science, against the background of the dual threat of individual variability and historical mutability. It was at once an epistemological, moral, and social vision of the kind of community required to safeguard knowledge and the kind of stripped-down, thinking self required to create such a community. Certain features of Einstein's formulation were no doubt all his own, but others—the religious metaphors, the community of minds past and present, the distillation of the self into pure thought—bear the clear imprint of a particular time and place. Above all, the emphasis on that which cannot change—the friends who cannot be lost, the "reliable" paradise—evokes a moment of perceived crisis in science, when the price of scientific progress seemed to be truth itself. If truth was not forever, was it worthy of the name? The paradise beyond the personal was to be a bulwark of permanence in a world in which, as Marx memorably put it, "all that is solid melts into air."

Conclusion: Objectivity as Invariance

At the heart of this vision of the paradise beyond the personal was the dream of invariance under transformations of all kinds, an idea also central to Einstein's understanding of the singularly ill-named theory of relativity. It is still at the heart of what many philosophers and scientists understand as objectivity. In a recent formulation, philosopher Robert Nozick defined an objective fact as "one that is invariant under (all) admissible transformations";[40] the title of the book in which this definition appears rings all the changes on the theme: *Invariances: The Structure of the Objective World.* Only structures, it is believed, survive the vicissitudes of many minds (human, angelic, Martian), of many worlds (physical, chemical, biological), and above all, of the many theories that litter the history of science. By X-raying the object of knowledge into structures, some semblance of eternal truth is preserved—or at least, that is the hope.

And the subjects of knowledge, the thinking beings who discern the structures? German mathematician and philosopher Hermann Weyl tried to capture the connections among objectivity, subjectivity, and invariance in the group theoretic language of invariance under "arbitrary linear coordinate transformations."[41] What the language of invariance under transformation obscures is a feature that leaps out from earlier visions of the paradise beyond the personal, namely that the knower must be many-headed and all but immortal. The knower is not the individual, not even a consensus among individuals, but the community. On this account, the condition of knowledge is the surrender of individuality, which late 19th-century scientists interpreted as the surrender of self *tout court.* It is not an accident that Poincaré's image of scientific stoneworkers evokes the anonymous medieval artisans who labored over centuries to build cathedrals. Peirce experienced this surrender as a sacrifice; Einstein, as a liberation. But both agreed that it was the precondition for scientific knowledge.

This is a conception of objectivity based not on self-restraint, but on the dissolution of self. Nietzsche, a sharp and acerbic observer of late 19th-century *Wissenschaftlichkeit,* had precious little patience for the ascetic pretensions of scientific objectivity based on self-denial, ridiculing its practitioners as a "race of eunuchs." But he made an exception for an objectivity he deemed a "positive quality": not passive indifference, but a "loving immersion [*Versenktsein*] in the empirical data," common to great artists and scientists alike.[42] Instead of the Venn diagram intersection of what a group of individuals have in common—intersubjectivity—this is a vision of fusion of subject and object. The knower is no longer the individual, nor even the community, however cosmic, but the world itself.

3

EINSTEIN'S JEWISH IDENTITY

Hanoch Gutfreund

"My Strongest Human Bond"

In 1952, David ben Gurion, the Prime Minister of the infant State of Israel, proposed to Albert Einstein to succeed Chaim Weizmann, who had just died, as the next President. In a letter to ben Gurion, Einstein conveyed his regret for not being able to accept the offer: "I am deeply moved by the offer from our State of Israel and I am saddened and ashamed that I cannot accept it," and he added, "I am the more distressed because my relationship to the Jewish people has become *my strongest human bond* ever since I became fully aware of our precarious situation among the nations of the world."[1]

What was the nature of this bond? What was its origin? How was did it evolve, and how was it demonstrated throughout his lifetime? These are the questions on which I will focus.

The Discovery of Being a Jew

Einstein's biographers describe the brief episode in his childhood years in Germany, when his assimilated parents, who did not lead a Jewish life, hired a young man to instruct the twelve-year-old Albert in the basics of Judaism. This generated strong religious feelings in little Albert, who—to the dismay of his parents—demanded that they begin to observe the rules of Jewish religious conduct. This did not last long. Albert's tutor had made a mistake. He had given him books in popular science, which diverted his interest from religion to science to the extent that Albert even refused to celebrate his Bar Mitzva. Years later, in his autobiographical notes,[2] Einstein referred to this short phase in his life as "the religious paradise of youth." The next episode in his biography with a significant Jewish component occurred twenty years later, in 1911–12 in Prague, where he met a group of young Jewish intellectuals at the university and was impressed by the old Jewish cemetery. To teach at the university, he had to declare a religion and he changed the previous entry in his identity papers from "none" to "Jewish."

I do not want to speculate whether one can trace the origins of Einstein's Jewish identity, which was so prominent in later years, to those scattered experiences. I would rather suggest that we accept his own judgment about the years before 1914: "There was nothing that called forth any Jewish sentiments in me."[3] In a letter to Willy Hellpach, a professor of psychiatry and President of the State of Baden (a letter to which I shall return later), Einstein wrote, "When I came to Germany fifteen years ago I discovered for the first time that I was a Jew, and I owe this discovery more to Gentiles than Jews."[4]

His personal "discovery" of being Jewish evolved into his most enduring commitment outside of physics. As Abraham Pais, one of Einstein's biographers, has put it, "I am sure that Einstein's strongest source of identity, after science, was to be a Jew, increasingly as the years went by." What should be emphasized here is that Einstein's Jewish identity was the result of a conscious process, of a mature person between the ages of thirty-five and forty, based on observation and rational analysis. I would like to point out some elements and consequences of this process.

Why Do They Hate Jews?

This process itself was triggered by the anti-Semitism that Einstein encountered for the first time when he returned to Germany in 1914. Anti-Semitism in those days had particularly affected Jews who came to Germany escaping anti-Semitic persecution in Eastern Europe. In Germany after World War I, they became the scapegoat for all the problems of the country. Einstein stood up in public, with courage and conviction, against this phenomenon. He wrote a lot about anti-Semitism, trying to understand its nature and motives, and looking for ways to cope with it. He made a distinction between anti-Semitism in Russia, where it was carried by the masses, and anti-Semitism in Germany, where he identified a culture of anti-Semitism that developed among the intellectual elites and was based on "rational" arguments. He was particularly concerned about how this hostile attitude of the Gentiles damaged the Jewish self-image and undermined the confidence of the best of his fellow Jews. He felt that the effect of anti-Semitism was greatest during the Emancipation period, when assimilated Western Jews lost their identity as Jews and lost the sense of belonging to a community. The question "Why do they hate Jews?" concerned him a lot. In 1938, he published an article under this title in *Collier's Magazine*, which summarized his thoughts.[5]

In 1920, Einstein was invited by the organization of German Citizens of Jewish Faith to a meeting to discuss ways to combat anti-Semitism. In his response, Einstein was critical of the presumption in that invitation, and sarcastically critical of the very name of that organization and its implied message. For him combating anti-Semitism was first of all developing the self-respect and courage to adopt the sense of national belonging. According to Einstein, anti-Semitism is a psychological phenomenon that will always be there "as long as Jews and non-Jews are thrown together," and maybe it is thanks to "anti-Semitism that we are able to preserve our existence as a race."[6] The term "German Citizens of Jewish Faith" generated in Einstein a melancholy smile. There was nothing about him that could be described as "Jewish Faith," although he was a Jew and glad to belong to the Jewish people. According to Einstein, claiming that one is a "German of Jewish Faith" meant that one belonged to a religion and not to a people.

This was his prescription for a remedy for the devastating effects of anti-Semitism: develop a sense of personhood independent of any religious contents.

How Did I Become a Zionist?

This understanding of anti-Semitism led Einstein to the conclusion that "Jewish nationalism is today a necessity" and in 1919 he started to openly support Zionism. He became involved in the Zionist movement and represented its positions with passion and conviction. This decision seemed to be in contradiction with his socioethical doctrines against chauvinism and the spell of nationalism. His first attempt to cope with this apparent contradiction was his article in the *Jüdische Rundschau* in 1921, "How I Became a Zionist."[7] In the following years, he defended his decision to identify with Jewish nationalism in numerous articles and in public addresses. To him Jewish nationalism was not only a necessity but also an undisputable fact to which every Jew should relate: "I am not a national Jew in the sense that I demand the preservation of Jewish or any other nationality as an end in itself. I look upon Jewish nationality as a fact, and I think that every Jew ought to come to definite conclusions on Jewish questions on the basis of this fact. . . . That was my main motive for joining the Zionist movement."[8] Yet, he always warned against narrow nationalism and emphasized that he referred to a "nationalism whose aim is not power but dignity and health" or that "my Zionism does not exclude cosmopolitan views."[9]

Einstein's open letter to Willy Hellpach, which I have already quoted, was his answer to an article published in 1929 in the *Vossische Zeitung* in which

Professor Hellpach alleged that Jews cannot be proper German citizens if they embrace the Zionist idea of Jewish nationalism. To this Einstein responded: "If we did not have to live among intolerant, narrow-minded and violent people, I would be the first to throw over all nationalism in favor of universal humanity."[10] The same arguments repeated themselves years later, in 1945 in the United States. This time the criticism of Zionism came from a Jewish organization, the American Council for Judaism, which claimed that Zionism enhanced anti-Semitism by casting doubt on the loyal citizenship of American Jews. Again, Einstein responded in public statements and in correspondence with individuals: "Antisemitism and Jew-baiting did flourish very well before Zionism was born. . . . The Gentiles had always found 'reason' to justify their behavior," and he added, "I dislike nationalism very much, even Jewish nationalism. But our national solidarity is forced upon us by the hostile attitude towards ourselves and not by the aggressive feelings which we connect with the word nationalism."[11]

DISCOVERING THE JEWISH PEOPLE

Einstein's discovery of being a Jew was followed by his discovery of the Jewish people. During his travels around the world, he met Jewish communities that were very different from the one he was familiar with in Berlin. These encounters played a role in shaping his Jewish identity.

In 1921, he joined Chaim Weizmann, President of the Zionist Organization, and a group of Zionist leaders on a six-week tour of Jewish communities in the United States. Years later he wrote about that trip:

> It was in America that I first discovered the Jewish people. I have seen any number of Jews, but the Jewish people I had never met either in Berlin or elsewhere in Germany. This Jewish people, which I found in America, came from Russia, Poland and Eastern Europe generally. . . . I found these people extraordinarily ready for self-sacrifice and practically creative. They have, for instance, managed in a short time to secure the future of the projected University in Jerusalem, at any rate so far as the Medical Faculty is concerned.[12]

Einstein's encounter with Sephardi Jews of Iraqi origin in Hong Kong led him to the following observation, expressed in his travel diary: "I am now convinced that the Jewish race has preserved its purity rather well in the last 1500 years, since the Jews originating from the Euphrates-Tigris countries resemble us very much. There is a very vivid feeling of shared belonging."[13]

The third such encounter with a Jewish community that is worth mentioning in this context of "discovering the Jewish people" occurred during his twelve-day visit in Palestine in 1923. Einstein could not identify with the kind of orthodoxy that he saw at the Wailing Wall. His reaction was: "black robed men, loudly praying, facing the wall and swaying their bodies back and forth, a pitiful sight of men with a past but no present."[14] He discovered a different world when he toured the country, visiting Tel-Aviv and Jewish settlements in the north. He then wrote in his travel diary, referring to his "tribal companions": "An incredibly active people, our Jews."[15] He was moved at a reception in a school: "I consider this the greatest day of my life. This is a great day. The day of liberation of the Jewish soul and it has been accomplished through the Zionist movement."[16]

A Cultural Center or a Political State?

Einstein's attitude toward Jewish nationalism was based on his perception of the legacy of Judaism. He did not identify with Judaism as a religion, but as a cultural heritage. According to Einstein, by divesting it from its religious attire, one is left with fundamental values, specifically with two characteristic features—the democratic ideal of social justice and the high respect for intellectual striving: "The pursuit of knowledge for its own sake, an almost fanatical love of justice and the desire for personal independence—these are the features of the Jewish tradition that make me thank the stars that I belong to it."[17] These are the features of Judaism that Einstein includes in his answer to the question "Just what is a Jew?"

The revival of Jewish nationhood in the land where this Jewish heritage first appeared on the historical arena was for Einstein a necessary framework for these values to come into play in modern time. This was the main role of Zionism. For the persecuted Jews of Eastern Europe, Zionism provided a national home. In Einstein's view this was an important, but not the most important, goal: "Palestine is not primarily a place of refuge for the Jews of Eastern Europe, but the reawakening corporate spirit of the whole Jewish nation."[18] Or, in other words, "Palestine will be a center of culture for all Jews, a refuge for the most grievously oppressed, a field of action for the best among us, a unifying ideal, and a means of attaining inward health for the Jews of the whole world."[19]

For many years, Einstein did not identify with the mainstream of political Zionism. He advocated the establishment in Palestine of a national home for the Jewish people in the form of a cultural-spiritual center, rather than a state with borders and an army. He thought that such an entity would not be compatible with the Jewish tradition. He believed that this goal could be achieved with the cooperation and agreement of the Arab population in Palestine, and he made personal efforts to explore the possibility of such an agreement. When Einstein testified in 1946 before the Anglo-American Committee, he still advocated the notion of a national home and expressed his views against a state with all the elements of political sovereignty. His views changed gradually in the late 1940s when he became disappointed with the attitude of the Arab League, and he welcomed the establishment of the State of Israel in 1948. And in 1948, when the infant state was still struggling for its existence, Einstein declared in a radio broadcast for the United Jewish Appeal, "It is inconceivable what consequences Israel and the entire Jewish people would face, should a lack of determination and commitment on our part bring about a collapse of the undertaking."[20]

In correspondence and arguments with leaders of the Zionist movement, and later with leaders of the State of Israel, Einstein frequently focused on attitudes toward the Arab population. In 1929, he wrote to Weizmann, "Should we be unable to find a way to honest cooperation and honest pacts with the Arabs, then we have learned absolutely nothing during our 2000 years of suffering and deserve all that will come upon us."[21] On an other occasion he said, "We, the Arabs and ourselves have to agree on the main outlines of an advantageous partnership, which shall satisfy the needs of both nations. A just solution to this problem is a goal no less important than the work of construction itself."[22] In his view the elimination of irrational psychological barriers between Jews and Arabs is an essential condition for Zionism to achieve its practical aim. Toward the end of his life, when Israel was already established and struggling for its existence, he wrote to Zvi Lurie, one of the heads of the Jewish Agency: "The most important aspect of our policy must be our ever present, manifest desire to institute complete equality for the Arab citizens living in our midst—the attitude we adopt toward the Arab minority will provide the real test of our moral standards as a people."[23] Despite his criticism, he never withdrew his support from the State of Israel or from the Zionist movement.

THE HEBREW UNIVERSITY

Einstein's most prominent involvement in the Zionist program was his commitment to the establishment of the Hebrew University in Jerusalem. He perceived the university as the essential arena in which the Jewish value of pursuing truth through a tradition of learning would be revived in modern times. His first major effort to promote the university project was the 1921 trip to the United States. To appreciate how important this was to him, I should mention that, to make this trip, Einstein gave up an invitation to participate at the Solvay Conference, which was about to take place in April 1921. The Solvay Conferences constituted the most prestigious stage on which the great ideas and advances in physics were presented and discussed. Einstein was the only German physicist invited to the first such gathering after World War I. Yet when Kurt Blumenfeld, a leading figure in the German Zionist

Organization, on behalf of Chaim Weizmann, invited Einstein to join the Zionist mission to the United States, he immediately agreed. In a letter to his friend, Dutch physicist Hendrik A. Lorentz, Einstein thanked him for arranging the invitation to participate at the Solvay Conference, and apologized for rejecting it because "the Zionists are planning to establish a university in Jerusalem and they believe that my trip to the US might influence rich Jews to contribute more generously. As much as it sounds strange they are probably right." He felt that it was his duty to help: "this initiative appeals to my heart and, as a Jew, I consider it my duty to contribute to its success as much as I can."[24] Lorentz responded by expressing his understanding for Einstein's decision and wishing him success.

Einstein's colleague and friend, the prominent German Jewish scientist Fritz Haber, was not so encouraging. He criticized Einstein's decision to go to the United States, invoking patriotic arguments in the name of his debt to German science and his loyalty to his German colleagues. He even used the argument that Einstein's decision might generate anti-Semitic sentiments against German Jews. Einstein was undeterred. He explained that his decision stemmed from his obligation "to help my morally oppressed ancestral companions as much as is in my power,"[25] and he categorically rejected any allegations of disloyalty to his German colleagues.

In 1922, on a stopover to Japan, Einstein raised funds for the Hebrew University in Singapore. Addressing the Jewish community there, he said: "One may ask—why do we need a Jewish University? Science is international but its success is based on institutions owned by nations. Up to now as individuals we have helped as much as possible in the interest of culture and it would be only fair to ourselves if we now, as a people, add to culture through the medium of our own institutions."[26]

Einstein inspired the entire Jewish world and many in the academic world with his vision of the Hebrew University. In the mission statement that he published at the inauguration of the university on April 1, 1925, he wrote: "A university is a place where the universality of the human spirit manifests itself," and he expressed the wish that "our University will develop speedily into a great spiritual center which will evoke the respect of cultural mankind the world over." In that document he once again expressed his attitude toward Jewish nationalism: "Jewish nationalism is today a necessity because only through a consolidation of our national life can we eliminate those conflicts from which the Jews suffer today." As he did so many times in the past, Einstein almost apologized for Jewish nationalism and expressed this hope: "May the time soon come when this nationalism will have become so thoroughly a matter of course that it will no longer be necessary for us to give it a special emphasis."[27]

In 1950, Albert Einstein wrote his last will and testament. In that document, he proclaimed "my final object is that any property (whether it consist of original manuscripts, or literary rights or property owned by my estate, or the proceeds from the disposition of any such property or rights) shall . . . pass to the HEBREW UNIVERSITY and become its property absolutely, to be thereafter retained or disposed of by it as it may deem to be in its best interests."[28] Einstein's personal papers, his literary estate, and his private library make up the Albert Einstein Archives at the Hebrew University, which constitute a cultural asset of supreme importance to humankind. Its holdings are unique; they consist of numerous manuscripts, prolific correspondence, and a large variety of additional material about Einstein. The material in the archives sheds light on the multifaceted aspects of Einstein's scientific work, his political activities, and his private life.

Concluding Remarks

Einstein's world view was based on universal social ethics and cosmopolitan views. Yet his Jewish identity and the role he played in the Jewish public arena reflect a great deal of particularism. The views he expressed in this context and the causes he supported show that he perceived Judaism and Zionism, the State of Israel and the Hebrew University, as frameworks and institutions in which universal values are embodied. This, according to Einstein, is their legacy, their justification, and their goal.

I would like to conclude by quoting Sir Isaiah Berlin's lecture at the Conference in Jerusalem in 1979 marking the centenary of Einstein's birth:

> Einstein, who tolerated no deviation from human decency, . . . believed in this movement [Zionism] and in this State [Israel] and stood by it to the end of his life, however critical he was at times of particular men or policies—this fact is perhaps among the highest testimonials on which any state and any movement in this century can pride itself. Unswerving support of an utterly good man . . . may not by itself be enough to justify a doctrine or policy, but neither can it be dismissed; it counts for something; in this case for a great deal.[29]

4

EINSTEIN AND GOD

Yehuda Elkana

THE RICHNESS OF EINSTEIN'S WORK and thought is such that literally almost all of it is relevant for our times.

The broad issue of Einstein and religion has been much discussed during the years, and recently beautifully treated by Max Jammer,[1] so I shall not touch on that topic at all. I want to concentrate on one small aspect of this huge oeuvre: Einstein's interest in thinking—in human thinking, in his own thinking, and God's thinking, in particular.

Like every human being, genius or not, Einstein is rich in paradoxes and contradictions. Most of them, in the specific context in which they occur, can be explained, or rather explained away, but the sheer number of them is impressive and thought provoking. One almost gets the impression when rereading his writings from this point of view that Einstein embraces contradictions. To put it differently: unlike most of modern science since at least Newton (probably not before, however), which is nondialectic and emphasizes a quest for black-and-white, yes-or-no answers to the scientific questions posed, Einstein is a thoroughly dialectical thinker. To make it plain what is meant by dialectical thinking: a readiness to admit that the way you formulate your question (which is the context) will influence what you consider the one unequivocal answer. As a self-exemplifying cue, Einstein himself fought a life-long battle, criticizing any statistical interpretation to be accepted as a final theory, and thus Quantum Mechanics, looking for strict determinism in Nature (nature with capital "N"), and for "Nature" very often simply reading "God."

Einstein had two major quests in life, though he often saw them as actually one: (1) to understand objective, real nature, which exists independent of human influence or knowledge, yet hides jealously its mysterious secrets; (2) to understand how people in general, how he himself, and above all, how God thinks. The two questions are linked by his commitment to the deep belief that mysterious, secretive Nature had been planned—thought out—by God, and thus in attempting to understand nature, we attempt to decipher God's thinking; moreover we (or at least he, Albert Einstein) can in principle enter God's mind and read his thoughts and plans, and discover what constraints apply even to God's thinking.

That is my thesis for this work. We are facing a wonderful psychological paradox: the almost unbearable tension between the deep humility Einstein, the scientist, feels toward Nature and the boundless hubris of presuming that he can enter God's mind and possibly answer his famous questions: Could God have created a different universe? Does God play dice? Is the dear God malicious or subtle? among many more.

Let us illustrate these ideas in Einstein's own words, and interpret each in its context.

(4.1) My religiosity consists of a humble admiration of the infinitely superior spirit that reveals itself in the little that we can comprehend of the knowable world. That deeply emotional conviction of the presence of a superior reasoning power, which is revealed in the incomprehensible universe, forms my idea of God.[2]

(4.2) . . . it suffices to stand in awe of the structure of the world, insofar as it allows our inadequate senses to appreciate it.[3]

(4.3) Nature conceals her secrets because she is sublime, not because she is a trickster.[4]

(4.4) The eternal mystery of the world is its comprehensibility. . . . The fact that it is comprehensible is a miracle.[5]

(4.5) One cannot help but be in awe when one contemplates the mysteries of eternity, of life, of the marvelous structure of reality. It is enough if one tries to comprehend only a little of this mystery every day.[6]

One could add many more in the same vein. It all may sound on first reading as clear expressions of deep humility toward nature and its mysteries. On further analysis the problems arise: if we *can* comprehend only parts of the "knowable world," in what sense is it "knowable"? And by whom?

The universe is incomprehensible to us, and thus it is a tenet of our faith that it was created by a superior reasoning power, namely God. Yet, its mystery is its comprehensibility. But if our "inadequate senses" can only partially comprehend it, how is it comprehensible? And again, by whom? By the superior power that must have created it on rational lines? But if it is incomprehensible to us, the mystery of its comprehensibility—that is, its rationality—is open only to God. Does that mean that rationality itself is not a characteristic of humans, but only of God? This is certainly not Spinoza's view, to which Einstein constantly refers.

In quotation 4.3, could we simply exchange "Nature" for "God"? My point is not to show pettily how illogical Einstein was, but rather to show how deeply involved he was in an emotional attitude toward Nature = God: standing in awe, admiring, admitting the mystery of its comprehensibility, knowing how small human powers are to comprehend it all, and yet working incessantly trying to do just that. This is neither less nor more rational than a love affair. Indeed, Einstein uses the same comparison when speaking of Faraday:

(4.6) This man loved mysterious Nature as a lover loves his distant beloved.[7]

One might look for an explanation like the claim that with age, and accumulating frustration of not being able to formulate a satisfactory unified field theory, the belief in the mystery and incomprehensibility of nature prevails, whereas in earlier years, in the midst of unprecedented success, the other view had upper hand, for example:

(4.7) My comprehension of God comes from the deeply felt conviction of a superior intelligence that reveals itself in the knowable world. In common terms, one can describe it as "pantheistic" (Spinoza).[8]

Compare this quote with quotation 4.1! Sometimes these contradictory views appear in the same sentence, as in quotation 4.1. By 1952, we find one of the most simply humble statements ever:

(4.8) My feeling is religious insofar as I am imbued with the consciousness of the insufficiency of the human mind to understand more deeply the harmony of the universe which we try to formulate as "laws of nature."[9]

Let us now move to the other, opposite attitude: Einstein and God, more or less on equal footing, and Einstein aspires—and finds it in principle not impossible—to read God's thoughts, or rather to think himself into God's mode of thinking and planning in order to find out God's intentions.

Already around 1920, in a letter to Berlin student Esther Salaman, we find this:

(4.9) I want to know how God created this world. I am not interested in this or that phenomenon, in the spectrum of this or that element. I want to know his thoughts. The rest are details.[10]

Then again in 1929:

(4.10) I believe in Spinoza's God who reveals himself in the harmony [that is again comprehensibility on rational lines] of all that exists, but not in a God who concerns himself with the fate and actions of human beings.[11]

And once again, speaking with his assistant Banesh Hoffmann in the late 1930s:

> (4.11) When I am judging a theory, I ask myself whether, if I were God, I would have arranged the world in such a way.[12]

To Ernst Strauss, another assistant, in the late forties:

> (4.12) What really interests me is whether God could have created the world any differently; in other words, whether the demand for logical simplicity leaves any freedom at all.[13]

The whole series of quotations about God not playing dice came from Einstein's opposition to a statistical interpretation of quantum mechanics—that is, his opposition to the idea that the world might be comprehensible only in probabilistic and not fully deterministic terms—and it started in an early letter to Max Born in 1926:

> (4.13) Quantum mechanics is very worthy of regard. But an inner voice tells me that this is not yet the right track. The theory yields much, but it hardly brings us closer to the Old One's secrets. I, in any case, am convinced that He is not playing dice.[14]

It is the same idea that God is not malicious:

> (4.14) The Lord God is subtle, but malicious he is not. [*Raffiniert ist der Herrgott, aber boshaft ist er nicht.*][15]

In this source, there is a note scribbled in Einstein's handwriting in which God is replaced by Nature: "Nature conceals her secrets because she is sublime, not because she is a trickster."[16]

In 1942, to his student-colleague, Cornelius Lanczos, he still uses this metaphor:

> (4.15) It is hard to sneak a look at God's cards. But that he would choose to play dice with the world . . . is something I cannot believe for a single moment.[17]

There is an unchecked anecdote that—if true—is very important, because it shows that his peers took some offence at this attitude. Niels Bohr supposedly responded to one of Einstein's remarks with, "Stop telling God what to do."[18]

As a last example of Einstein's "conversations" with God, Einstein said to James Franck:

(4.16) I can, if worst comes to worst, still realize that God may have created a
world in which there are no natural laws. In short: chaos. But that
there should be statistical laws with definite solutions, i.e., laws that
compel God to throw dice in each individual case, I find highly dis-
agreeable.[19]

It is interesting to note that Einstein must have been very conscious of his
audience. It cannot be accidental that all his public statements emphasize the
line of humbly admiring nature and being struck by awe when realizing Nature's (that is, God's) rationality. All his remarks about thinking himself into
God's mind occur in private letters or were said in personal conversations with
his students. This is not meant to deny a strong public relations mentality on
his part, but rather a sensitivity of how people would interpret him if he exposed publicly his inner thoughts; at the same time, he considers these private
thoughts to be a major factor of his creativity and tries to impart them to his
students.

That Einstein was quite clearly aware of the importance of his contributions—in consultation with God or not—we can see very early. Once
again, since it borders on hubris, he notes only in a private communication:

(4.17) I cannot find the time to write because I am occupied with truly great
things. Day and night I rack my brain in an effort to penetrate more
deeply into the things that I gradually discovered in the past two years
and that represent an unprecedented advance in the fundamental
problems of physics.[20]

Nowadays much can be heard about the so-called anthropic principle. The
name was "suggested by Brandon Carter. One way to put [what it means] is
that the reason why the physical constants have the values they do is that
if they did not, we should not be here to measure them." This is a quotation
from the magisterial overview of physics and its history by Gerald Holton and
Stephen Brush, who further explain:

So far, theoretical physicists have not been able to explain why the dimensionless constants have the values they do. Perhaps they are random, not determined by any law of nature. But Wheeler and other cosmologists have shown
that if these constants were very much different from their actual values in our
universe, the formation of planetary systems and the evolution of higher forms
of life might be impossible.[21]

Not surprisingly this had occurred to Einstein, too, way before there was
experimental evidence or appropriate exact calculations that enabled Wheeler's
work. In a 1930 conversation with Rabindranath Tagore, he said:

(4.18)   There are two different conceptions about the nature of the universe;
(1) the world as a unity dependent on humanity; (2) the world as a
reality is independent of the human factor.[22]

Most physicists tend to disregard the anthropic principle, such as Holton
and Bush describe above, as a scientific statement. Interestingly, Steven
Weinberg—certainly not a mystic—relates to it thoughtfully: he considers it
a last resort but does not pooh-pooh it as nonsense:

> If this version of the anthropic principle were true, there would be a kind of cir-
> cularity built into nature and one would then I suppose have to say that there is
> no fundamental level—that the arrows of explanation go round in circles. I
> think most physicists would regard the anthropic principle as a disappointing
> last resort. We'll just have to see.[23]

Elsewhere in the same book, he is even more positive:

> We too may have to discover new sorts of hypotheses, which may at first seem
> to us as uncongenial as Einstein's symmetry principles seemed to Lorentz. . . .
> [T]he so-called anthropic principle states that the laws of nature must allow for
> the appearance of living beings capable of studying the laws of nature. This
> principle is certainly not widely accepted today, though it provides what is so far
> the only way we have of solving the problem of a large vacuum energy density.
> (Too large a vacuum energy density would, depending on its sign, either prevent
> galaxies from forming, or end the Big Bang too early for life to evolve).[24]

Einstein certainly considered his late (and as it turned out, unsuccessful)
efforts to find an all-embracing unified field theory to be in a way jointly undertaken by himself and God. At one point, when he briefly believed he had
succeeded, he said:

(4.19)   I believe that this is the God-given generalization of general relativity
theory. Unfortunately, the Devil comes into play, since one cannot
solve the [new] equations.[25]

In the same vein:

(4.20)   Had God been satisfied with inertial systems, he would not have created gravitation.[26]

And when the failures continued, and fewer and fewer of the serious physicists showed enthusiastic readiness to continue the search, Einstein somewhat
bitterly said:

(4.21)   The unified field theory has been put into retirement. It is so difficult to
employ mathematically that I have not been able to verify it somehow,

in spite of all my efforts. This state of affairs will no doubt last many more years, mostly because physicists have little understanding of logico-philosophical arguments. [27]

Einstein might be surprised at how much progress has been made in recent years toward a possible unified field theory, especially by the creators of string theory. Yet his remark about logico-philosophical arguments was deeply appropriate, because string theory is exactly that type of argument.[28]

And again, two years later:

(4.22) That no one can make a definite statement about [the unified field theory's] confirmation or non-confirmation results from the fact that there are no methods of affirming anything with respect to solutions that do not yield to the particularities of such a complicated nonlinear system of equations. It is even possible that no one will ever know.[29]

The picture that emerges is a complex, sometimes inconsistent, dialectical relationship between two attitudes standing in awe of the creation. Nature, or God, created a wonderful rational world, which sometimes looks comprehensible, and sometimes hides its secrets and leaves the scientist wondering with amazement at the incomprehensible. The belief in this divine order is cosmic religion. It is dialectical in the deep sense, when Einstein declares that:

(4.23) The eternal mystery of the world is its comprehensibility.[30]

Whether it is all comprehensible or not, in any case the world is rational. It is rational insofar as it is "comprehensible to reason":

(4.24) The supreme task of the physicist is to arrive at those universal elementary laws from which the cosmos can be built up by pure deduction. There is no logical path to these laws; only intuition, resting on sympathetic understanding of experience, can reach them.[31]

The world is rational, it can be built up by pure deduction from universal laws, which we can reach by intuitive leaps. These cannot be taught—there is no method for finding them. Rationality and deduction are formal, logical, and thus mechanically available intellectual tools. It is not about these that Einstein would like to think himself into God's mind; it must therefore be the divine thought that involves those great intuitive leaps, which then help us find and formulate the universal laws.

What does Einstein think about thinking? What is the type of thinking that yields intuitive leaps and allows one, or rather Einstein (and his peers, if such exist), to penetrate God's mind, and is at the same time rational like nature itself and capable (by the way) of unerring logical deductions? Is this type of thinking covered by any of our psychological theories? Is it one of Howard Gardner's "multiple intelligences"? Could it be an additional thema among Holton's "science-making [Maxwell's term] themata"? There is one important study on Einstein's thinking, by Gerald Holton,[32] which does not, however, refer specifically to Einstein's belief that he can think himself into God's mind. As far as I know, most other analyses of Einstein's thinking have been along psychoanalytic rather than cognitive lines, and among them, two are notable: Eric Erikson's study on psychoanalytic aspects, and Roman Jakobson's, which correlates the psychology of thinking with linguistics in a very interesting manner.[33]

But let us look again to Einstein's own words:

(4.25) I very rarely think in words at all. A thought comes and I may try to express it in words afterwards.[34]

What does that mean for our query? Does it mean that when Einstein investigates how God would have made decisions—or "think" about what kind of world to create—that God, too, would have projected nonverbal images, pictures, appearances . . . of possible worlds? Intriguing idea—we shall never know.

In his autobiographical notes,[35] Einstein talks about the development of his "thought-world"—*Gedankenwelt* (translated as "mental world")—and he revisits the idea briefly elsewhere:

(4.26) The development of that mental world is a continual flight from "wonder."[36]

He repeatedly made the point that the seeker for knowledge "wonders" as long as he or she admires the unknown. With achieved knowledge the wonder diminishes and slowly disappears. Interesting to recall that the etymology of "to theorize" is "to wonder."

Indeed, a few lines before the above quoted sentence, we find:

(4.27) I have no doubt that our thinking goes on for the most part without the use of signs (words), and, furthermore largely unconsciously. For how, otherwise, should it happen that sometimes we "wonder" quite spontaneously about some experience? This "wondering" appears to occur when an experience comes into conflict with a world of concepts that is already sufficiently fixed within us.[37]

Applying this to Einstein's own work we should understand that he had "experienced" time and space, or light, in a way that clashed with much of the classical physics he knew at the time, and he started to wonder—not in signs, but by some other means, and probably unconsciously. Perhaps his—or our—luck was that when this process started, the accepted concepts were *not* so deeply fixed within his mind. If he now tried to understand how God thought about the world, he surely would have deduced that the conventional concepts—i.e., the then-accepted physics—were not those in terms of which God had thought, and indeed created, the world. This is surely not even close to what great though conventional physicists meant by "god-like" knowledge. Max Planck famously referred once to "god-like" knowledge as knowing all that is knowable, in a deterministic world.

Klaus Meyer-Abich recounts that, in a discussion of the Bohr-Heisenberg interpretation of quantum mechanics in the late 1930s, Planck rejoined to Bohr, "But you must admit, Bohr, that a god-like eye could know both the particle's position and momentum." To which Bohr replied, "I do not think that this is a question of what a god-like eye can know, but of what you mean by 'knowing.'" And indeed, Meyer-Abich elucidates that "what a god-like eye would observe the world to be is a fairly good description of the kind of knowledge sought by a classical physicist."[38]

Mathematical thinking is clear: it is deduction of phenomena from principles by mathematical tools. Einstein refers admiringly to Newton when saying this:

> (4.28)   Newton was the first to succeed in finding a clearly formulated basis from which he could deduce a wide field of phenomena by means of mathematical thinking—logically, quantitatively, and in harmony with experience.[39]

. . . which is what is meant by the notoriously famous saying:

> (4.29)   All of science is nothing more than the refinement of everyday thinking.[40]

Whose everyday thinking? Einstein's, or of any serious scientist, or any lay person? And what kind of thinking is it? Is it different from the mathematical thinking of Newton, who made deductions from principles to everyday phenomena? Is it the rational procedure by which we tend to develop conventional science, since nature is rational and logic is universal and immutable? Or, perhaps, is it the uniquely Einsteinian everyday thinking, which leads from experiencing the world toward new concepts and theories by intuitive leaps? Statements like that can be attributed either to the humble, nature-admiring public Einstein or to the Einstein in "dialogue" with God about inventing/creating the rational world. In the context of the 1936 paper, we read that in these times:

(4.30)   the very foundations of physics itself have become problematic . . .
         when experience forces us to seek a newer and more solid founda-
         tion. . . . In looking for a new foundation, he [the physicist] must try
         to make clear in his own mind just how far the concepts which he uses
         are justified, and are necessities.[41]

But as we learned elsewhere and repeatedly, precisely this process of clarifying the mind to check whether the old concepts are satisfactory in order to
draw from experience the appropriate theories can be achieved only by intuitive leaps. There is no royal road, no method. And the next paragraph starts
with "All of science is nothing more but refinement of everyday thinking."
Thus we are forced to draw the conclusion that this refinement is the pecu-
liarly God-guided use of free imagination. The task, then, of the ever-reflective
Einstein—or, as he claims, of every physicist—is to consider critically "the
problem of analyzing the nature of everyday thinking."[42] This is made even
more problematic if we continue to read carefully and analyze what is said
further in this very seminal writing.

In an aside, in parentheses, we read:

(4.31)   [thinking] is . . . operations with concepts, and the creation and use
         of definite functional relations between them, and the coordination of
         sense experiences to these concepts.[43]

Thinking can put in order our sense experiences by means of concepts, and
this leaves us in awe: "The eternal mystery of the world is its comprehensibility."

These concepts—ordering tools—are reached only by intuitive leaps:

(4.32)   In my opinion nothing can be said concerning the manner in which
         the concepts are to be made and connected and how we are to coordi-
         nate them to the experiences.[44]

And, somewhat confusingly:

(4.33)   The connection of the elementary concepts of everyday thinking with
         complexes of sense experiences can only be comprehended intuitively
         and it is unadaptable to scientifically logical fixation.[45]

Where does it leave us? I am incapable of dealing with this complex of
ideas without resorting here to an old work-horse of analysis; first-order and
second-order concepts. First-order concepts are part of everyday thinking, but
even these we cannot find by any automatic method; these emerge when experiencing the world, representing them to ourselves with recurring pictures,
and juxtaposing the sense experiences to conventionally accepted concepts of

the accepted science. Then we have formed our concepts of everyday thinking. Next we coordinate these newly found first-order concepts with complexes of sense experiences. This connection can be comprehended only intuitively, by leaps of imagination, through which our new insights—our second-order concepts—are created. This process is an uncontrollable god-like process. Here are the real problems that Einstein confronts in science, and here he consults God by trying to understand, to think himself into God's mind:

> (4.34) The totality of these connections—none of which is expressible in notional terms—is the only thing that differentiates the great building which is science from a logical but empty scheme of concepts, presumably those first-order concepts which are part of everyday thinking.[46]

Alternatively, we drop the distinction between two types of concepts, leave the concept of "concept" for everyday thinking, and draw the conclusion that when Einstein consults God to find out how God thinks, it is a "dialogue" conducted not in any "notional terms" but probably in pictures and other nonverbal signs. In other words, Einstein and God think alike in nonverbal terms.

What follows now is a clear analysis of what ordinary science is about. It is a first stage of the contemplation of nature, in terms of what were so far called concepts of everyday thinking and now are called primary concepts, the ones that are intuitively connected with typical complexes of sense experiences:

> (4.35) Science concerns the totality of the primary concepts, i.e., concepts directly connected with sense experiences. And theorems connecting them. In its first stage of development, science does not contain anything else. Our everyday thinking is satisfied on the whole with this level. Such a state of affairs cannot, however satisfy a spirit which is really scientifically minded.[47]

Now the struggle continues: after having created a "first layer" of the system, we now create a second layer of greater logical unity by having its own elementary concepts (concepts of the second order),

> (4.36) only those which are no longer directly connected with complexes of sense experiences. . . . Thus the story goes on until we have arrived at a system of the greatest conceivable unity, and of the greatest poverty of concepts of the logical foundations, which are still compatible with the observation made by our senses. We do not know whether or not this ambition will ever result in a definite system.[48]

One must wonder if this extremely convoluted, not really consistent, attempt at explaining his own thinking in a public document would not have become much, much simpler if Einstein allowed himself to state publicly that, at some point, after he accounted for all the complexes of direct sense experiences and tried to bring them in harmony with conventional knowledge, and failed, then he turned to God and asked how God would have thought about the problem that emerged. If this consultation is fruitful, Einstein reaps the leaps of imagination by which, out of the blue, he connects the complex of sense experiences in a completely novel and unexpected way.

One last point to make: Hubris or not, Einstein considered himself part of a small, select company of laborers in the building of the genuine cathedral of science. This finds its expression in the celebrated address delivered in Berlin in 1918 on the occasion of Max Planck's sixtieth birthday. This time I have to quote at somewhat greater length:

> (4.37) In the temple of science are many mansions, and various indeed are they that dwell therein and the motives that led them thither. Many take to science out of a joyful sense of superior intellectual power; science is their own special sport to which they look for vivid experience and the satisfaction of ambition; many others are to be found in the temple who have offered the products of their brains on this altar for purely utilitarian purposes. Were an angel of the lord to come and drive all the people belonging to these two categories out of the temple, the assemblage would be seriously depleted, but there would still be some men, of both present and past times, left inside. . . . I am quite aware that we have just now light-heartedly expelled in imagination many excellent men who are largely, perhaps chiefly responsible for the building of the temple of science; and in many cases our angel would find it a pretty ticklish job to decide. But of one thing I am sure: if the types we have just expelled were the only types there were, the temple would never have come to be, any more than a forest can grow which consists of nothing but creepers. For these people any sphere of human activity will do, if it comes to a point; whether they become engineers, officers, tradesmen, or scientists depends on circumstances.[49]

There follows a lengthy, poetic description of those in the third category—those motivated by the wish "to escape from everyday life with its painful crudity and hopeless dreariness, from the fetters of one's own desires. A finely tempered nature longs to escape from personal life into the world of objective perception and thought. . . ."[50]

One wonders whether a person whose strongest wish is to escape the "merely personal" is indeed a "finely tempered nature." Is the association Bach's *Wohltemperiertes Klavier*? On the positive side, as Einstein puts it, is the much more important general trait of men that they try to make sense of the world:

> (4.38) Man tries to make for himself in the fashion that suits him best a simplified and intelligible picture of the world; he then tries to some extent to substitute this cosmos of his for the world of experience, and thus to overcome it. This is what the painter, the poet, the speculative philosopher, and the natural scientist do, each in his own fashion. Each makes his cosmos and its construction the pivot of his emotional life.[51]

Clearly we are now describing a group of people, among them Einstein. Actually, what Einstein here says about such men is verbatim what he repeatedly said about himself. And this Einstein is seemingly part of a select company of present and past seekers—perhaps prophets—without whom the temple of science would not have been built. And all of these, each in his own way—again like the prophets—communicate with God in order to fathom God's thoughts.

Let me conclude with one last quotation, which will bring us back from the unfathomable sublime black box to one of Einstein's perhaps most important insights—very much down to earth, very appropriate, and indeed of "science-making" quality. Alas, though often quoted, its origination with Einstein cannot be confirmed. Alice Calaprice puts it in the category of "attributed to Einstein":

> (4.39) We cannot solve the problems we have created in our world by thinking the way we thought when we created them.[52]

5

EINSTEIN'S UNINTENDED LEGACY: THE CRITIQUE OF COMMON-SENSE REALISM AND POST-MODERN POLITICS

Yaron Ezrahi

THE HISTORY OF SCHOLARLY or scientific critique of common beliefs is of course as long as the history of science. Molière captured an important aspect of this critique when a doctor in one of his plays says to a patient, who surprisingly recovered following an unorthodox treatment: "Sir, it is better to die according to the rules than to live in contradiction to the faculty of medicine." Still, of course, the Enlightenment's ideal of closing the gap between science and popular beliefs has probably been one of the most resilient themes in our culture.

An example can be found in the December 2004 issue of the *New York Review of Books*. In his review of the New York Library's exhibition *The Newtonian Moment*, the Princeton historian Anthony Grafton sees a special link between the American democratic enterprise and Newton, who

> redefined the study of nature by insisting that it must rest not on hypotheses but on evidence. . . . We Americans trace our origins, spiritual and intellectual legacy to the heralds of the Newtonian movement: writers and doers like Benjamin Franklin. The creators of the United States couched their arguments for its independence and their visions of its constitution in the Newtonian language of reason, nature's laws and factual evidence.

Grafton sees in our time a dangerous erosion of this legacy. He refers in particular to a (now famous) statement made by a top aide to President George W. Bush in which he rejected what he called the "reality-based community," referring to people who believe in the relevance of factual analysis to policy making. "That's not the way the world really works anymore," he asserted. "We are an empire now, and when we act, we create our own reality." In reminding the public, Grafton writes, that the men who made America belonged to the "reality-based community" of their day, the New York Public Library and Professor Feingold, the curator of the exhibition, offer the best kind of civic education.

I would like to reflect on the almost rhetorical question, why Newton and not Einstein? Why one would regard Newton's legacy so important to contemporary democratic civic education, and not Einstein's? How come in 2005 we would link democracy and civic education to a physicist who lived about three hundred years ago, with Newton as an icon of the respect for reality and factual evidence, and not with Einstein and 20th- or 21st-century physics? Surely contemporary physicists cannot be regarded as lacking respect for "factual evidence."

At least part of the answer to this question may help discern why the relevance of science to the culture of democratic politics has eroded in our time, why it is the Newtonian moment that is cherished by those committed to the democratic form of government, and not the much more recent and no less revolutionary Einsteinian moment. Whereas Newton's physics and its mathematical foundations remained elusive to laypeople during his lifetime and after, the early democratic political imagination nevertheless found in Newtonian science an invaluable resource for democratic political worldmaking.

This relates to the role of modern science in the modernization of common sense and in the rise of common-sense realism as the epistemological frame of modern democratic politics. Historically, the 18th-century Enlightenment had associated the rise of modern science with the rise of modern democratic ideology and politics. As Louis Dumont argued, perhaps the most important political implication of the rise of modern individualism was antihierarchical sensibility, which led to the idea that the only legitimate political order is that which is built from the bottom up rather than from the top down.[1]

This revolutionary reversal of political and legal theory created a powerful demand for cultural forms that would enable equal and discrete individuals to reach agreements and cooperate in exposing the "lies" upon which the Old Regime was founded, and then build an alternative and more legitimate order. It is not surprising that democratic thinkers such as Locke and Montesquieu and later figures like Condorcet, Thomas Paine, John Stuart Mill, and John Dewey regarded Newtonian empiricism as a valuable resource of the emerging democratic culture. Science, or rather scientism since the 18th century, carried the great promise of serving the elimination of the errors and prejudices that were regarded as the source of disputes and violent conflicts. As Bishop Sprat already anticipated in 1667, the new science has exemplified a successful novel participatory epistemology that can underlie the novel participatory politics of democracy. Very characteristically, Thomas Jefferson was convinced that the assertion that the king of England is a despotic ruler is something that can be proved if the relevant "facts be submitted to a candid world."[2] The principal spokesmen of the democratic revolutions in America and France relied on the idea that the book of nature is a script full of clues and instructions for the architecture of the modern democratic state. Once nature was established as a source of social and political norms, the study of nature and, in particular, any claims about the universal accessibility or inaccessibility of nature to common-sense knowledge became politically loaded.

Against this background, it is not surprising that the doubts raised about the reliability of common-sense realism by the very 20th-century physicist who revised Newton could have provoked anxious interest in wide circles; that less then two hundred years after the American Declaration of Independence linked the laws of nature with self-evident truths, Einstein's reference to the "plebian illusion of naïve realism, according to which things 'are' as they are perceived by us through our senses,"[3] would appear to problematize a constitutive premise of democratic political culture. What was at stake in that discussion was not the epistemological or the scientific, but the political fact that popular realism is a condition for the power of citizens to claim the knowledge and authority without which they cannot assess and criticize their government nor make sense of their decisions on election day. Traditionally, science appeared to guarantee the possibility and the value of the distinction between facts and fictions just as the gold standard guaranteed the value of paper currency.

W.V.O. Quine argued in his famous *Two Dogmas of Empiricism* that, like the Homeric view of the world, the scientific view of the world as a physical object is but a myth. Still, he goes on to observe that "the myth of physical objects is epistemologically superior to most in that it has proved more efficacious than other myths as a device for working a manageable structure into the flux of experience."[4] I doubt that Quine considered in this connection the efficacy of the myth of the world both as a publicly accessible physical object and as a resource for democratic political worldmaking. But this was clearly an important reason for why the Newtonian moment and the democratic moment have formed such a symbiosis since the 18th century. The history of modern social science and particularly of political science indicates the extent to which the example of Newtonian physics, especially in its popular versions, has been transferred to the spheres of society and politics, suggestively implying that society and politics can be described and explained in terms of observable facts and objective events. This attitude was closely linked with the development of a popular democratic political imagination that envisioned politics and government as potentially transparent and therefore accountable. The norms and rules of empiricism deployed in society and politics were supposed to correct false ideas and establish common sense on more solid grounds. The commitment, embedded in the emerging democratic order, to common-sense realism, to the capacity of the public to distinguish facts from fictions, constituted the epistemological rationale for empowering democratic citizens and the decentralization of political power. It forced democratic governments to submit themselves to continual tests of transparency.

The scientific approach, as advocates of democratic politics like Jefferson, Paine, and Condorcet understood it, had a crucial role in rationalizing and legitimating common-sense realism as the foundation of such important democratic aspirations as achieving consensus among a multitude of discrete individuals and enabling a government by elections. The notion that politics can be simplified by reducing it to "simple facts and plain arguments" and that the authority of science backs it up was a central premise of the new democratic culture and particularly of the trust in the capacity of discrete individuals to evolve a political order from the bottom up without recourse to hierarchical authorities. In late 19th-century America, for instance, Henry Rowland, the head of Johns Hopkins research laboratory, insisted that students of experimental physics will become better democratic citizens because they "stand face to face with nature; they must learn to test their knowledge constantly and thus see for themselves the sad results of vague speculation."[5] Regarding scientific education as an integral part of the democratic education of the citizens, a way to secure their ability to be free, turned out to be a powerful public justification for the obligation of the state to secure universal education and support the universities.

In this political context, empiricism and induction have acquired the status of moral-political principles, which legitimate the results of their application even when they do not scrupulously adhere to intellectual standards. The moral, political, and even legal authority of judgments based on public factual evidence and on both direct and virtual witnessing[6] has supported the very rise of an independent democratic press committed to "separating facts from opinions" and capable of exposing arbitrary uses of governmental powers. Hence, especially against the background of Fascism, any move to detach the connection of the elementary concepts of everyday thinking and sense experience from scientific standards of validity was bound to appear as freeing public opinion from necessary intellectual constraints and criticism. But this is precisely what has occurred in the course of the 20th century, as public opinion has emancipated itself from the original grounds of its democratic authority and repeatedly defied and defeated public policy choices that respect scientific evidence and analysis. This development has been demonstrated most clearly in cases where lay opinion prevailed even in the face of contradictory scientific opinion. The persistence of the belief in the existence of unidentified flying objects despite the Condon report of the National Academy of Science, or the decision of an American jury not to convict O. J. Simpson despite compelling scientific evidence can illustrate the point.[7]

The decline in the authority of science in public affairs has apparently been more the consequence of changes in the structure of democratic culture than of internal scientific revisions of established theories and methodologies. But the influence of the latter has not been negligible. If scientists like Joseph Priestley and Michael Faraday could convince their lay audience that science evolves in the world of the common experience, thus discarding the invisible causalities invoked by alchemists, magicians, and other esoteric groups, contemporary scientists can no longer argue with the same facility that current scientific claims can be visibly demonstrable to the lay public. Now not only magic and alchemy but even common-sense realism (which largely replaces them), the very epistemological sphere of democratic politics, is characteristically regarded by scientists as but "a collection of prejudices," in Einstein's words. The sciences have long departed from the common visual and experiential domain they shared with laymen, which lent them authority and presence as critics. Gone are the days when the scientists of the Academia del Cimento recruited folks from the street to check the accuracy of their own observations, or when The Royal Society of London had lay corresponding fellows across the world sending to London "samples" of nature accompanied by their observations.

The pace of the progress of science has been much faster than the pace of the transformation in the premises and conventions of democratic politics. The gap between common-sense realism and scientific notions of reality was bound to grow as 18th-century concepts and images of nature, science, and evidence were frozen into the foundations of modern democratic conceptions of politics. Scientists who are still ignoring this gap and the vital role of the simplifications of common sense in sustaining faith in the possibility, although not always the practice, of participatory democratic politics have come to exemplify what I would like to call *the fallacy of misplaced rationality*. They are missing the role of scientifically baseless common-sense realism as the core of the *moral epistemology of democratic politics* and its normative working frame; they overlook the latent functions of "naïve realism," of what Einstein denigrated as the "plebeian illusion," as a means to justify the decentralization of power, checking its arbitrary uses and holding governments, at least ritualistically, accountable to the public. The tragic paradox of postmodern politics is that public opinion has been able to fly high on the moral and political fuel of the Enlightenment concept of the informed public while discarding its content; that it could keep its shell while denying its substance.

In some ways science never had a chance. This development was inevitable. Thomas Hobbes already recognized the idea that a communal notion of reality rests primarily on the reciprocal moral commitments of lay witnesses rather than on cognitive standards. In *Leviathan*, he observes that when two or more men know of one and the same fact, they are said to be conscious of it one to another; which is as much as to know it together. And because such are the fittest witnesses of the facts of one another, or of a third; it was, and ever will be reputed a very Evill [evil] act, for any man to speak against his Conscience; or to corrupt or force another so to do.[8]

Clearly the moral cement of knowing together was no less and probably more important to him than the truth of what is known.

Albert Einstein was actually aware of the social interactive process through which humans evolve their concept of the objective world, but he seemed to disregard its normative status and its crucial regulatory, albeit latent, functions in the political context. Considering the social perceptual process by which subjective sense experiences of something like lightening is transformed from a personal experience into an objective event, Einstein observed that such a transformation occurs when a person who has the experience "it is lightening," which appears first as a personal experience, is also experiencing other people behaving in a way that suggests to him they, too, have the same experience "it is lightening." In other words, Einstein attributes to what appears to people as behavioral signals of having *socially* simultaneous individual experiences of the world, the power to transform a personal event into what is imagined as an "objective event."[9] But whereas his argument is made in terms of the social psychology of perception, it is perfectly compatible with Hobbes' moral account of the implicit commitment of individuals who experience an event together, not to deny the fact of what they have witnessed together. Einstein apparently did not think that, in the sociopolitical context, the operation of such common-sense realism can or should be normatively autonomous in relation to science. He insisted on the role of science in the "refinement of [and] in considering critically . . . the problem of every day thinking."[10] He was aware of the gaps between common-sense notions of the world and scientific ones, but he seemed more concerned about protecting science from the prejudices of common sense than protecting the stability of common-sense realism from the impact of the revolutionary shifts in science. As we understand it better today, this "assortment of prejudices" called common sense is a cultural system that constitutes a vital organ of democratic institutions like the press, the parliament, and public discourse.

Of course, the sense of sharing a physical event like lightening may not be morally significant. But reciprocal individual confirmations of experiencing together, of witnessing, such social or political events as massacres, executions, coerced displacements and elections are, of course, morally and politically loaded. Obviously, in such cases, denials, too, which would disrupt the sociopsychological process of objectifying such events, would violate the Hobbesian imperative and constitute acts against one's own conscience. Further, one can infer from this position the moral and political significance of historical memory, the readiness to acknowledge the occurrence of past events as a basis for such acts as peace and reconciliation.

A democracy is not founded merely on a set of commitments to principles like freedom and equality, but also on a complementary set of explicit and tacit commitments to the procedures by which decisions about facts and values are collectively legitimated. In politics, there is a strong relation between the moral and the causal orders. They are actually two sides of the same ideological coin. In politics as in law, what constitutes evidence is therefore a matter of both fact and policy. In both politics and law, fictions are no less useful and necessary in organizing and regulating social life and political discourse than theoretical entities are in the generation of scientific theories and the organization of scientific discourse. In all such cases, their ontological status is either beside the point or secondary to other tests of adequacy.

The idea that objective reality is inscribed on the sensually perceivable surface of our experience is such a necessary democratic political fiction. This common-sense reality plays a crucial role in linking the moral-political and the causal orders, such that causality and responsibility can be publicly related. Hence, any disruption of the credibility of common-sense notions of causality, and especially of the authority of the procedures that define what is an "objective" fact or event, is bound to have political and metapolitical repercussions. In politics, common-sense reality as a simultaneous intersubjective experience does not merely fix the terms of public discourse. It is not only the very foundation of democratic journalism as well as, at least partly, its very product; it is also constitutive of the public itself as a democratic political agency. Moreover, in the age of mass communication, not only eyewitness visualization, but also the mere potential of simultaneous eyewitness visualizations, achieved by sheer acts of publicity, legitimates social and political claims. John Hartley had made the apt observation that journalism does not only share with Newtonian science the premise that the world is knowable, that observation is a means to the advancement of knowledge of nature, and that scientific standards can and should be extended to include also the knowledge and understanding of society. Journalism has added to this its commitment "to generalizing (i.e., making generally available to the general public) these militant scientific realisms (laws of order) through the forms of its own methodology of eyewitness visualization; its credibility depends not only on veracity but also on popularity"—that is, on ubiqity.[11]

But the record of the interrelation between contemporary mass communications and politics reveals that the journalistic commitment to ubiquity, to diffusion, can eclipse its commitment to accurate representations of the social and the political world, that in our culture repeated publicity in the mass media can lend a report, at least for a short time, the status of a trustworthy mirror of external objects and events without the backing up of certified facts and analysis. Politicians and advertisers have been very effective in exploiting this fact to influence public perceptions. This means, of course, that the sociopsychological mechanism, which Einstein discerned as underlying people's notions of objective physical events, also operates in full force in the political sphere. As was illustrated during the presidency of George W. Bush, this can lead even a relatively enlightened public opinion to support a dubious war. This does not mean that scientists should relinquish their responsibilities to warn and enlighten the public; only that scientists are now radically disempowered in alerting the public to the gaps between what is ordinarily experienced and what is happening.

By and large the public as an aggregate of discrete individuals is not aware that viewing the same live television broadcast of an event, the same news report or documentary, is always viewing through the particular angle of a specific camera, situated in only one of many possible spots by a particular camera person with a specific perspective. The technology of visual communication facilitates the development and sustenance of the illusion that discrete individuals are seeing the same objects or events because of these objects' or events' objective existence. They are therefore oblivious to the fact that the sense of sameness, of the objectivity of the things seen, is not the result of the qualities of specific objects or scenes or of their ontological status, which allow such an aggregate of separate viewers to experience the same "reality," but as W.J.T. Mitchell has repeatedly observed, the result of the technological-cultural capability of the camera as a mechanism to render a vision from only one perspective and appear as a neutral or natural intersubjective vision of objective things and events.[12]

This kind of artificial, technologically made-up "seeing the same thing" cannot claim to carry the weight of sameness as a result of the convergence of a multiplicity of natural experiences by a multitude of observers situated in different places. But its power derives from successfully mimicking such a convergence. The extent of the illusion created by the tendency to naturalize what is actually a particular artificially mediated vision from a singular and often-biased perspective, is well illustrated by the case of what Mitchell calls the "televisual spectacle" produced by the Rodney King beating. Mitchell observes, "the first attempt to prosecute the police officers who beat Rodney King was hampered by its naïve trust and reliance on the transparency of the video, its 'self evident' authority. The second prosecution succeeded largely because the prosecutors realized the need to supplement the videotape with

King's own testimony."[13] In a court of law, it was clear that what was taken as an objective natural seeing of the event by means of a video was no more than one particular out of several possible and not necessarily compatible representations.

We can conclude, perhaps unfortunately, that the faith of thinkers like Leibnitz, Locke, and Condorcet that public knowledge based upon, or corrected and certified by, science could induce consensus and diminish civil conflicts was never warrantable. Like most other Enlightenment thinkers and ideologues, with the obvious exception of Vico and Rousseau, they considered but did not assign enough weight to the powerful influence of emotions, interactive sociopsychological perceptions, developments in communications technology, and collective imaginaries on the cognitive maps of modern individuals as well as on political behavior and institutions. The Enlightenment was excessively optimistic in using Newtonian physics and the simplified understanding of how it works as a model for how citizens can and should relate to sociopolitical realities.

Twenty-five years ago, in my paper in the centennial Einstein symposium in Jerusalem entitled *Einstein and the Light of Reason*, I held that the gap between Einstein's physics and common-sense realism shook lay confidence in a world of objective public facts and objective events guarded by scientific reason. Now I have come to believe that the debates about reality and objectivity provoked by relativity theory and quantum mechanics, and later by the much-maligned works of historians and sociologists of science often classified as postmodern, were only the smallest among the forces that have been influencing the growing public distrust in Enlightenment notions of realism and public facts, and in their capacity to expose and constrain arbitrary exercise of political power. I would like to suggest that the record of the last twenty-five years since the Einstein centennial highlights the fact that the contemporary decline of this faith, as well as the erosion of common-sense realism in contemporary culture, have more to do with the impact of the omnipresence of the camera in our society and politics, and with the rise of what can be called radical postmodern normative perspectivism and the culture of subjectivism, than with the impact of the emancipation of scientific notions of reality from what Van Fraassen has called the "manifest image of the world."[14] The shift of contemporary physics away from the domain of the ordinary lay sense experience of material reality has been important but not crucial for these developments.

In many respects, those scientists who wage war against the supposed slayers of scientific reason, objectivity, and empiricism among the historians and sociologists of the sciences have been barking up the wrong tree. What has made the "reality-based community," about whose marginalization Professor

Grafton appears to worry, so vulnerable? The very erosion in the common-sense conventional dichotomy between fact and fiction has apparently been induced by the uncanny effects of the photographic or the iconic turn in our culture, the deleterious effects of the photographic mediation of politics on public trust in the reliability of long-standing explicit and implicit procedures for certifying and objectifying political acts and events. The camera, as an instrument that combines the documentary and the fantastic while blurring their boundaries, has radically transformed political phenomenology in modern teledemocracies.[15] As Rafael Sanchez writes, "Television now contaminates everyday appearances with undecidability."[16]

Given this development, even Newtonian science, as a part of democratic civic education, may be helpless. John Hartley has made the apt observation that "attempts to see contemporary photographic and 'taking' [of] pictures in the light of traditional notions of public and private, reality and illusion (truth and seduction), are doomed to failure. Pictures no longer mean what such binarism says."[17] In the modern electronic media environment, the concepts of "the public" and "public space" have undergone changes that require a new post-Newtonian and post-Jeffersonian vocabulary for talking about citizenship, political participation, transparency and accountability, checking arbitrary governmental power, and the role of science in civic affairs. In the final analysis, the public is no more than an abstraction, or a collective imagining supported by methodologically problematic polls, often-controversial periodic election results, and tendentious projections by the mass media. Contemporary publics have become increasingly unsure of what constitutes their own reality, their own accurate self-representation, and what constitute trustworthy pictures of the world. Hence much of what is going on depends on rituals of legitimation and the suspension of disbelief upon which they rest.

So if even the Newtonian moment has now questionable relevance for how democratic politics can handle the relations among freedom, reality, and power, what about the future relevance of the Einsteinian moment? Does the decline of the Enlightenment open the possibility of a new relationship between physics and politics? I promise to give my considered answer to this question in the 150th anniversary of Einstein's famous papers in the conference scheduled for 2055. The outline of that presentation may very well include the following three points:

1. Einstein as an icon of the powers of a single individual mind to revolutionize the understanding of the cosmos is one of the most precious enduring expressions of the value of free individual thinking and of its capacity to resist group-mind and coerced opinion. As such the Einsteinian movement can only add strength to the Newtonian movement. Both

Newton and Einstein symbolize the power of the individual to reach truths, and therefore the possibility that the majority may be wrong and that individual freedom is an irreplaceable social asset.

2. Einsteinian and post-Einsteinian physics can generate politically significant metaphors that could orient democratic citizens to appreciate the notion that, far from sanctioning relativism, contemporary science endorses the multiplication of objectivities and allows for acknowledged ambiguities to replace dogmatic notions of external objective nature by more elastic interactive concepts that facilitate a more "negotiable" concept of nature.[18]

3. Einsteinian and post-Einsteinian physics, while exemplifying the importance of bold abstractions in modern science, do not entirely discard the role of observation and visualization in science. On the contrary, they can provide the cultural support for a much more sophisticated public visual culture. Among other things, such an advanced visual culture may encourage a more skeptical public to be alert to the delusions inherent in naturalizing the machine-produced television pictures of events, objects, people, and delusions, which have created a dangerous gap between what democratic citizens think they see and what is actually happening.

6

SUBVERSIVE EINSTEIN

Susan Neiman

UNLIKE THE DISTINGUISHED EINSTEIN SCHOLARS who appear in this volume, my relation to Einstein was purely accidental. I didn't choose him; you might say he chose me when I took over the direction of an institute founded to nourish Einstein's heritage as a public intellectual. I might claim that before 2000, my attention was simply elsewhere, and I felt no more than the distant respect that people in the humanities feel for Newton or Kepler. But I'd rather confess it at the outset: when I began to work at the Einstein Forum, I didn't even *like* Einstein. This was something I hardly admitted to myself, because it wasn't until later that I understood the reason why: I don't like *Luftmenschen*—and Jews don't have saints. (I used to think *Luftmensch* was a German word, but it's one of the few Yiddish words that sounds German but isn't. It means someone who has his head in the clouds, or seems to live on air. In practice, of course, since none can live on air, it means someone who keeps his head in the clouds and lets someone else worry about putting food on the table.)

Was Einstein a *Luftmensch*? Here's part of the *Time Magazine* article declaring him person of the century: "Einstein was the iconic 20th century scientist, the bumbling professor with the German accent, a comic cliché in a thousand films. . . . Yet he was unfathomably profound, the genius among geniuses who discovered, merely by thinking about it, that the universe was not as it seemed."[1] Perhaps only in America could a German accent count as a sign of bumbling. But even that aside, here are all the clichés: comic and faintly ridiculous; a sweetheart perhaps, but who would chose to follow him? If the price of what *Time* calls "profundity" is to become a comic cliché, wouldn't you rather be superficial?

Now you may not expect much of *Time*, but what about *Die Zeit*? The material in one of its recent issues devoted to Einstein is predictably written in a more ponderous voice than the American quickie, but its content is surprisingly similar. Einstein is described as an eternal child. At first we're told of a *wise* naivety, but the wisdom drops out quickly, for soon he's described as a sad fool, *ein trauriger Narr*.[2] Nor is the cliché confined to the news media. J. Robert Oppenheimer called him childlike and wholly without sophistication, Isaiah Berlin called him innocent, and even Fritz Stern writes that his views were "well-meant in the usual sense, but lacking a certain closeness to reality." Now, only clowns can be both comic clichés and sad fools; so all in all, the picture of Einstein as clown is alive and well on both sides of the Atlantic. But not even circuses lead me to have much patience for clowns.

The other side of the clown is the saint, and they have much in common. When Einstein isn't regarded as comic, he's enveloped in an air of solemnity he couldn't escape whether he put his tongue in or out of his cheek. St. Albert, we know, was in favor of peace and helped little children with their math homework. But even these can be given a faintly ridiculous spin. Who, precisely, is not in favor of peace? Einstein's unofficial canonization provokes worse things than ridicule: where there are saints, many will suspect clay feet and hunt for what is sententiously called "the man behind the myth." Einstein's engagement for peace is countered with the false claim that he fathered the atomic bomb, while his fondness for strangers' children is countered with the slightly more plausible claim that he rather neglected his own. Sainthood provokes demasking—one reason why Einstein, who realized he was being canonized in his lifetime, did what he could to make it a matter of irony.

I suspect I speak for other non-Einstein specialists when I say that neither the air of piety nor the touch of derision that surrounds the man was remotely appealing, so I took detours around him wherever I could. But when I came to the Einstein Forum, I had to approach him somehow, and the first thing that surprised me were the pictures. Well into his sixties, every photo shows immediately that this was a man of extraordinary power, with the sort of charisma that drew the attention of men and women, small American children and aging European aristocrats, and virtually everyone else in between. *Luftmenschen* attract pity and a bemused sort of affection, but they're never charismatic. And saints may be luminous, but they really aren't sexy. Einstein was clearly both. So what evidence do we have for the bumbling professor?

He hated formal clothing and wore his hair long, possible grounds for thinking someone out of touch with reality in the 1940s and 1950s— remember the Beatles?—but his own attitude about this was anything but silly. (One of his better remarks was uttered in Caputh, where his second wife Elsa, always more enamoured of bourgeois convention than he was, fussed at him to get dressed before a delegation of dignitaries arrived. Einstein fired back, "If they want to see me, here I am. If they want to see my clothes, you can open the closet.")[3] His biographers revel in stories of forgetfulness: Einstein forgot the key to his apartment and had to wake his landlord after his first wedding, which, says Fölsing, shows *something*—without telling us what. (Isn't that the day when one is supposed to have other things on one's mind?) Others make something of the fact that he left a suitcase on a train. (Wouldn't the objects that have been sacrificed to train systems by those who are reading this volume suffice to stock a department store, or at the very least an umbrella shop?)

Did Einstein pretend to be a *Luftmensch*? It's a pose that can be useful. There are smart men who assume it just to avoid doing the dishes. If Einstein did play at being more unworldly than he was to preserve himself from too much worldliness, he had an awful lot to avoid. Einstein is called the first intellectual superstar, but no intellectual before or since, and possibly no superstar, was ever treated like this. Recall what happened on his first trip to America in 1921: Reporters stormed the ship when it docked, an impromptu parade, Stars and Stripes as well as blue and white Magen Davids lined the streets of New York to greet him, and crowds reacted similarly elsewhere. And all this was *before* the Nobel Prize. After it, the circus got larger: On his next visit to America, nine years later, he addressed a mass meeting to celebrate Hanukah in Madison Square Garden, was handed the keys to the city by New York's mayor, gave advice to Rockefeller about his educational foundation, and was escorted by Charlie Chaplin to the world premiere of *City Lights* in Hollywood. (There the crowd not only rioted but shrieked.) According to the German consul at the time, "Einstein's personality, without any clearly recognizable reason, triggers outbursts of a kind of mass hysteria, not only among . . . romantic dreamers . . . but also among relatively level-headed circles."[4] One final story: When I. F. Stone received a check from Einstein to subscribe to his newsletter, he called and asked if he might simply frame the check. His secretary sighed, for it was a common request, which made it extremely difficult to keep his bank account balanced. She appealed to Stone to cash the check, and promised to send it to him once it was cancelled. This is a sort of superstardom that Mick Jagger, or even Dylan, may not match. Confronted with *this much world* wherever he went, Einstein might be forgiven most any pose he took up that would allow him to withdraw from it once in awhile.

But the more I have learned about Einstein, the more I suspect that he wasn't a *Luftmensch* at all, genuine or fake. Rather, I've come to believe that the myth of Einstein as *Luftmensch* is one that was invented to cope with him. Einstein himself said the one good piece of news in the otherwise intolerable cult surrounding his person was the fact that "it is a welcome phenomenon in our supposedly materialist time that it makes heroes of men whose goals lie exclusively in the spiritual and moral domain."[5] Alas, that cult is marked by ambivalence. For by turning Einstein into a saint or a fool, we can pretend to celebrate the intellectual while actually undermining it.

To question this apotheosis, there is all sorts of evidence that he had his feet on the ground: from the testimony that he made an excellent patent officer—a profession he seems to have enjoyed, and hardly a calling for *Luft-menschen*—to his extraordinarily savvy self-awareness. Einstein knew exactly how he was perceived and used, and made use of it when he wanted. He often complained that he was liked, but not understood; one person who understood him quite well was David Ben Gurion. It's well known that Einstein was asked to become president of the State of Israel; less well known is Ben Gurion's remark after the invitation was made: "Tell me what to do if he says yes! I had to offer him the post because it's impossible not to, but if he accepts it we are in for trouble."[6]

It's the business of public intellectuals to speak truth to power, and as one whose task was consolidating the fledgling State of Israel, Ben Gurion knew what he'd have on his hands if Einstein were his partner. The first record we have of the way Einstein made people in power nervous stems from a teacher's remark that, although he didn't do anything wrong, the way he sat in the back of the class and smiled violated the feeling of respect a teacher needs from his class. Einstein radiated not only charisma, but a distinct and steady antiauthoritarian aura from the time he was young. He was not conventionally religious; Spinoza was one of his heroes. But no less than the 17th-century philosopher, Einstein's works exude a sense of awe toward the Creation. Unlike those skeptics whose goal is to admire nothing, Einstein's antiauthoritarian standpoint never degenerated into the purely negative. As I will shortly sketch, his was a Kantian idealism: never merely skeptical for its own sake, his refusal to accept received wisdom was always in service of an ideal.

To think about what it means to serve ideals without forgetting the real, let's recall some high points in his career as subversive. Einstein's work as a public intellectual can be divided into four distinct causes: his work against war, against political repression, against racism, and in favor of socialism. In all but the last case, his position was quite similar, and it was both clear and complicated. He insisted on the rule of law and on strengthening international law, but when law itself failed, he championed nonviolent but decided civil disobedience. The volume of material he wrote shows that this was very clearly work, not merely a hobby. The first time he put his reputation into political service was shortly after the outbreak of World War I. It's one thing to be in favor of peace today, when even those who valorize war do it with Orwellian language praising peace that no one is really fooled by, but most of us have learned to swallow. It was another in 1914, when the heroic language with which boys marched off to die had not even been called into question. Most of Europe marched with them. While scientists like Planck and Haber, and artists like Reinhardt and Liebermann, were urging support for the war effort, Einstein was one of only four German intellectuals, out of a hundred approached, who was willing to sign a manifesto at the beginning of the war calling for its ending and for the creation of a united Europe. He went on to join the harassed League of the New Fatherland and was so outspoken that French pacifist Romain Rolland wrote: "Einstein is incredibly free in his judgments on Germany, where he lives. No German enjoys that freedom."[7]

At the time, even less bold actions put people at risk. The civilized and aristocratic Bertrand Russell, in civilized and aristocratic England, spent time in jail for pacifism during World War I. And everywhere he went, Einstein was subject to pressures Russell never knew. There were so few Jewish professors at the time he became one that a letter testifying he was free of unpleasant Jewish qualities was seen as necessary for his first professorship. Though his commitment to the religion ended when he was twelve, Einstein never dreamed of hiding his origins, which could spell doom for anyone who wasn't willing to blend into the background. His courage on this score should be underlined: this was a time when many Jews still took conversion as the price of entry into the local culture, and even those who didn't, leapt to military service and displays of patriotism to show what good Germans (or French, or Americans) they'd become. Einstein's refusal to blend into the background hardly escaped notice. A Berliner who offered a reward to anyone who managed to kill Einstein was merely sentenced to a fine—reason enough for Einstein to leave Berlin briefly after the murder of Rathenau, with whom he'd been friendly. But though he confessed that the assassination left him on edge, he was soon back to provoking whenever he thought it was needed.

Though he had been one of the few pacifists during World War I, he enraged those who remained unremitting pacifists later with his support for any step in the war effort that would serve to defeat the Nazis: after World War II was over, he returned to support everything that could be done to defuse the cold war. Addressing the U.N. General Assembly in 1947, Einstein urged the United Nations to "strengthen its moral authority by bold decisions," and in clear and concrete terms took America to task, over and over, for its share of responsibility for the Cold War. While insisting on the need to strengthen international institutions, he was hardly naïve about their limits. Where they prove ineffective he urged defiance, for he argued that the judgment at Nürnberg confirmed what he held to be self evident: Where the law is immoral, we have the duty to follow our conscience instead.

Einstein's instinct for political repression was born under the spell of the Nazis, but it was never confined to them. It doesn't take a rocket scientist to criticize the Nazis, but keeping eyes open for other dangers took both more acumen and more courage. Unfortunately, the need for vigilance did not cease when Einstein reached the safety of American shores, though he praised the country for a democratic spirit second to none. But his troubles with the United States started even before he emigrated. In 1932 he became target of a group called the Women Patriots, described in their own newspaper in 1918 as antisuffragists waging an unceasing war against feminism and socialism. Having lost the battle against women's suffrage, they turned all their attention on socialism, whose world leader, they claimed, was Einstein—worse than Stalin himself. Some of their charges:

> Albert Einstein . . . advocates "acts of rebellion" against the basic principle of all organized government. . . . He advocates "conflict with public authority"; admits that his "attitude is revolutionary" . . . he teaches and leads and organizes a movement for unlawful "individual resistance" and "acts of rebellion" against officers of the United States in times of war.[8]

The charges go on for sixteen single-spaced pages. They led to interrogation by the U.S. Consul in Berlin, to whom Einstein pointed out that he hadn't asked to go to America but was invited there, and would cancel his trip if the visa wasn't delivered in twenty-four hours. His wife then relayed Einstein's words to the *New York Times*: "Wouldn't it be funny if they won't let me in? The whole world would be laughing at America."[9] The combination of chutzpah and media savvy at work here, neither of them qualities much in evidence with *Luftmenschen*, impressed the State Department, too, for he was issued a visa immediately. Though a group of Women Patriots tried to prevent him from getting off his ship in California, they did indeed appear ridiculous. A year later Einstein was offered unconditional citizenship through a special act of Congress, citing his qualities as a genius, a humanitarian, a lover of the United States, and an admirer of its Constitution—all of which were true. Since he rejected special treatment in speeding up the process, formal citizenship didn't come until later, but his right to stay in the country was thereby assured. His safety, however, was not, and he was urged to blend quietly into the background in just the same way smart Jews were urged to do in Wilhelminian Germany. Arriving in Princeton, he was met with this letter from the director of the Institute for Advanced Study: "I have conferred with the local authorities . . . and the national government in Washington, and they have all given me the advice . . . that your safety in America depends upon silence and refraining from attendance at public functions. . . . You and your wife will be thoroughly welcome at Princeton, but in the long run your safety will depend on your discretion."[10]

It was advice that Einstein ignored. He spoke and wrote to virtually everybody he could stand, about virtually everything he cared about, in virtually every format, large or small, and as anticommunist hysteria swept over America, he became increasingly outspoken. Though he was one of the few leftist intellectuals never tempted by communism, he thought anticommunism posed a far greater danger for America. Einstein's engagement during the McCarthy period took several forms. One was simply supporting prominent people who were under siege, appearing for a photo-op, for example, with Henry Wallace and Paul Robeson at a time when supporting either was taking a risk. After Robeson, whom he much admired, had his passport revoked, Einstein wrote to Wallace describing America as half-fascistic, and to Queen Elizabeth of Belgium expressing his fear and sadness that America reminded him of Germany in the 1930s. By that time McCarthyism was in full swing, and the House Un-American Activities Committee had compiled lists of subversive organizations. Anyone associated with any of them was ipso facto suspected of treason. Einstein was connected with thirty-three.

Even more impressive than his willingness to support people and organizations who were already in the limelight was his willingness to support those who weren't. He used his influence to help young people who were threatened for refusing to serve in the army or to go along with the HUAC. The letter that he wrote to a young teacher who refused to testify, William Frauenglass, was published on the front page of the *New York Times*. Unfortunately, the letter has again become so timely that it should be quoted at length:

> Dear Mr. Frauenglass,
> The problem with which the intellectuals of this country are confronted is very serious. The reactionary politicians have managed to instil suspicion of all intellectual efforts into the public by dangling before their eyes a danger from without. Having succeeded so far, they are now proceeding to suppress the freedom of teaching and to deprive of their positions all those who do not prove submissive, i.e., to starve them. What ought the minority of intellectuals to do against this evil? Frankly, I can only see the revolutionary way of non-cooperation in the sense of Gandhi's. Every intellectual who is called before one of the committees ought to refuse to testify, i.e., he must be prepared for jail and economic ruin, in short, for the sacrifice of his personal welfare in the interest of the cultural welfare of this country. This refusal to testify must be based on the assertion that it is shameful for a blameless citizen to submit to such an inquisition and that this kind of inquisition violates the spirit of the constitution. If enough people are ready to take this grave step they will be successful. If not, then the intellectuals of this country deserve nothing better than the slavery which is intended for them.[11]

Would the U.S. government really jail the world's most famous scientist at the age of seventy-three? The fear may seem exaggerated, but Einstein had reason to worry about going to jail in the days before this letter was published. Two years earlier, the older W.E.B. DuBois had been brought to court in handcuffs, and one of Einstein's closest friends had just been denied a passport. Einstein was the target of a flood of hate mail; one right-wing commentator urged the Senate to ban refugees altogether so that America would not "get another Einstein"; and even liberal sources like the *New York Times* and the *Washington Post* called his letter "extreme" and "unwise." The letter made international headlines and greatly strengthened the morale of those brave young teachers who refused to testify. A few of them found their way to Einstein's door, where they discovered he was not only willing to tell them to prepare for economic ruin in the interests of their country, but to try to prevent it, and he spent some time working to help those who had been fired find new jobs. He was not himself molested; try as he might, J. Edgar Hoover could never find anything that was *criminally* subversive. But it's important to recall one reason why the popular view of Einstein as father of the bomb is false: even if he had wanted to work on the Manhattan Project, he couldn't get a security clearance. Alerted to his socialist convictions in the early 1930s, the FBI director opened an investigation that didn't succeed in deporting him, but did prevent the world's greatest scientist from working on classified defense projects.

The third of Einstein's political concerns, racism, was also nurtured in the womb of the anti-Semitism he and other Jews experienced, but his intolerance for mistreatment of minorities was never confined to mistreatment of Jews. Einstein's deep commitment to the fledgling State of Israel was always accompanied by universalist concerns; in 1929, in a famous letter to Chaim Weizmann, he warned that without better policy toward the Arabs, "we will not have learned anything from our 2,000 year old ordeal."[12] His hatred of racism showed itself in the same mixture of public and private behavior he showed in other cases. Shortly after arriving in Princeton, for example, upon hearing that singer Marian Anderson played to a sold-out concert but was denied a room at the Nassau Inn, he promptly invited her to stay at his house, beginning a friendship that lasted for the rest of his life. Einstein often compared American treatment of African Americans to German treatment of the Jews. Those who find this extreme should know that in 1946, racist violence in the United States killed fifty-six black people, most of them veterans returning from service in World War II. Far more appalling than this is the fact that an anti-lynching campaign, headed by Robeson and others, was initially unsuccessful. One might assume that lynching was murder, and therefore already considered to be criminal, but the FBI found the group's left-wing origins more threatening than lynching itself, and a multiracial delegation failed to convince Truman, who argued that the Cold War was no time for passing controversial laws. Einstein was too ill to join the delegation, who took a letter he wrote for Truman, but he continued to work with Robeson as long as he could.

Finally, a few words about Einstein's commitment to socialism must be mentioned. Einstein has been described as a temperamental socialist. Despite the best efforts of his second wife, he had virtually no interest in consumer goods, and his convictions were shaped accordingly. What's particularly unusual was the way he consistently managed to maintain his own set of convictions without bowing to *any* set of dogmas. His intellectual and practical courage were shown in stories like this one. When Einstein received word in the chaos of 1918 that a group of revolutionary students and soldiers had taken over the University of Berlin and imprisoned the rector and several professors, he went with two friends to the Reichstag. There, armed revolutionaries denied them entrance until somebody recognized Einstein. They were astonished when he argued against their attempt to create a program in which only socialist doctrines would be taught. Though well known as a socialist, he was appalled at the breach of intellectual freedom the revolutionaries proposed. They nonetheless allowed him to reach the newly elected Social Democratic president, whom Einstein persuaded to write the note that released the imprisoned academics.

Yet his essay "Why Socialism"—written in 1949, hardly an opportune time—goes so far as to say that "the economic anarchy of capitalist society as it exists today is, in my opinion, the real source of evil."[13] Socialism, for him, is the reasonable response to a crisis of value: he thought the present situation to be a result of the lack of connection between the individual and society. That break itself is furthered by the fact that the media are so thoroughly controlled by economic interests that individual citizens cannot use the political rights they have, while fear of unemployment makes them docile and tame. Einstein doesn't pretend to offer original political theory here or elsewhere. Rather, what he gives are standard arguments for the idea that capitalism cripples the lives of those who seem to benefit by it, as well as those who don't. What's unusual in Einstein's arguments are, first, his unequivocal rejection of Soviet-style communism: "No purpose is so high that unworthy methods in achieving it can be justified." And second, his equally un-Stalinist claim that socialism can never be scientific. Einstein's socialism was a moral commitment, the only one he thought could give life meaning.

My survey has been brief, but I hope it's been sufficient to suggest that the popular picture of the good-hearted bumbler is sorely in need of revision. Einstein was a man who knew exactly what he was doing in the world. This is not Don Quixote, well meaning but looking backward; if some of his views seemed *weltfremd* at the time, the world has been catching up to him ever since.

In a valuable essay that also helps to undermine the picture of Einstein as unworldly, Yehuda Elkana pointed to continuities between the epistemology behind Einstein's scientific and moral views. It may be helpful to add that both of them are extremely Kantian. Though we know that Einstein first read the *Critique of Pure Reason* at the age of thirteen, standard discussion of Einstein and Kant concentrates on space and time. At least as worthy of further scholarship would be Einstein's own remark in the Schilpp volume: "I did not grow up in the Kantian tradition, but came to understand the truly valuable which is to be found in his doctrine . . . quite late. It is contained in the sentence: 'the real is not given to us, but is put to us (by way of a riddle).'"[14]

To view reality as a riddle that is put to us is to question statements like Fritz Stern's, which I quoted in beginning, and with them the picture of Einstein as far from reality. Such statements assume that we know what reality is: what is certain and what is not, what can be known and what can only be dreamed or intuited, what is given to us from objects outside ourselves and what we contribute to their structure, what can be confirmed by experience and what calls experience into question. To view reality as a riddle is to ask all these questions and more. Kant's major reason for doing so was to call attention to the difference between the way the world is and the way the world should be. The first is the object of science, the second of ethics, and we confuse them at our peril. Those whose only reality is what we experience leave no room for experience to be changed by ideals of justice and progress that challenge the authority of experience itself. Yet those whose lives are guided by ideals without regard to experience are in danger of becoming merely utopian, or even totalitarian. Both in science and in ethics, Einstein was aware of the risks of tradition-bound empiricism as well as of foolish idealism. More than anything else he was a Kantian idealist, with a commitment to maintaining ideals that are not derived from experience, but that shape it. While maintaining a clear-eyed view of the way the world is, he never forgot the way it should be, and always acted according to the latter. Did this make him unrealistic? Telling someone to be more realistic is a way of saying, Decrease your expectations of the world. Einstein never did.

When we look at the positions he took, what's astonishing is how often he was right. To recapitulate: his reputation skyrocketed after 1919 not only because observations confirmed his bold theoretical speculations, but also because experience confirmed his lonely opposition to World War I. The Weimar government was eager to use him to represent the new republic, for he was one of the few German public figures who'd opposed the war from the start, and that position was suddenly in demand. Einstein had nothing against being useful. He continued to argue for clear and complex positions—supporting any means to stop the Nazis, whether a united front at a time when most of the left found cooperation impossible, or a race to build the bomb before the Nazis did—while never losing sight of longer-term goals as soon as the war was over. His fears about the Israeli-Palestinian conflict were extraordinarily prescient. Albrecht Fölsing calls his proposals for a secret council of Jews and Arabs "of course both arbitrary and unrealistic."[15] What such a council could have accomplished under the British Mandate is hard to say, but in past decades, secret meetings between Israelis and Palestinians have been major steps toward peace. Even George W. Bush, who found the demand for a strong United Nations to be obsolete in 2003, has since been forced to acknowledge its necessity, at least in principle. And those who find Einstein's call for world government to be Quixotic should know that global governance has become not just a buzzword but the focus of attention by such great utopian institutions as the Deutsche Bank.

Einstein's instinct for the right position, far in advance of his time, has been so clear in the cases of racism and political repression that it would be tempting to claim that reality has been running after him ever since, until the Bush administration turned the word "terrorist" into the sort of epithet "communist" was for McCarthy. Progress may be possible, but it is hardly inevitable. Backsliding is always an option, and it would not have surprised Einstein: his universalism always recognized universal possibilities for failure. Socialism may seem the one issue on which history supports the picture of Einstein as a sad fool. But the real, no-longer-existing socialism that's been consigned to the dust heap of history was never the socialism Einstein supported. His anti-Stalinist warnings were as clear as his conviction that unrestricted capitalism could not provide humankind with either the justice or the meaning it deserves. As for the latter, the jury's still out. Not two decades have passed since the break-up of the U.S.S.R. and all of Eastern Europe in 1989, but the view that unrestrained capitalism is the solution to global development is already opposed by many of its earlier partisans. The more we're confronted with the reality of globalization, the more realistic social democracy may come to appear.

In fact, there's only one important question where I think he turned out to be wrong: his universalist convictions weren't strong enough to withstand the spectacle of Nazi Germany, to which so many of his colleagues capitulated. He never forgave Germany, nor believed it could be the home of a decent and democratic society. Of course, forgiveness is a private matter, and Einstein was one of many emigrants who refused to set foot on German soil after the Holocaust. These are private decisions. But it is worth questioning Einstein's judgment that the Germans would never be able to reject the militarism and authoritarianism that led them, and the world, into its greatest war. It is hard to say he was wrong without at least a shiver of trepidation: we all know what happened to Weimar. Yet the 100th anniversary of relativity theory that occasioned the Einstein Year turned out to coincide with the 60th anniversary of the end of World War II. Empiricist that he was, I believe Einstein would have been impressed by the strength of contemporary German democracy, and by the depth of its attempts to come to terms with the moral ruins that once made up its landscape. Through those attempts, the Federal Republic earned the right to take Einstein as a model in the year-long state-sponsored celebrations devoted to his legacy. Had he known, he might even have agreed to an appearance.

For let us be clear: what the Einstein Year sought was not a man behind a myth. A year of festivities, and however many millions the bill ultimately totalled, would not be devoted to celebrating a man. What was sought is a model, and we should ask which ones are available. Neither the saint nor the *Luftmensch* is borne out by the facts, so why are those models so widely accepted?

People like *Luftmenschen*, and they've liked them for a long time. The image of Thales, called the world's first philosopher, cannot be proven, but it's comforting to think that intellectuals have their heads in the clouds and stumble into the well before their feet. It allows us to feel that having ideas, and ideals, is vaguely ridiculous, and ultimately futile. If those who have the courage to think further than the rest of us are made to look like sad or silly fools, who will be tempted to follow them? *Luftmenschen* are useful for the same reason saints are useful: we don't want to be the one, and we know we're not the other. Turning Einstein into either is a way to tame him and to ensure that his life cannot challenge us into asking questions about our own.

In fact, I think Einstein is something much rarer: he's a genuine Enlightenment hero. Heroes make us uneasy. We're often too quick to diagnose that unease and to dismiss the notion entirely: the word sounds bombastic at worst, kitschy at best, and false notions of heroism have surely wreaked any amount of havoc in any number of cultures. But heroes make us uncomfortable because they represent ideals we *could* follow, demands we *could* make: to increase our expectations of our own lives, to become a little less certain about what's realistic, and a little more daring about which pieces of the world might be changed.

For the question is not merely whether the world caught up to him, but whether Einstein's own efforts had anything to do with the process. *Luftmenschen* are by definition ineffective, and saints (or their body parts) generally work miracles after their deaths. Einstein himself asked what a handful of intellectual workers can affect in politics, and his answer in a 1946 radio address was hopeful: ". . . but they can see to it that concise information about the situation and possibilities for successful action be made widely available. By spreading enlightenment they can prevent able statesmen from being impeded in their work by general prejudice and reactionary opinion."[16]

Telling good news today always gets you in trouble. For any reason you give them to be hopeful, any bright teenager can give you ten reasons for despair. But if we don't remember the good news that's been made, we're unlikely to have the will to make any good news of our own. Let us take a moment to remember the post-Hiroshima world with one quote from another Nobel laureate. In his 1950 address in Stockholm, William Faulkner declared, "There are no problems of the spirit. There is only the question: when will I be blown up?"[17] Now the nuclear danger is anything but past; Mohammed el Barradei argues it has never been so great. But our best hope of combating it lies in continuing the work that Einstein and others began. It was tireless, it was repetitive, it teetered between pathos and boredom, and it played no small role in keeping us from the brink.

I began by saying that when I came to the Einstein Forum, I didn't really like him; and there were moments during the hoopla surrounding the Einstein Year that my skepticism only grew. But the more I've had to do with him, the more I've changed my mind, so that hearing recordings of his radio speeches finally put me in danger of joining the Einstein cult myself. Einstein had guts, good sense, and humor. He was subtle where you need to be subtle and clear where you need to be clear. He knew how to attend to the details, and when to keep his eye on the ideals that go beyond them. He maintained an extraordinary balance between modesty and chutzpah, and the claims that each of them make on our lives, much in the sense of Kant's most famous quote:

> Two things fill the mind with awe and wonder the more often and more steadily we reflect upon them: the starry heavens above me and the moral law within me. . . . The former view of a countless multitude of worlds annihilates my importance as an animal creature, which must give back to the planet (a mere speck in the universe) the matter from which it came, the matter which is for a little time provided with vital force, we know not how. The latter, on the contrary, infinitely raises my worth as that of an intelligence by my personality, in which the moral law reveals a life independent of the whole world of sense . . . a destination which is not restricted to the conditions and limits of this life but reaches into the infinite.[18]

Few lives reflect this passage as much as Einstein's did. It's a privilege to be able to honor him.

7

EINSTEIN AND NUCLEAR WEAPONS

Silvan S. Schweber

INTRODUCTION

Quantum mechanics was responsible for a restructuring of the physical sciences that was as consequential as that brought about by the conceptual developments encompassed by the Scientific Revolution. It is one of the characteristics of the 20th century that the insights and understanding provided by quantum mechanics were soon thereafter marshaled for destructive ends. The design of the first atomic bomb was based on the conceptual tools provided by quantum mechanics: the uranium bomb was never tested before its use on Hiroshima. Similarly, the understanding of the structure and properties of metals and semiconductors given by quantum mechanics made possible the design of the first transistor, and like nuclear weapons, computers have transformed our world. Though readily conceding its extraordinary success, Einstein never accepted quantum mechanics as *the* theory demarcating the path to a more fundamental understanding of the physical world. Thus in a 1945 interview he stated: "The quantum theory is without a doubt a useful theory, but it does not reach to the bottom of things. I never believed that it constitutes the true conception of nature."[1]

The period from 1927 to 1933, during which quantum mechanics was established as the dynamics governing the atomic and nuclear domains, was a time of great change for Einstein. By virtue of his stand with respect to quantum mechanics and his views regarding the unification of gravitation and electrodynamics at the classical level, he became a marginal influence in setting the intellectual agenda of the physics community during his late years. And when in October 1933 he became an emigrant in the United States, he also lost the use of his mother tongue, German, as the means of communication with colleagues, being unable to express himself in English with ease at the time. Nonetheless, his stature as the greatest scientist since Newton, his modesty, his courage, and his forthrightness made all his statements important and newsworthy, and they often became transformed—even the political ones—into moral pronouncements.

Einstein was well aware of his unique position and was always ready to use it in calling attention to the dangers facing humankind or to the ills done to and the suffering of individuals. From after World War I until his death, as the evils of fascism, Nazism, Stalinism, and after World War II, American militarism, became apparent and the threat of war mounted, he became ever more vociferous, but he remained circumspect about the causes he would support and the help he would extend.[2]

After his formulation of general relativity in 1915, Einstein's research agenda was molded by his conviction that nature could be described by deterministic laws that did not relate solely to possibilities and their changes, but to the temporal changes of an exterior real world. Underlying his quest for a unified, nonlinear, field theory of gravitation and electromagnetism was the hope that it would yield a deterministic description of nature that would encompass the explanation of microscopic phenomena given by quantum mechanics. He was constant in his belief that this was the right path. He was likewise constant in his views about war, the causes of war, and the consequent ills of humankind. For him World War II, *qua* war, was quantitatively but not qualitatively different from World War I. What had been *qualitatively* different was the collective inhumanity the Germans demonstrated by their organized murder of six million of his brethren, of six million Jews. For him atomic bombs were but a *quantitative* change in the scale of destructiveness that weapons and war could wreak.

In an interview with Raymond Gram Swing in the fall of 1945, Einstein indicated that the release of atomic energy did not create a new problem, but the increased destructiveness of atomic weapons made the necessity of solving the problem of avoiding war more urgent and more imperative.[3] Their control, even their elimination, were not the solution to how to avert wars, if for no other reason than other technological means such as biological weapons would be developed. The solution to the problem was not arms reduction or arms limitation, but the elimination of war through the establishment of a world government with military powers to enforce the keeping of peace, together with a world court that would arbitrate conflicts between nation-states and whose rulings would be binding on and abided by the member states.[4] Einstein had expressed identical views during the 1930s, and his stance did not change after World War II.[5]

Much has been written on the subject of Einstein and nuclear weapons. Otto Nathan and Heinz Norden's fairly comprehensive *Einstein on Peace* is a valuable source for many of Einstein's statements on nuclear weapons,[6] as are articles on the subject by Paul Doty and by Bernard Feld in the volume edited by Gerald Holton and Yehuda Elkana that contains the papers presented at the 1979 Einstein Centennial Symposium in Jerusalem.[7] Much material is also available on the web.[8]

Here I will re-examine Einstein's participation in the development of nuclear weapons to indicate that his involvement at the political level was more extensive than usually related and greater than he himself depicted. I shall then briefly review Einstein's reaction to the use of atomic bombs over Hiroshima and Nagasaki, his subsequent efforts to ban nuclear weapons, and his relentless call for a supranational government. An epilogue connects his theorizing in physics to that in politics.

Einstein and the Atomic Bomb

The economic and academic context in Europe after World War I was such that only the very best young theoretical physicists could hope for a professorial position, and only the most outstanding among them gained entrance into the inner circle around Neils Bohr, Einstein, Max von Laue, Max Planck, and Arnold Sommerfeld. One such young man was Leo Szilard—a member of the group of brilliant young Hungarians that included Michael Polyani, John von Neumann, and Eugene Wigner. They had come to Berlin in the early 1920s to study engineering, but became captivated by science and mathematics. Szilard was not quite as proficient in mathematics as his two dazzling friends, von Neumann and Wigner, but for his Ph.D. dissertation, he had submitted a path-breaking analysis of the relation between entropy and information, whose importance was generally recognized only much later.

During the 1930s Szilard made seminal contributions to nuclear physics. Very soon after the discovery of the neutron in 1932, sensitized by H. G. Wells's *The World Set Free*,[9] a book he read as a youth not long after its publication in 1914 in which the devastation from an atomic bomb had led humankind to renounce warfare, Szilard became obsessed with the possibility of a nuclear chain reaction and, in fact, patented the idea. After the discovery of fission by Otto Hahn and Fritz Strassmann in the winter of 1938 and the interpretation of that phenomenon by Otto Frisch and Lise Meitner, he immediately realized that the imagined chain reaction was now a real possibility, should a uranium nucleus also emit neutrons when undergoing fission.[10] In early spring of 1939, Carl Anderson, Enrico Fermi, and Szilard, working together at Columbia, were the first to measure the average number of neutrons released during fission and to establish the viability of a chain reaction. Szilard and Fermi shortly thereafter designed the first nuclear reactor.[11]

After it had been established that on average 2.5 neutrons were emitted in the fission of U^{235}, Szilard became deeply troubled about the dangers that an atomic bomb—now a real possibility—would pose if developed by Nazi Germany. Recall that these experiments were being carried out at a time when the world knew the Germans were threatening war, which broke out a few months later.

As is well known, Szilard, after conferring with Eugene Wigner and Edward Teller, enlisted the help of Einstein. Initially Szilard, Wigner, and Teller had thought of obtaining Einstein's help to have the Queen of Belgium—a personal friend of Einstein—use her influence to prevent the sale of uranium ores from the Belgian Congo to Germany and for the United States to buy the available raw materials. However, the mathematician Oswald Veblen, a colleague of Wigner in Princeton, had convinced him that the U.S. government ought to be informed of the implication of the discovery by Hahn and Stassmann and of Szilard's findings, that is, of the possibility of creating nuclear weaponry and of construction of nuclear reactors for the propulsion of ships and submarines. Szilard, who together with Fermi had already approached the U.S. Navy for support and had been rebuffed, was skeptical. However, they agreed that the letter to be sent to the Queen should be submitted to the U.S. State Department for approval.

On Wednesday, July 12, 1939, Wigner met Szilard early in the morning at the King's Crown Hotel, located next to Columbia University in Manhattan, to drive to Peconic on Long Island to see Einstein, who was spending his summer vacation there at the home of a friend.

After Szilard and Wigner found Einstein, and Szilard told him about his experiments establishing secondary neutron emission in fission and of his calculations indicating the possibility of a chain reaction in a uranium pile moderated by graphite, Einstein exclaimed: "That never occurred to me! [*Daran habe ich gar nicht gedacht*]." Einstein readily agreed to have a letter sent under his signature to a Belgian cabinet minister, and he dictated a draft of it. It was agreed that Szilard and Wigner would edit it and send it to him for his signature. Wigner did so and sent the draft of his letter to Szilard.

In the meanwhile Szilard received a letter from Gustav Stopler, a friend whom he had consulted about these matters, indicating that he had discussed the problems with Dr. Alexander Sachs, a vice-president of the investment firm Lehman Corporation, who had direct access to President Franklin Roosevelt[12] and that Sachs wanted to talk to him. A day or two after his receipt of Stopler's letter, Szilard saw Sachs. At their meeting Sachs convinced Szilard that Einstein's letter should be addressed to Roosevelt to inform him of the dramatic gains—military and economic—should the fission process be mastered and successfully controlled to yield bombs and reactors, and to alert him of the danger that Nazi Germany might develop nuclear weapons first.

As Wigner had gone West on vacation, Szilard took it upon himself to alter the previous draft into a letter to be submitted to Roosevelt, with Sachs as a possible intermediary. He mailed the draft to Einstein, who in turn indicated that he would like Szilard to come again to Peconic to revise the letter. This time it was Edward Teller, who that summer was teaching at Columbia, who drove Szilard to Long Island to see Einstein; this on Wednesday, August 2. Einstein again dictated a draft, this time to Teller, and asked Szilard to convert it into shorter and longer versions. Einstein would decide which one to send upon receiving the two versions. As the three of them felt themselves not experienced enough to decide who would be the best person to transmit the letter to Roosevelt, upon his return to New York Szilard phoned Sachs, who suggested three further names as possible couriers to Roosevelt: the financier and elder statesman Bernard Baruch, M.I.T. president Karl T. Compton, and the aviator Charles Lindbergh. Given Lindbergh's Nazi sympathies, this last suggestion was somewhat odd.

Szilard thereafter penned the longer and shorter versions of the letter and sent them to Einstein. Einstein indicated that he preferred the more detailed letter, but suggested that he make it somewhat more "straightforward" for "it always gives one pause for thought when a person wants to do something too smartly." Szilard did so and it was this version, dated August 2, that was transmitted on October 12 to Roosevelt by Sachs. Szilard also wrote a technical addendum to the letter in which he indicated that the research in nuclear physics over the past five years suggested that "a nuclear chain reaction could be maintained under certain well defined conditions in a large mass of uranium" even though "[i]t still remains to prove this conclusion by actually setting up such a chain reaction in a large-scale experiment."

The delay in Einstein's letter reaching Roosevelt[13] resulted from Hitler's invasion of Poland on September 1, the subsequent declaration of war by France and Great Britain, and the consequent reassessment of U.S. policy in the face of these developments.

The content of the letter Einstein's letter is well known. Let me only focus on its main points:

1. The recent work of Fermi and Szilard implied that "uranium may be turned into a new and important source of energy in the immediate future"; furthermore, that it appeared almost certain that a chain reaction could be induced in a large mass of uranium. The new phenomenon could also lead to the construction of extremely powerful bombs. The situation called for watchfulness and, if necessary, "for quick action on the part of the Administration."

2. In view of this situation, it would be desirable to have permanent contact maintained between the Administration and the physicists working on chain reactions in the United States. One way of achieving this could be to have a person Roosevelt trusted serve in an unofficial capacity (a) to help speed up experimental work by providing governmental funds or to help get such funds from private sources; and (b) to keep the appropriate governmental departments informed of further developments and make recommendations for governmental actions.

3. There were reliable indications that Germany was actively pursuing these lines of research; and furthermore, that it had "stopped the sale of uranium from the Czechoslovakian mines which she [Germany] has taken over."

Upon hearing Sachs, Roosevelt asked General Edwin M. Watson, his military aide and executive secretary, to act as his liaison. He also instructed Watson to introduce Sachs to Lyman Briggs, the director of the National Bureau of Standards, and to constitute, on the President's behalf and chaired by Briggs, a committee of representatives from the Armed Services to consider the ideas and memoranda submitted by Sachs. Roosevelt also suggested to Sachs that he stay through the following day to meet Briggs. The choice of Briggs was unfortunate for he was an "inarticulate and unimpressive man."[14]

At their meeting on October 12, Sachs and Briggs agreed to include nongovernmental representatives as well, with Sachs as presidential representative and representatives from the Army and Navy to be selected by Briggs in consultation with Watson. The committee convened on October 21 with Lt. Col. Keith F. Adamson and Commander Gilbert C. Hoover as the representatives for the services, and Wigner, Teller, Fermi, and Szilard as the nongovernmental representatives. The discussions at the meeting were not friendly, with the military representatives of the opinion that those interested in the political-military implications were premature in "converting a mere potential into an actual result of research."[15] Nonetheless, on November 1, Briggs submitted on behalf of the committee a report to Roosevelt that recognized the implied military and naval applications, though it characterized them as only possibilities, and recommended that adequate support for a thorough investigation of the subject should be provided.[16] Subsequently the sum of $6,000 was made available to the Columbia investigators.

The report also recommended the enlargement of the committee "to provide for the support and coordination of these investigations in different universities" and suggested that M.I.T. president Compton, Sachs, Einstein, and George Pegram, the dean of science at Columbia, be invited to join.

The Sachs correspondence in the Einstein Archives documents Einstein's considerable involvement in the activities of the Briggs Committee until mid-June 1940, at which time the committee, which until then had overseen the atomic energy program, was placed under the aegis of Vannevar Bush's National Defense Research Committee (NDRC). Both Sachs' and Einstein's connections with governmental atomic energy matters ended with that changeover.

Until June 1940, Sachs kept Einstein informed of all his governmental activities, as did Szilard and Wigner of their scientific research. Sachs' and Szilard's reports that the scope and pace of the work was unsatisfactory prompted a suggestion that Einstein prepare another review of the situation for submission to Roosevelt. Einstein did write such a letter, as drafted by Szilard, to Sachs, in which he called attention to the intensification of investigations on uranium in Germany, "which research is being carried out in great secrecy,"[17] and to the fact that the German government had assembled a group of physicists and chemists from the Kaiser Wilhelm Institutes for Physics and for Chemistry to work on uranium problems, placing it under the leadership of Carl F. von Weizsäcker. He added that "should you think it advisable to relay this information to the President, please consider yourself free to do so." Sachs did transmit Einstein's letter to Roosevelt. In reply Roosevelt instructed Watson once again to convene a conference that would include the members of the Briggs Committee "at a time convenient for you and Dr. Einstein."[18] Watson thereafter wrote Sachs asking him for suggestions of additional academic scientists to be invited to the conference, adding that "perhaps Dr. Einstein would have some suggestions to offer as to the attendance of the other professors."[19] After conferring with Einstein and Szilard, Sachs wrote Watson to recommend that Einstein and Fermi also be invited and to suggest that the conference be held on April 29.[20] However, after meeting Einstein in Princeton, Sachs wrote that "it became clear that indisposition on account of a cold and the shyness which makes Dr. Einstein recoil from participating in large groups would prevent his attendance."[21]

Einstein himself wrote to Briggs to suggest that Fermi be invited to the conference and to indicate his interest in Sachs' suggestion "that the special Advisory Committee submit names of persons to serve as a Board of Trustees for a non-profit organization which, with the approval of the Government Committee, should secure from governmental or private sources, or both, the necessary funds for carrying out the work."[22] The conference took place on April 27, 1939.

Einstein was not involved in any of the technical developments of atomic energy and atomic bombs after June 1940, except for some work on diffusion processes that he did not associate with the bomb.[23] He was involved in two other episodes in the wartime history of the atomic bomb, and both had serious consequences for Niels Bohr. The first occurred when Bohr first came to the United States in the winter 1943 and the second a year later. In public statements after World War II, Einstein conveyed the impression that his only connection to the atomic bomb project was writing his letter of August 1939 to Roosevelt; however, the facts are different. Although he did not know any of the technical details of the enterprise, it seems clear that he was aware that important developments were taking place. Undoubtedly, whoever had specific knowledge about the Manhattan Project would have been very guarded and judicious in conversations with Einstein, nonetheless allusions must have been dropped. This must have been the case when Bohr suddenly appeared in Princeton on December 22, 1943, shortly after his escape from occupied Denmark. Bohr had become a member of the British team working at Los Alamos. When he heard from Chadwick of the vast Manhattan Project, he immediately realized that the creation of an atomic bomb would bring fundamental changes in the world. Thereafter, working out the implications of atomic weaponry and preventing an arms race between the U.S. and the U.S.S.R. after the war became his primary concern. At Los Alamos he profoundly influenced J. Robert Oppenheimer regarding the political implications of the possession and use of atomic bombs and had many discussions with him and Fermi on how these problems should be addressed.

Nothing much is known of the conversations that Bohr had with Einstein in December 1943. In some undated notes from the war, Bohr stated simply that he had tea with Einstein, Herman Weyl, Carl Ludwig Siegel, Ostwald Veblen, James Alexander, Wolfgang Pauli, and others at the Institute for Advanced Study on Wednesday, December 23, 1943. However, Einstein must have gleaned some hints from Szilard because he evidently told Bohr at the tea that he was glad that Bohr had come to help with the atomic bomb project because the "American Army was making a frightful mess of the uranium work and no doubt he [Bohr] would be a able to put this right."[24] And a letter from Wallace Akers to M. W. Perrin of January 27, 1944, contains this comment:

> With regard to his own [Bohr's] moves, he is now satisfied that it is quite impossible for him to go to Princeton, to the Institute of Advanced Studies. It seems that he had a devastating experience when he first went there, on his arrival here; because Einstein, who is a very old friend of his greeted him in a crowded room with the statement that he [Einstein] was delighted that Bohr had come to America as he had heard from Szilard that the American Army was making a frightful mess of the uranium work and no doubt he [Bohr] would be able to put this right.[25]

The second episode occurred in the fall of 1944. Otto Stern, who was a consultant to the Met Lab in Chicago[26] and whose friendship with Einstein dated back to their association in Prague and Zurich, then told Einstein about the likely success of the bomb project. They must also have discussed some of the postwar implications of the existence of atomic weapons. These talks with Stern had deeply troubled Einstein. With the successful building and operation of the reactors at Hanford in the summer of 1944, the Met Lab, whose responsibility had been the design of these reactors, found itself under much less pressure, and informal talks concerning how to use the bombs and the consequences of the existence of such bombs were taking place.[27] Szilard in particular had prompted many such discussions. Stern, who had just been awarded a Nobel prize for the Stern-Gerlach experiment, must have been privy to some of these discussions.

Otto Stern had another conversation with Einstein in early December 1944. On the day after Stern's visit on December 11, Einstein wrote a letter to Bohr, care of the Danish embassy in Washington, DC,[28] in which he referred to atomic bombs euphemistically, calling them "weapons with technological means." Reflecting the discussions taking place at the Chicago Met Lab, a very anxious ("recht alarmiert") Stern had, in his meeting with Einstein, predicted that after the war in all countries

> there will be a continuation of the secret arms-race of weapons with technological means, which inevitably will lead to preventive wars (veritable wars of annihilation, worse in the loss of life than the present one). And since the politicians don't know of these possibilities they are unaware of the magnitude of the threat. Every effort must be made to avert such a development.[29]

Einstein went on to say that he shared Stern's view of the situation, but that in the past he and Stern couldn't see any joint effort that could help. "But yesterday, when Stern was here again, it seemed to us that there might be a way—even though trifling—that could be successful." There are, Einstein wrote, in the relevant, important countries quite influential scientists to whom the politicians would listen and who could make them aware of the enormity of the danger.

> [Of such eminent scientists] with international contact there is you [Bohr], A. Compton here, Lindemann in England, Kapitza and Joffe in Russia, etc. The idea is to induce them to jointly bring pressure to bear on the political leaders in their country, in order to achieve an internationalization of military power—a road that was some considerable time ago rejected as being too adventurous. But this radical step with all its far-reaching political prerequisites concerning supranational government seems the only alternative to a secret technological arms-race.[30]

Stern and Einstein had agreed to have Einstein apprise Bohr of their plan and to enlist him in its implementation. In his letter Einstein admonished

Bohr not to say immediately "impossible," but to wait a few days till he had gotten used to this strange idea. And if he found something worthwhile in it, "if only with a 0.001 chance" of success, he entreated him to get together with Stern (in Pittsburgh) and Frank (in Chicago) for joint consideration.

The establishment of an effective supranational government and the internationalization of military power had been the basic components of Einstein's solution to the problem of war since the early 1930s, and were to remain the basic components of Einstein's views of any solution concerning the threat posed by nuclear weapons.

For Bohr the immediate goal in the face of the existence of atomic weapons was more limited. He wanted wide-ranging discussions of atomic issues among the United States, Great Britain, and the Soviet Union before any tests of the bombs were carried out, and he had urged that the Soviet Union be told about the progress being made toward the building of atomic bombs. The Soviet Union was then the indisputable ally of the United States in the war against Nazi Germany. Bohr had come to the United States to try to convince Roosevelt of the necessity of such an approach. He had drafted a document prior to meeting Roosevelt in August 1944 that outlined his ultimate aim, namely to seek

> an initiative aiming at forestalling a fateful competition about a formidable weapon, [an initiative which] should serve to uproot any cause of distrust between the powers on whose harmonious collaboration the fate of coming generations will depend. . . . Of course, the responsible statesmen alone can have the insight in the actual possibilities.

Earlier that year, a few weeks before D-Day, Bohr had met Churchill with disastrous consequences, as he could not get his points across to an inattentive, otherwise preoccupied Churchill.[31] Although Bohr thought that the meeting with Roosevelt went well, in fact, Roosevelt had been upset because Justice Felix Frankfurter, who had been influential in arranging the meeting between Bohr and Roosevelt, had become privy to atomic secrets that he shouldn't have known about. At their meeting in Hyde Park following the Quebec conference in early September 1944, Churchill convinced Roosevelt that the bomb project should be kept strictly secret until further notice. And they agreed that "inquiries should be made regarding the activities of Professor Bohr and steps should be taken to ensure that he was not responsible for any leakage of information, particularly to the Russians."[32]

It was in this context that Bohr received Einstein's December 12 letter informing him of his talks with Otto Stern.

Because Bohr had made reasonable and well thought out suggestions to men who wielded considerable power in Great Britain and the United States— Anderson, Frederick Lindemann, Frankfurter, Bush—and had been rebuked, he thought Einstein's proposal surely unrealistic and having no chance of being taken seriously. Furthermore, Bohr probably feared that Einstein might himself write to Pyotr Kapitza and Abram Joffe and thus jeopardize the secrecy surrounding the Manhattan Project—an act that could readily be interpreted as treason. Moreover, the Soviet scientists that Einstein had suggested for Bohr to contact, Kapitza in particular, could hardly have been more ill-chosen. Earlier that year Kapitza had invited Bohr and his family to settle in Russia, and Bohr had graciously declined! Given his views of the Soviet Union and of communism, these exchanges between Bohr and Kapitza were grounds for Churchill's apprehension concerning Bohr.

Upon receiving Einstein's letter, Bohr consulted with Frankfurter,[33] who also considered himself a friend of Einstein. It was Frankfurter who had made possible Bohr's interview with Roosevelt and had facilitated the one with Churchill. Bohr then rushed to Princeton on December 22, and in a lengthy get-together persuaded Einstein to remain silent. He wrote a report of his meeting and handed it to security officials in Washington; the typed copy is among Bohr's papers. It opened with a statement that he had visited Einstein and had indicated to him that "it would be quite illegitimate and might have the most deplorable consequences if anyone who was brought into confidence about the matter concerned, on his own hands should take steps of the kind suggested." Bohr continued:

> Confidentially B [Bohr] could, however inform X [Einstein] that the responsible statesmen in America and England were fully aware of the scope of the technical development, and that their attention had been called to the dangers to world security as well as to the unique opportunity for furthering a harmonious relationship between nations, which the great scientific advance involves. In response X assured B that he quite realised the situation and would not only abstain from any action himself, but would also—without any reference to his confidential conversation with B—impress on the friends with whom he had talked about the matter, the undesirability of all discussions which might complicate the delicate task of the statesmen.[34]

On December 26, Einstein wrote Stern a diplomatic letter stating that "a leaden cloud of secrecy" had descended upon him because of his letter to Bohr, so that he could only inform him that they were not the first to consider the situation. He has the impression that the issues will be given serious attention, and "that the matter will be best served, if for the time being one does not speak of it, and especially that one does not promote it in any way, because at the present moment public attention hinges on this. Such a nebulous way of talking is difficult for me, but this time I cannot change it."[35]

It is thus clear that as of December 1944, Einstein knew a fair amount about developments of nuclear weapons and about attempts to influence American and British statesmen regarding nuclear weapons policy. It is against this background that Einstein's March meeting with Szilard and Einstein's March 22 letter to Roosevelt to introduce Szilard should be assessed. Einstein's statement that he is writing the letter "in spite of the fact that I do not know the substance of the considerations and recommendations which Dr. Szilard proposes to submit to you" is to be taken literally. He did not know what Szilard had written and meant to protect Szilard. He wanted to be explicit about the fact that he was aware that the terms of secrecy under which Szilard was working "do not permit him to give me information about his work," and that there had not been any breach of secrecy. "However," Einstein could state, "I understand that he now is greatly concerned about the lack of adequate contact between scientists who are doing this work and those members of your cabinet who are responsible for formulating policy."[36]

In retrospect, as significant as had been the diplomatic and political activities of Szilard, Einstein, Wigner, Teller, and Sachs, the bomb program in the United States effectively started upon receipt of the British MAUD report[37] in September 1941, which detailed the Frisch-Peierls calculations of the critical mass of U^{235} needed for a uranium bomb and their estimation of the feasibility of separating the U^{235} isotope from "natural" uranium. It is only after studying this document that James Conant, who was in charge of the NDRC atomic energy program, became convinced that an atomic bomb could be produced in time to alter the course of the war. Thereafter, in mid-October 1941, Bush recommended to Roosevelt to go ahead with the project to build a bomb. This was a few weeks before Pearl Harbor. The timing was crucial. It would not have been possible to obtain the top priority ranking for the project after Pearl Harbor.

One further matter merits comment concerning this initial phase of Einstein's involvement with atomic bombs. On several occasions after the war, Einstein was criticized for having "participated" in the bomb project. His answer was almost always the same:[38]

> I have never done research having any bearing upon the production of the atomic bomb. My sole contribution in this field was that in 1905, I established the relationship between mass and energy, a truth about the physical world of a very general nature, whose possible connection with the military potential was completely foreign to my thoughts. My only contribution with respect to the atomic bomb was that, in 1939, I signed a letter to President Roosevelt in which I called attention to the existing possibility of producing such a bomb and the dangers that the Germans might make use of that possibility. I consider this my duty because there were definite indications that the Germans were working on such a project.[39]

To this statement should be added the comment Einstein made to Linus Pauling during his visit in Princeton on November 16, 1954: "I made one great mistake in my life—when I signed the letter to President Roosevelt recommending that atomic bombs be made. But there was some justification—the danger the Germans would make them."[40]

AFTER HIROSHIMA AND NAGASAKI

Helen Dukas, Einstein's secretary, heard the news of the bombing of Hiroshima over the radio on August 6 and told Einstein what had happened. His response was *O weh!*—a wrenching *cri-de-coeur* that is not readily translated.

A letter to Robert M. Hutchins, the president of the University of Chicago, dated September 10, 1945, contains one of the first reactions of Einstein to Hiroshima and Nagasaki:

> As long as nations demand unrestricted sovereignty we shall undoubtedly be faced with . . . wars fought with bigger and technologically more advanced weapons. The most important task of intellectuals is to make this clear to the general public and to emphasize over and over again the need to establish a well organized world government. They must advocate the abolition of armaments and of military secrecy by nations.[41]

The question was, what kind of world government? In the spring of 1945, the Allies had met in San Francisco for two months to negotiate the charter of the United Nations, which was ratified in early June. The charter created a security council with permanent seats for the United States, the Soviet Union, China, and the United Kingdom, each wielding veto power over council decisions.

In the summer of 1945, Emery Reves, an economist, political writer, and successful publisher[42] whom Einstein had known for a number of years,[43] published a book entitled *The Anatomy of Peace* to very favorable reviews.[44] Emery Reves (1904–1981) was a remarkable man. He was born in a small village in southern Hungary to a Jewish family supported by the father's wood and grain business. His father's original name was Rosenbaum, but was changed to Revesz so that he could do business with the Hungarian government. A very talented and accomplished musician, Emery almost became a concert pianist. In 1922 he left Hungary to study at the universities of Berlin and Paris and eventually earned a doctorate in political economy at the University of Zurich, writing a dissertation on Walter Rathenau. His two best friends in Zurich were fellow Hungarians John von Neumann and William Fellner. Fellner later became a distinguished economist. Revesz was fluent in nine languages and, during the Weimar Republic, worked in Berlin as a journalist, an activity he had begun in Zurich by interviewing all the prominent people passing through the city.

In Berlin Revesz got the idea "that the world needs an organization that can arrange publications in the press of articles on world issues, articles by Englishmen and Germans in the French Press, by Frenchmen and Germans in the English press, and so on." He there founded the Co-Operation Press Service for International Understanding in 1930, the first international media company, which eventually serviced 400 newspapers in 60 countries. He and the Press Service left Berlin for Paris in April 1933 when it became clear that the Nazis were going to arrest him and close the Co-Operation Press Service. He first met Winston Churchill in 1937 and enlisted him as one of the writers for the Press Service; other contributors included Austen Chamberlain, Clement Attlee, Anthony Eden, Leon Blum, Paul Reynaud, Eduard Benes, and Einstein. Revesz' response to Hitler's bombastic speeches was to disseminate Churchill's replies in the newspapers of Europe the next day, which restored Churchill to public awareness and began his political comeback. Churchill got Revesz a position working for the British Ministry of Information, and in February 1940, Revesz became a British citizen.

In June 1940, just before the Germans occupied Paris, Revesz fled to London, where Churchill put him in charge of propaganda to the United States and other neutral countries. He was wounded in a London air raid in September 1940. In January 1941, on his own initiative, Revesz went to New York to help with the public relations campaign designed to enlist the United States' support of Britain in its battle with Germany. While in New York, he changed his name to Emery Reves and met fashion model Wendy Russell, who became his companion and later his wife. After the war, Reves made Churchill and himself very rich by negotiating the American royalties for Churchill's *The Second World War* and by obtaining for himself all foreign language rights of the volumes. From the proceeds the Reveses bought La Pausa, the Riviera home near Monte Carlo of Coco Chanel, the founder of the Chanel perfume business. Churchill often stayed there for weeks at a time.

To understand the impact of Reves' *Anatomy of Peace*, the context of its publication should be recalled. In early 1945 Harvard historian Crane Brinton, in a book entitled *The United States and Britain*, had tried to demonstrate the impracticality of world government at the time. Reves in his book eloquently defended the opposite view. In August 1945, Reves mailed a copy of his book to Einstein with a statement that he was particularly indebted to him, as he felt "that without [Einstein's] philosophical outlook this book could not have been written."[45] Reves further indicated that, since its publication, the book had "provoked rather extraordinary reactions. Columns and columns are devoted to its discussion in the American press."[46] Supreme Court Justice Owen

Roberts had read the book "twice in two days" and had informed Reves that "it expresses entirely his convictions." Roberts had drafted an open letter warning the American people not to believe that the ratification of the U.N. Charter would bring peace, but that the thesis in Reves' *The Anatomy of Peace*—"that peace is law and that war between nations can be stopped only if a legal order is established, standing higher than the present nation states"[47]— must be recognized. Three senators, including J. William Fulbright, had endorsed the letter, and Reves asked if Einstein would sign it.

Two days later Einstein answered, telling Reves that he had read his book

carefully and finished it in 24 hours. I agree with you wholeheartedly in every essential point and I admire sincerely the clarity of your exposition of the most important problem of our time. . . . I believe Justice Roberts' action is of the greatest value in the matter of enlightenment of public opinion in this country. I find the text of his open letter excellent and convincing and am gladly willing to sign it.[48]

On October 10, 1945, the *New York Times*, the *Washington Post*, and some fifty other newspapers published the *Open Letter to the American People* signed by Einstein and eighteen other leading public figures. The letter contained an explicit endorsement of Reves' *The Anatomy of Peace*, for "it expresses clearly and simply" what the signers had been thinking. The letter urged "American men and women to read the book, to think about its conclusions, to discuss it with neighbors and friends privately and publicly."[49] Later that year Reves excitedly wrote Churchill:

Events happening around my book almost daily have forced me to postpone my trip [to London] from week to week. It is sweeping this country in an extraordinary way. It became a textbook in several colleges, sermons are made on it in many churches, students in Harvard, Yale and Columbia began to form groups to spread it and new editions of 10,000 copies disappear in two days. . . . The *Reader's Digest* is now organizing for January and February discussions of *The Anatomy of Peace* in 15,000 American clubs and discussion groups, with three speakers in each place presenting the three parts of the book. This is unprecedented in American publishing history and may have unpredictable effects.[50]

In late summer 1945, Reves forwarded Einstein the statement that a group of Oak Ridge scientists had issued. In it they had recommended that an international agency, a world security council, be made the sole custodian of nuclear power in the world, as well as all scientific and technical knowledge relating to it, and have the right to complete, detailed, periodic inspections of all scientific, technical, industrial, and military installations in the world. In addition to the Oak Ridge statement, Reves included a letter to Einstein in which he argued that the recommendations of the scientists were "completely fallacious and may have tragic consequences if the people are given to believe that on the basis of these suggestions peace between the nations is possible and an atomic war between nation-states is made impossible." Reves continued:

The question of preventing the use of revolutionary weapons such as the atomic bomb is nothing new. Whenever a new weapon was invented—whether . . . gunpowder, dynamite, the machine gun, the tank, gas, the submarine or the bomber—people were always afraid the new weapon would mean the final destruction of civilization and tried to "outlaw" or "control" it. All these attempts failed because war is essentially a political, a social problem and not a technical one.

There is only one way to prevent an atomic war and that is to prevent War. Once war breaks out, it is certain that every nation will use every conceivable and available weapon to defend themselves and to achieve victory. Analyzing the wars of history . . . I think it is possible to isolate the virus of war and to define the one and only condition in human society that produces war. This is the non-integrated coexistence of sovereign powers units in contact. . . . Peace is law. Peace between warring sovereign social units . . . can be achieved only by the integration of these conflicting units into a higher sovereignty. Peace between the nation-states in the 20th century can be achieved only by the transfer of parts of the people's sovereignty into a higher system of government—legislative, judiciary and executive—regulating the relations of man in the international field; by the creation of a world government having direct relations with the individual citizen.

Attempts to maintain peace between nation-states by a league structure such as the San Francisco Organization in which sovereignty continues to reside in the individual members . . . are pitifully outdated and bound to fail. . . . There is only one way and one way alone to make the United States secure from an attack by atomic bombs. The method is the same that today makes the states of New York and California (non producers of atomic bombs) safe from annihilation by the states of Tennessee and New Mexico (producers of atomic bombs).[51]

No group of people today have such influence on the public as do the nuclear physicists. Their responsibility in making political suggestions is tremendous. . . . [52]

I have quoted at length from the letter because it succinctly summarizes the conclusions of Reves' book, and it encapsulates Einstein's position after the war.

Let me briefly indicate Einstein's views in 1947 by presenting an interview he had with William Golden,[53] carried out on behalf of the Secretary of State James Byrnes in June 1947. In it Einstein told Golden that he believed:

a) that the world is heading for an atomic war. The American monopoly cannot be maintained for more than a few years. Russia will surely develop an atomic bomb. With each side having the atomic bomb, the premium on surprise attack is greater than ever before. In case of war when the two sides have the bomb, one or the other will surely use it, either "from nerves or fear if not from policy. The devastation that will result from atomic warfare will be fully up to the [i.e., will match] popular conception."

b) that the United Nations has been ineffectual and cannot control the situation.

c) that the only way of averting an atomic war within a few years is through an effective supranational world government to which military power will be transferred. All countries should be invited to join; however, if Russia does not join, then proceed without them. If the world government is strong, Russia will join eventually. Nonmembers of the world government would be invited to send observers to its council so that they could assure themselves that they weren't being plotted against.

d) that delegates to the world organization should be elected directly by the citizens of the member nations, and not appointed by the national governments of the member states. (AEA 32-447-1)

Golden noted that though Einstein didn't insist on it, he indicated that "perhaps the number of votes of each country should be proportional to the number of professional men (or college graduates or some other such standard of intellectual hegemony) rather than the total population." Golden commented further that

it pains [Einstein] to see the development of a spirit of militarism in the United States which follows from its experience in the last war. The American people are tending to become like the Germans—not like the Nazis but those of the Kaiser. He says that Americans are beginning to feel that the only way to avoid war is through a Pax Americana, a benevolent world domination by the United States. He pointed out that history shows this to be impossible and the certain precursor of war and grief. There can be no lasting enforced peace. The benevolent despot becomes a tyrant or a weakling; either way the structure crumbles.

The German people have been ruined by their military spirit which stems from Bismarck.[54]

He made these views public in an article entitled "The Military Mentality," published in *The American Scholar*, the quarterly of Phi Beta Kappa, in the summer of 1947.

Hydrogen Bombs

During the fall of 1949, following the detonation of "Joe 1"—the first Soviet atomic bomb—on August 29, important deliberations took place within government circles to determine what the American response to this perceived threat should be. Although the feasibility of fusion weapons had been considered as early as 1942, there were as yet no known methods to trigger and sustain a fusion process that could translate into a effective and deliverable weapon. At a meeting in late October 1949, the General Advisory Committee to the AEC, which Oppenheimer chaired, recommended unanimously that the United States not embark on a crash program to build an H-bomb, a weapon whose destructive power would be a thousand times greater than the Hiroshima or Nagasaki bombs and could be made larger at will.[55] Nonetheless on January 31, 1950, President Harry Truman ordered the AEC to do so.

Shortly before Truman's decision, Einstein was petitioned to take a stand on the matter. Abraham J. Muste,[56] the secretary of the Fellowship of Reconciliation, a Christian pacifist organization whose activities Einstein had supported in the past, wrote him asking him to

> throw the full weight of your influence against that decision, no matter what it might cost in reputation or even life to do so. The very least this could mean, would be to say that negotiations with Russia must now be seriously taken up, and that pending these negotiations, and regardless of their outcome, there must be no production of hydrogen bombs, or other super weapons of mass destruction by the United States. It would be necessary to call on your fellow scientists and technicians to join you in making that proposal. . . .
>
> Mere words of protest or resolutions advocating negotiations with Russia, the establishment of international control, etc. are useless in this situation. The only thing that will count, that may bring the people and our political and military leaders to their senses, would be for you and your leading colleagues to take public action indicating your refusal to have anything more to do with the manufacture of weapons of mass destruction, calling upon all scientists and technicians to take the same position, those now involved to resign their posts.[57]

Einstein answered Muste a few days later, telling him that he recognized the sincerity and seriousness of his intentions, but that his request was based on false assumptions. Since he, Einstein, had never participated in any work of a military nature, and except for signing a letter to Roosevelt "in which attention was drawn to the fact that the possibility existed to make [an atomic] bomb, [which he considered to be his duty] because there were definite indications that the Germans were working on such a project," he had never done any work for the atomic bomb. "It would, therefore, be quite ridiculous if I should make a declaration refusing to participate in armament work."

Moreover, Einstein went on to tell Muste that he did not believe that his proposal to refrain from making the hydrogen bomb goes to the heart of the matter. The fact is that those people who have the actual power in this country and in the Soviet Union do not have the intention to avoid the "cold war": "Both sides are using the 'cold war' for their internal political goals—regardless of consequences." In the United States, the process started immediately after Roosevelt's death, and the people in power succeeded in deceiving, intimidating, and "fanaticizing" the public. "I see no effective way how the small group of reasonable and well meaning people could stop this disastrous course of events."

Einstein concluded his letter by stating that he never shirked from expressing his opinion freely and has always considered it his duty to do so. "However, the voice of an individual is powerless against the shouting of the masses—this has always been so."[58]

Muste replied on January 26, taking Einstein to task. He told Einstein that he was well aware that Einstein hadn't participated in any of the technical aspects of the atomic bomb, and that there was no likelihood that he would be asked to work on the H-bomb. Muste explained that he had written because he thought that Einstein believed that "scientific conscience and fidelity to the interests of science" require scientific and technical men and women generally to take a conscientious stand against any involvement in the production of such weapons "whether by direct work or simply by failure to take a public stand on ground of conscience."

Muste went on to ask whether it was not as much Einstein's responsibility and privilege to intervene now with the highest political and military authorities to point out "that it is both unwise and evil, under present conditions, to go for the productions of super atomic bombs" as it was in 1939, under those conditions, to advise Roosevelt of the possibility of making atomic bombs. Furthermore it seemed to Muste that

> Quite apart from the immediate political effect of his action or inaction, the individual has the responsibility before God and future generations as well as his own, to witness for what he believes to be right and true, even though he has to stand alone and his is a voice crying in the wilderness.

Einstein did not reply to the letter. Muste made a further plea in a telegram to Einstein, in which he was joined by several clergymen. They appealed to Einstein to use his great influence to secure the delay of the hydrogen bomb decision "pending thorough public discussion. People must have opportunities to ponder and discuss this life and death issue if the United States is to remain a democratic nation."[59] Einstein replied the same day:

> I received your telegram. But the way you propose seems to me quite ineffective. As long as competitive armament prevails it is not possible to stop the process in one place. The only possible way out is an honest attempt to work for a reasonable agreement with Soviet Russia and beyond this for security on a supra-national basis.[60]

Einstein's refusal to sign Muste's appeal wasn't because he was reluctant to make his views public. He did so on Eleanor Roosevelt's radio program in February 1950. Efficacy was the issue. Einstein was unswerving in his refusal to sign his name to appeals that he believed would prove to be ineffective.

More perplexing was Einstein's refusal in late March 1954 to sign an appeal to President Eisenhower calling on him to order an immediate suspension of the H-bomb tests scheduled to take place in the Pacific in April. The March 1 Bikini tests had resulted in severe radiation burns to military personnel, to the inhabitants of one of the nearby atolls, and to the crew of *The Lucky Dragon*, the Japanese vessel that had been fishing some 80 miles from the nuclear explosion and had been coated with radioactive ash. Once again, Einstein declined, stating that he appreciated the motive of Muste's action, but that he could not participate in the effort:

> Such a small scale enterprise of a few private persons has not the slightest influence on the behavior of people who have already decided and have de facto no freedom to change their attitude. Only powerful political agencies can influence the course of events. I find it not reasonable to do something only to satisfy one's personal urge.[61]

That Einstein was fully aware of the tragic consequences of the Bikini tests is indicated by the little equation he put on top of his copy of the pamphlet entitled "STOP THE BOMB: An attempt to the Reason of the American People." The pamphlet had been sent to him by Muste to support his request to have Einstein sign the appeal to Eisenhower to discontinue testing fusion weapons in the Pacific. The pamphlet described what had happened to the crew of the *Lucky Dragon* and to the fish they had caught. The equation read:

$$A.E.C. = Atomic\ Extermination\ Conspiracy^{62}$$

Similarly, perhaps upon reading the statements that Admiral Lewis Strauss had made in defense of the tests and in denying that any of the inhabitants of the Marshall Island atolls, or any of the military personnel, or any of the Japanese fishermen had been exposed to a hazardous level of radiation,

Einstein composed the following aphorism: "The patriotic lie—that unfailing weapon of political scoundrels."[63]

INDIVIDUAL VS. COLLECTIVE STANDS

Before turning to Einstein's involvement with the Russell Manifesto, I would like to point to a feature of Einstein's stance when participating in collective vs. individual political action. In the collective case, he was rather sensitive to who are the other members of the collective, even when he fully agreed with them on the particular issues that have brought them together. Thus, in his answers to Muste when he refused to affix his signature to Muste's appeals, the reason he gave was not that he disagreed with the position taken by Muste, but rather that the appeal would be ineffective given who the signatories were. There are other similar instances in which Einstein declined to join because he thought the petition ineffective or impracticable. When asked to participate in an appeal to scientists to refuse to work on developments of nuclear power because of its possible evil uses, he declined to so because the answers to the critical questions were negative:[64] Would any action by a group as small as the one they were contemplating have any decisive influence? Would the physicists and engineers necessarily follow the suggested course of action?

Yet he was willing to sign a letter to warn the American people that the United Nations is a "tragic illusion unless we are ready to take further steps necessary to organize peace" and aim at a "Federal Constitution of the World and a world wide legal order." Perhaps the proposal was rendered impractical because it was drafted shortly after the end of World War II, with the memory and pictures of Hiroshima and Nagasaki still fresh in people's minds. But I believe the fact that the cosigners were *eminent* people—Justice Owen Roberts, Senator J. W. Fulbright, Albert Lasker, Thomas Mann, Mark van Doren, Garner Cowles, Mortimer J. Adler, Louis Finkelstein—was also a factor.

Einstein was more willing to affix his name to statements when the cosigners were distinguished scientists and more ready to participate in collective activities with other scientists. Thus at the urging of Szilard in May 1946, Einstein agreed to serve as chairman of the Emergency Committee of Atomic Scientists. The initial mission of the committee was "to advance the use of atomic energy in ways beneficial to mankind," to diffuse knowledge and information about atomic energy, and to promote the general understanding of its consequences. The committee consisted originally of Einstein, Szilard, Hans Bethe, Thorfin Hogness, Harold Urey, and Victor Weisskopf, with Selig Hecht, Philip Morse, and Linus Pauling joining later. Einstein took a fairly active part in the affairs of the committee and allowed his name to be used rather freely for fund-raising purposes.

We shall see that the issue of the eminence of the cosigners of the Russell Manifesto was a factor in the minds of both Bertrand Russell and Einstein.

Einstein's sensitivity to signing documents with others contrasts sharply with his readiness to take a stand as an individual against the infringement of civil liberties brought about by the "military mentality" that was taking a hold in the United States after World War II. Einstein was outspoken and passionate in his defense of civil liberties. In his book *The Einstein File*,[65] Fred Jerome cites numerous examples of Einstein's courageous stand against the excesses of House and Senate committees investigating "fellow travelers." He became prominently featured in the news for protesting the "inquisitions" of the House Un-American Activities Committee and of Senator Joseph McCarthy, Senator William Jenner, and the Senate Internal Security Subcommittee.[66]

THE RUSSELL-EINSTEIN MANIFESTO

It was a reaction to the Bikini hydrogen bomb tests that made Bertrand Russell suggest once again that all fissionable raw materials be owned by an international authority. Russell's concerns with nuclear weapons had started during World War II when he became aware that work was being done to develop fission bombs. Shortly after the bombing of Hiroshima and Nagasaki, Russell made a speech in the House of Lords in which he warned that nuclear weapons based on the fusion mechanism could be made much more destructive than fission bombs, and that in time fission bombs would become much cheaper to manufacture. He was convinced that the Russians would soon have their own atomic bombs and recommended that nuclear weapons be placed under international control. He had supported the Baruch Plan for an International Atomic Development Authority. In fact, he considered the situation when the Soviet Union and other nations would be in possession of nuclear weapons so perilous that, in late 1948 when the United States still was the only nuclear power, he advocated that the United States force the Soviets to accept nuclear disarmament, even by threatening immediately to go to war against it if necessary.[67] Russell's anti-Communism became more moderate after the death of Joseph Stalin. McCarthyism, American nuclear policy, and in particular, the Bikini test in 1954 slowly led Russell to believe that United States was a greater threat to world peace, and more likely to start a war in which nuclear weapons would be used, than the Soviet Union.[68]

Most of Russell's efforts after World War II were spent on writing and lecturing about the danger posed by nuclear weapons, advocating world government, being actively involved in peace negotiations, and during the Vietnam conflict, engaging in civil disobedience to protest the war policies. Like Einstein, he believed that world government was the only alternative to the tragedy and ruin that would be brought about by a nuclear war, and that a world government was the prerequisite for the prevention of such a war. He thought that the problem preventing the establishment of a world government was that nations were not yet willing to give it enough power to be effective. Yet war was inevitable as long as different sovereign states tried to settle their disagreements by the use of armed force.

On December 23, 1954, Russell gave a radio address over the BBC on "Man's Peril from the Hydrogen Bomb," which evoked a very favorable response in Great Britain. He indicated that he was speaking not as a Briton or a European but as a human being and stressed that: "There lies before us, if we choose, continued progress in happiness, knowledge and wisdom. Shall we instead choose death because we can not forget our quarrels? I appeal as a human being to human beings. Remember your humanity and forget the rest. If you can do so the way lies open to a new Paradise; if you cannot, nothing lies before you but universal death."[69]

Russell followed this address by drafting a statement for eminent scientists to sign. On February 11, 1955, he sent the following letter to Einstein:

> In common with every other thinking person, I am profoundly disquieted by the armament race in nuclear weapons. You have on various occasions given expressions to feelings and opinions with which I am in close agreement. I think that eminent men of science ought to do something dramatic to bring home to the public and governments the disasters that may occur. Do you think it would be possible to get, say, six men of the very highest scientific repute, headed by yourself, to make a very solemn statement about the imperative necessity of avoiding war? These men should be so diverse in their politics that any statement signed by all of them would be obviously free from pro-Communist or anti-Communist bias.[70]

Einstein answered Russell quickly, telling him that he agreed with every word of his letter. He supported the notion of a public declaration "signed by a small number of people whose scientific attainment (scientific in the widest sense) have gained them international stature and whose declarations will not lose any effectiveness on account of their political affiliations. . . . The neutral countries ought to be well represented. It is vital to include Niels Bohr." He also indicated he would write to some colleagues to obtain names of people in the United States and behind the Iron Curtain who might be willing to sign.[71] Russell agreed, and Einstein asked Bohr to get in touch with Russell. However, Bohr ultimately did not sign the Manifesto. Russell's last letter was dated April 5, 1955, and contained a draft of the Manifesto. A few days before he died, Einstein wrote Russell that he "gladly [is] willing to sign your excellent statement."[72]

CHAPTER SEVEN

The Russell-Einstein Manifesto[73] was issued in London on July 9, 1955; its content closely parallels Russell's BBC Broadcast statement of December 1955. In addition to Russell and Einstein, it was signed by nine other scientists: Max Born (Germany), Percy W. Bridgman (USA), Leopold Infeld (Poland), Frédéric Joliot-Curie (France), Hermann J. Muller (USA), Linus Pauling (USA), Cecil F. Powell (UK), Joseph Rotblat (UK), and Hideki Yukawa (Japan). The Manifesto was an appeal to undertake the abolition of war through a commitment to arms reduction, the first step of which would be the forsaking of nuclear weaponry. It called on scientists to assume their special social responsibilities and inform the public of the technological threats, particularly the nuclear threats, confronting humanity. Since nuclear weapons threaten the future of mankind, mankind must put aside its differences and address this paramount problem. However, it stressed that the renunciation of nuclear and other modern technological weapons is not a solution to the peril; war as an institution must be abolished.

Thus as one of his last public statements, Einstein lent his voice to call for the renunciation of nuclear weapons as a first step toward banning war. The Resolution of the Manifesto succinctly stated what he had struggled for all his life:

> In view of the fact that in any future world war nuclear weapons will certainly be employed, and that such weapons threaten the continued existence of mankind, we urge the Governments of the world to realize, and to acknowledge publicly, that their purpose cannot be furthered by a world war, and we urge them, consequently, to find peaceful means for the settlement of all matters of dispute between them.

One response to the Manifesto was the creation of the Pugwash conferences. From the first in 1957 to the most recent in 2004, the Pugwash conferences have constituted the most sustained attempt by the scientific community to address the threat of nuclear war on an international basis. They "bring together, from around the world, influential scholars and public figures concerned with reducing the danger of armed conflict and seeking cooperative solutions for global problems." They have been consequential and fruitful. Their importance was recognized and rewarded with the Nobel Prize for Peace in 1995.[74] The citation from the Nobel Committee read:

> The Pugwash Conferences are founded in the desire to see all nuclear arms destroyed and ultimately, in a vision of other solutions to international disputes than war. . . . It is the Committee's hope that the award of the Nobel Peace Prize for 1995 to Rotblat[75] and to Pugwash will encourage world leaders to intensify their efforts to rid the world of nuclear weapons.

Epilogue

There is, to my mind, a striking parallel between Einstein's theorizing in physics—in particular, his struggle to unify general relativity, i.e., gravitation and electromagnetism, and the hopes he had for the ensuing theory—and his views regarding world government. Einstein's position regarding a world government, a world court, and the particular features of these organizations mirrors the one he enunciated regarding "constructive theories" and "principle-theories" in physics:[76]

> [Constructive theories] attempt to build up a picture of the more complex phenomena out of the materials of a relatively simple formal scheme from which they start out. . . .
>
> Principle-theories employ the analytic, not the synthetic method. The elements which form their basis and starting point are not hypothetically constructed but empirically discovered ones, general characteristics of natural processes, principles that give rise to mathematically formulated criteria which the separate processes or the theoretical representations of them have to satisfy.[77]

For Einstein, the establishment of a world government and of a world court were the starting point and the basis for any proposal to prevent war. They formed the axiomatic basis for any model delineating particular institutions and specific mechanisms to eliminate wars. Their nonexistence in the past and the concomitant constant occurrence of war was the empirical proof of their necessity. Their stipulation—the charter that would establish them and delineate their broad powers—articulated the analogue of a principle-theory. Einstein's varying, detailed suggestions for their implementation—details about membership, governance, mechanisms for arbitration—were the parallel to constructive theories.

Acknowledgments

I am indebted to Finn Aasserud, Snait Gissis, Paul Forman, Gerald Holton, Victor McElheny, and Skuli Sigurdsson for very helpful comments. I would like to express my appreciation to Ms. Chaya Becker, Ms. Barbara Wolff, and Dr. Roni Grosz at the Einstein Archives at the Hebrew University Library in Jerusalem for their assistance in helping me obtain relevant materials. I would also like to thank them for permission to quote from these materials.

For Further Reading

Goudsmit, S. A. *Alsos*, rpt. ed. Los Angeles, CA: Tomash Publishers, 1989 [originally published in 1947].

Heisenberg, W. *Encounters with Einstein.* Princeton, NJ: Princeton University Press, 1989.

Hoffman, B., and H. Dukas. *Albert Einstein: Creator and Rebel.* New York: The Viking Press, 1972.

Holton, G. *Victories and Vexations in Science.* Cambridge, MA: Harvard University Press, 2005.

Masters, D., and K. Way. *One World or None.* New York: McGraw Hill, 1946.

Reves, E. *A Democratic Manifesto.* New York: Random House, 1942.

Rhodes, R. *The Making of the Atomic Bomb.* New York: Simon and Schuster, 1988.

Rhodes, R. *Dark Sun.* New York: Simon and Schuster, 1995.

P A R T

II

ART AND WORLD

EINSTEIN AND 20TH-CENTURY ART: A ROMANCE OF MANY DIMENSIONS

Linda Dalrymple Henderson

A CENTURY AFTER EINSTEIN's *annus mirabilis* of 1905, much research remains to be done on the impact of Einstein and Relativity Theory on 20th-century art as a whole. For the scientist whom *Time* magazine declared the "Person of the Century" in December 1999 and whose effect on 21st-century culture is unquestioned, this situation may seem rather surprising.[1] In large part, it is the result of a misidentification—that of Einstein's theories with Pablo Picasso's Cubism—which began to be made widely in the 1940s and which came to dominate the question of Relativity's relationship to modern art. Typical of this view is painter Philip Courtenay's essay "Einstein and Art" for the volume *Einstein: The First Hundred Years*, published in 1980 to mark the centennial of the scientist's birth. "Cubism attempted to incorporate Einstein's fourth dimension to gain 'realism of conception,'" Courtenay asserts, concluding, "how much modern art was aided by Einstein's ideas is an open question; that it was aided is not."[2] As recently as 2004, Mary Acton's *Learning to Look at Modern Art* declared, "two years before Picasso painted *Les Demoiselles d'Avignon*, Einstein had published his Theory of Relativity. A central idea in the theory was that our view of the world cannot be understood in a purely three-dimensional way because of the existence of the fourth dimension, which is time."[3]

Such vague associations of Cubism and Relativity, usually on the grounds of references to the "fourth dimension" in Cubist literature, had first been promulgated widely in Sigfried Giedion's *Space, Time and Architecture* of 1941 and were given their fullest exposition in the writings of Paul Laporte in the late 1940s. For both Giedion and Laporte, connecting modern architecture and Picasso's Cubism to Einstein was a means to validate new forms of artistic expression and to argue for their grounding in culture at large.[4] Having observed this development firsthand, prominent art historian Meyer Schapiro reacted strongly to these claims, and at the Jerusalem Einstein Centennial Symposium in 1979, he delivered what Gerald Holton has described as "an extensive and devastating critique of the frequently proposed relation between modern physics and modern art."[5]

Schapiro's specific target was the purported Cubism-Relativity link. Unfortunately, he never reworked the talk into an essay for the 1982 publication of the conference proceedings, and thus his arguments reached a larger audience only in 2000, with the posthumous publication of his book *The Unity of Picasso's Art*.[6] In the meantime, other scholarship on this subject had begun to appear, including my 1983 book *The Fourth Dimension and Non-Euclidean Geometry in Modern Art*. That text pointed up the absence of accessible literature on Relativity Theory in France in the pre–World War I era and established, on the contrary, the Cubists' focus on the spatial "Fourth Dimension" that had been the subject of intense popular interest in the early decades of the century.[7]

Instead of the fourth dimension as time in the four-dimensional space-time continuum Minkowski had formulated in 1908 for Relativity Theory, Cubist painters and theorists were stimulated by the notion of a suprasensible fourth dimension of space that might hold a reality truer than that of visual perception. An outgrowth of the mid-19th-century development of *n*-dimensional geometry, the spatial fourth dimension had first been popularized widely in E. A. Abbott's 1884 *Flatland: A Romance of Many Dimensions by a Square*, a cautionary tale about refusing to believe that one's world was merely a section of the next higher dimensional space. A massive amount of popular writing on the subject followed, including a 1909 *Scientific American* essay contest on the question, "What is the fourth dimension?" with entries received from all over the world.[8]

Beginning with the Cubists in pre–World War I Paris, artists in almost every modern movement engaged the spatial fourth dimension in one way or another during the first three decades of the century. In both Picasso's *Portrait of Ambroise Vollard* (Fig. 8.1) and French geometer E. Jouffret's 1903 rendering of a "see-through" view of a four-dimensional solid, transparent, multiple views of an object as well as shifting, shaded triangular facets create an ambiguous space that cannot be read as three dimensional (Fig. 8.2). "I paint objects as I think them, not as I see them," Picasso declared, and—along with his engagement with the art of Cézanne and African sculpture (as well as the science discussed below)—contemporary interest in a higher dimension of space encouraged his increasingly conceptual approach to the visible world.[9] Cubist theorists Albert Gleizes and Jean Metzinger drew directly on Henri Poincaré's ideas on tactile and motor sensations in his 1902 *La Science et l'hypothèse*, in

FIGURE 8.1 Pablo Picasso, *Portrait of Ambroise Vollard*, 1909–10; oil on canvas. Pushkin State Museum of Fine Arts, Moscow. © 2006 Estate of Pablo Picasso/Artists Rights Society (ARS), New York.

which he asserted that "motor space would have as many dimensions as we have muscles" and suggested that one might represent a four-dimensional object by combining multiple perspectives of it.[10]

Whereas Schapiro had argued correctly against the Cubism-Relativity myth, his treatment of "science" in this period solely as Einstein and Relativity led him to argue against any sort of art-science connection in the early 20th century. In fact, Picasso and his fellow Cubist Georges Braque—as well as virtually all artists before the later 1910s—actually *were* responding to certain ideas in physics. However, it was the exhilarating discoveries of the 1890s redefining the layperson's understanding of matter and space (e.g., X-rays, radioactivity, the electron, and the Hertzian waves of wireless telegraphy)— and not Relativity Theory—that excited artists and writers in the first two decades of the new century.[11] In addition, the ether of space and its model of continuity and interpenetration had been embraced by the general public and were not to be dislodged easily, even after the popularization of Einstein's theories in the wake of the 1919 eclipse expedition that confirmed his prediction of the curvature of light by the mass of the sun.[12] Rather than Relativity Theory, then, Picasso's *Vollard* portrait gives visual form to the contemporary conception of space as suffused with ether and matter as transparent and

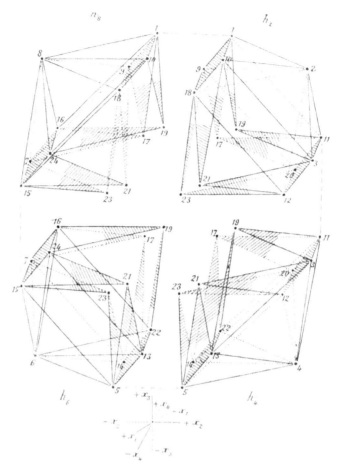

FIGURE 8.2 "*Perspective cavalière* of the Sixteen Fundamental Octahedrons of an Ikosatetrahedroid." From E. Jouffret, *Traité élémentaire de géométrie à quatre dimensions* (Paris, 1903), fig. 41.

continually dematerializing into the ether on the model of radioactivity. If Gustave Le Bon's 1905 bestseller *L'Evolution de la matière* was the primary French popularization of this view, American science writer Robert Kennedy Duncan captured its essence in his 1905 book *The New Knowledge*: "How much we ourselves are matter and how much ether is, in these days, a very moot question."[13]

Marcel Duchamp, Picasso's counterpart in the early 20th century, was the artist most fully engaged with both late Victorian ether physics and the geometrical fourth dimension—a prowess manifested in his extensive notes for his nine-foot-tall work on glass, *The Bride Stripped Bare by Her Bachelors, Even* of 1915–23 (Fig. 8.3).

FIGURE 8.3 Marcel Duchamp, *The Bride Stripped Bare by Her Bachelors, Even*, 1915–23; oil, varnish, lead foil, lead wire, and dust on glass panels encased in glass. Philadelphia Museum of Art, bequest of Katherine S. Dreier, 1952. © 2006 Artists Rights Society (ARS), New York/ADAGP, Paris/Succession Marcel Duchamp.

The *Large Glass*, as it is known, is a scientific and mathematical allegory of frustrated desire rooted in the insuperable contrasts Duchamp created between the realm of the Bride above and the domain of the Bachelors below. Here Duchamp's organic Bride hangs, gravity-free, in an ethereal, indeterminate space he defined as four-dimensional, forever beyond the reach of the three-dimensional, perspectival gravity-bound Bachelor Machine below.[14] The rising acclaim for Einstein and Relativity Theory after 1919, however, would alter the terrain for Duchamp and other early 20th-century artists, gradually displacing the popular spatial fourth dimension and inaugurating the conception of time as the fourth dimension that would characterize the public's view for much of the century. Only in the 1960s and 1970s would the spatial fourth dimension begin to be recovered in popular literature, and it would re-emerge full-blown only in the 1980s with the rise of string theory and computer graphics.[15]

Duchamp, who died in 1968, lived through this metamorphosis and, as a result, waited until 1966 to publish his playfully inventive *Large Glass* notes on four-dimensional geometry and space in his *White Box* or *A l'infinitif*, issued by the Cordier & Ekstrom Gallery in early 1967.[16] Duchamp had published facsimiles of others of his notes in two earlier artist's boxes, and Einstein's status as cultural hero by century's end was made clear in the deluxe publication of his 1912 manuscript on Special Relativity. Advertised as "the book of the century," the manuscript was published with a slipcase and marketed as if it, too, were a deluxe artist's book.[17] From the 21st century we can now observe the waxing and waning of the two competing traditions represented by these individuals and objects: the spatial fourth dimension engaged by Duchamp and others in the early 20th century versus Einstein and Relativity Theory.

From its earliest days, the popular fourth dimension had quickly acquired a variety of nongeometric associations—from mystical higher consciousness and infinity to science fiction usages—that made it attractive to a wide range of artists. Einsteinian Relativity, by contrast, represented a much more specifically scientific or mathematical source, which was also less immediately suggestive to the visual imagination of artists. Nonetheless, a good many artists took up the challenge of addressing Einstein and/or Relativity, and we can begin here to trace the shape of those responses during the 20th century and even propose an initial typology of reactions to them. Their varied form reflects, in part, the changing attitudes toward and understanding of Einstein and his physics during the course of the century. With Einstein standing as the single cultural icon of science for much of this period, any examination of his impact necessarily ranges beyond painting, sculpture, architecture, and experimental film to include the broader field of visual representations—cartoons, popular photographic images, book and magazine covers, and specific scientific illustrations—that served as vehicles for the art world's "romance of many dimensions" with the scientist and his theories.

The cover of the *Berliner Illustrirte Zeitung* for December 14, 1919 (Fig. 8.4), documents Einstein's sudden rise to celebrity that year, declaring, "a new great in world history: Albert Einstein, whose researches, signifying a complete revolution in our concepts of nature, are on a par with the insights of a Copernicus, a Kepler, and a Newton."

As might be expected, the first widespread artistic response to Einstein and his ideas occurred in his home city of Berlin, which during the early 1920s became a crossroads for the international artistic avant-garde. Soon after this cover photo appeared, Berlin Dadaist Hannah Höch incorporated it into her monumental (over a meter tall) photomontage, *Cut with the Dada Kitchen Knife through the Last Weimar Beer-Belly Cultural Epoch in Germany*, dated 1919–20 (Fig. 8.5).

In her commentary on postwar Weimar culture, Höch places Einstein in the upper left quadrant with other signs of revolution—literal and figurative—in opposition to the German military and Kaiser Wilhelm and in league with the Dadaists at lower right. Höch's disjunctive and cacophonous technique of photomontage shouts Dada's critique of values held sacred in art and culture—just as Einstein's Relativity Theory had undercut the absolutes of Newtonian science.[18] No knowledge of the new physics was necessary for

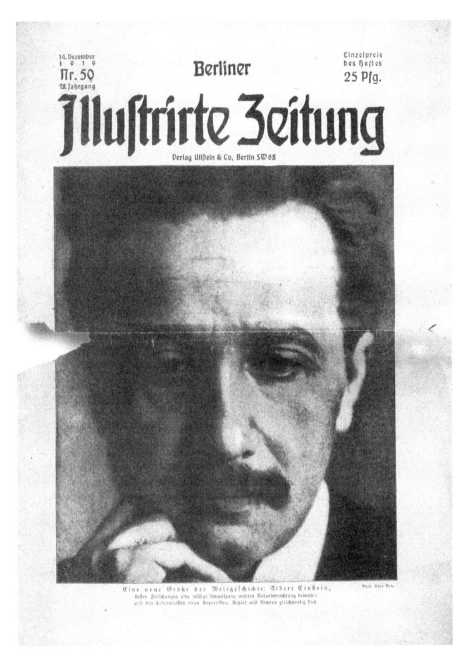

FIGURE 8.4   *Berliner Illustrirte Zeitung*, cover, December 14, 1919.

Höch's message—the face of Einstein sufficed for her purposes, as it would for a number of artists, designers, and cartoonists later in the century.

Predating Höch's exposure to Einstein, the first major attempt in Germany to embody any aspect of Relativity physics was actually Erich Mendelsohn's *Einstein Tower* (Fig. 8.6), built in 1920–21, but designed between 1917 and

FIGURE 8.5   Hannah Höch (1889–1978), *Cut with the Dada Kitchen Knife through the Last Weimar Beer-Belly Cultural Epoch in Germany*, 1919; collage, 114 × 90 cm., NG 57/61. Photo: Jörg P. Anders. Nationalgalerie, Staatliche Museen zu Berlin, Berlin, Germany, Bildarchiv Preussischer Kulturbesitz/Art Resource, New York. © ARS, New York.

FIGURE 8.6   Erich Mendelsohn, *Einstein Tower*, 1920–21. Potsdam, Germany.

1920. Mendelsohn was unique in having a direct connection to Einstein's theories well before his 1919 emergence as a celebrity—in this case through the astronomer Erwin Finlay Freundlich. Mendelsohn had met Freundlich in 1913, when he was already in contact with Einstein about conducting experimental observations to test General Relativity. Freundlich conceived the *Einstein Tower* project with this goal in mind, and as Kathleen James has documented, Mendelsohn's design was clearly stimulated by the astronomer's exposition of Einstein's theories.[19] Seeking to express the new awareness of the energies inherent in mass, the architect chose reinforced concrete (ultimately replaced with brick in key places) to create a "dynamic . . . and rhythmic condition" in architecture. Mendelsohn was also particularly struck by the discussion in Special Relativity of the contractions of form that would be observed at the speed of light, and such deformations of distance and time would subsequently become one of the visual signs of Relativity Theory.[20] Rooted in Jugendstil and Expressionist architectural styles, the *Einstein Tower*—like the Italian Futurism to which Mendelsohn also looked—interpreted science and technology in organic terms. Here his building surges forward as if its internal energies and formal distortions were those of muscles of the human body.

Mendelsohn's organic vision of Relativity Theory was soon to be replaced by an emphasis on geometric form in the context of the Berlin avant-garde's links to Dutch De Stijl and Russian Constructivism. Not until French Surrealism in the 1930s and 1940s would the organic again become a preferred language for Relativity-oriented art. Although in the 1940s the Surrealists' focus would be on imaginative renderings of the space-time continuum, as discussed below, Salvador Dalí's 1931 *The Persistence of Memory* (see Plate 1) engages the distortions of space and time in Special Relativity that had also struck Mendelsohn. Among the Surrealists, Dalí was one of the artists most engaged with science, and he drew on both psychoanalysis and recent developments in science to support his theory of Paranoic-Critical Activity intended to "contribute to a total discrediting of the world of reality."[21] Dalí's interest in Einstein and Special Relativity is clear in a seminal essay of 1930, in which he compares the paranoic "psychic dilation of ideas" to Einstein's "physical dilation of measures." In *The Persistence of Memory*, Dalí, inspired by a plate of melting cheese, gives visual form to the temporal and spatial distortions in Special Relativity, producing, as he later described it, "the soft, extravagant, and solitary paranoic-critical Camembert of time and space."[22]

Given the variety of these three initial images, it is useful to categorize such responses in order to begin to characterize the ways in which 20th-century artists have engaged Einstein and Relativity Theory. Höch's collage (Fig. 8.5) is the first of the genre that would dominate visual representations of Einstein and Relativity—the image of Einstein himself.[23] Formal distortion or specific contraction of forms in the work of artists aware of Einstein such as Mendelsohn or Dalí—or graphic designers of book covers like David Cassidy's 1995 *Einstein and Our World*—would be second on the initial list of eight approaches this essay will suggest.[24] That sign of Relativity, however, has been rarer than the primary artistic response to Einstein that emerged in the early 1920s—the incorporation of time into art. As noted earlier, the new definition of time as the fourth dimension was rooted in Minkowski's 1908 four-dimensional space-time continuum that provided a framework for the viewpoints of all observers after Einstein had made them relative in 1905.[25] Whereas Gottfried Ephraim Lessing in the 18th century defined painting as the spatial art and music as the temporal art, Einstein's and Minkowski's theories now encouraged artists to claim time as their domain as well. This infusion of time and motion into art, which stands as a third type of response, was by far the most prevalent among artists throughout the 20th century.

No other German artist enjoyed Mendelsohn's front-row seat on developing Relativity physics during the 1910s. After 1919, however, Einstein's presence in Berlin and popular fascination with his theories made the city a locus for avant-garde innovation in response to the newest science.[26] Moreover, because of the much greater scientific interest in Relativity Theory in Germany and Russia in the 1910s (in contrast to France, England, and the United States), German and Russian artists who gathered in Berlin were more likely than others to have heard something of Relativity Theory before the 1919 eclipse.[27] Indeed, the two major Russian artists who participated in the Berlin milieu, Naum Gabo and El Lissitzky, had both spent time studying in Germany: Gabo pursued medicine, science, and art history in Munich, and Lissitzky studied architectural engineering in Darmstadt before earning a diploma as "engineer/architect" in Moscow. Gabo later recalled first hearing of Relativity Theory in 1911 or 1912, when he was taking physics classes in Munich.[28] Rather than physics, though, it was Gabo's engineering skill that prepared him to create the first time-based work of art.

While still in Moscow in 1920, Gabo had produced his *Kinetic Construction (Standing Wave)* (Tate Gallery, London), a vertical steel rod mounted on a base containing an electromagnet and springs that set the rod into the vibratory pattern of a standing wave.[29] When Gabo exhibited the work in Berlin in 1922, he highlighted the sculpture's temporal quality with the title *Kinetic Construction (Time as a New Element of Plastic Art)*. Already in their *Realist Manifesto* of 1920, Gabo and his brother Antoine Pevsner had alluded to the newly popular space-time of Relativity Theory: "Space and time are re-born to us today. . . . The realization of our perceptions of the world in the forms of space and time is the aim of our pictorial and plastic art."[30] As the first sculpture in the history of art to incorporate motorized motion, Gabo's *Kinetic Construction* is a milestone in the development of the tradition of kinetic art that would reach its height in the 1950s and 1960s. In a 1957 interview, Gabo affirmed his continued commitment to the temporal fourth dimension: "Constructive sculpture . . . is four-dimensional in so far as we are striving to bring the element of time into it."[31]

By 1922 the Berlin studio of the Hungarian artist László Moholy-Nagy had become a central gathering point for avant-garde artistic discussion. In that year Moholy conceived the kinetic work he called a *Light Prop for an Electric Stage*, which he filmed in 1930 as it produced its moving patterns of reflected light (Fig. 8.7). The *Light-Space Modulator*, as the work came to be known after his death in 1946, would become the single most important icon of the "space-time" world the artist subsequently promoted in his books such as *The New Vision* (1928, 1946) and *Vision in Motion*, first published in 1947 and in print into the 1970s.[32] Instrumental in establishing the curriculum at the Bauhaus, Moholy actually met with Einstein in 1924 to discuss the possibility of his writing a popular book on Relativity for the Bauhausbuch series.

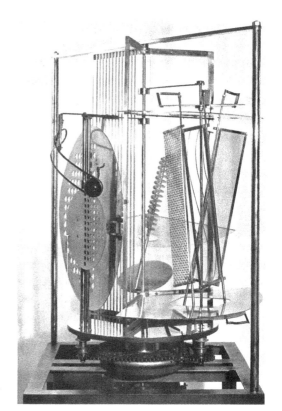

FIGURE 8.7   László Moholy-Nagy
(Bacsborsod, Hungary, 1895–1946,
Chicago, Ill., USA), *Light Prop for an
Electric Stage* (*Light-Space
Modulator*), 1930; aluminum, steel,
nickel-plated brass, other metals,
plastic, wood, and electric motor.
Photo: Junius Beebe. Courtesy of the
Busch-Reisinger Museum, Harvard
University Art Museums, Gift of Sibyl
Moholy-Nagy, BR56.5. © 2006 Artists
Rights Society (ARS), New York/VG Bild-
Kunst, Bonn.

Although Einstein did not write the book, he lent his name to the school's Circle of Friends.[33] For Moholy, Relativity Theory was emblematic of a fundamental cultural shift toward a more dynamic worldview to which artists must respond by replacing the static methods of the past with an art of motion and time.

The first artists to write extensively about the new importance of time to art were the Dutch De Stijl artist Theo van Doesburg and the Russian Constructivist El Lissitzky, who were both drawn to Berlin in this period. Having heard of the experiments in film by Hans Richter and Viking Eggeling, van Doesburg visited Berlin in December 1920 and ultimately stayed on in Berlin and Weimar from 1921 through early 1923.[34] In a lecture of 1921 or 1922 entitled *The Will to Style*, van Doesburg addressed the pioneering experiments of Eggeling and Richter:

> As a result of the scientific and technical widening of vision a new and important problem has arisen in painting and sculpture beside the problem of space, and that is the problem of time. [Noting the need to incorporate time in painting, sculpture and architecture, he suggests that the synthesis is already being

achieved in film.] Here . . . new artistic form is being created from the combining of the impetus of space and time (example: V. Eggeling and Hans Richter).

. . . Using film technique in the painting of pure form gives the art a new ability: the artistic solution of the dichotomy of static and dynamic, of spatial and temporal elements, a fitting solution to the artistic needs of our time.[35]

Eggeling and Richter had been so confident in the relevance of their new "universal language" of abstract forms in motion that in 1920 they had solicited from Einstein, among others, letters of support for their application to Universum-Film AG for technical assistance. The film studio granted their request, although there is no evidence that Einstein wrote such a letter.[36]

Van Doesburg, like a number of more established artists by the 1920s, had actually explored the spatial fourth dimension before Einstein's emergence in 1919, and those ideas remained with him as he contemplated the implications of Relativity Theory for art. After his initial enthusiasm for Eggeling's and Richter's direct incorporation of time in abstract film, van Doesburg, who died in 1931, spent much of the rest of his career seeking in his architecture and painting to merge the earlier, spatial fourth dimension with Relativity Theory's emphasis on time. His most successful efforts were in architecture, where drawings such as those in his 1924 *Color Construction in the Fourth*

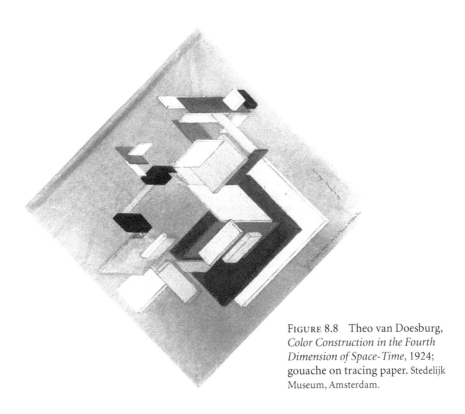

FIGURE 8.8   Theo van Doesburg, *Color Construction in the Fourth Dimension of Space-Time*, 1924; gouache on tracing paper. Stedelijk Museum, Amsterdam.

Dimension of Space-Time (Fig. 8.8) served as models for his designs for houses. Van Doesburg argued that architecture must break out of the traditional "box," and he compared the space of his new anticubic architecture to the hypercube of four-dimensional geometry.[37] He further added color to his buildings to emphasize the necessity of movement in time to viewing the new architecture. Along with Moholy-Nagy, van Doesburg was an important source for Sigfried Giedion's theory of modern architecture as the expression of Einstein's space-time world in his 1941 *Space, Time and Architecture*.[38]

Before considering the role of time and space-time in the work of van Doesburg's fellow visionary Lissitzky, a final aspect of the De Stijl artist's film theory reveals a fourth vehicle for artists responding to Einstein in this period and later: light—either alone or in relation to time and/or filmmaking. Echoing the argument he had made for an anticubic architecture, van Doesburg utilized the four-dimensional hypercube in a 1929 essay on film to suggest that the film projection surface should be broken open to create a new "light-space continuum."[39] Van Doesburg's terminology, in fact, may well have been a source for the later usage of *Light-Space Modulator* as the title for Moholy's *Light Prop*. Both Moholy's emphasis on the role of light in the new space-time world in the *New Vision* and Giedion's *Space, Time, and Architecture* were critical stimuli in the 1940s for American painter Irene Rice Pereira. Like van Doesburg, Pereira was grounded in the tradition of the spatial fourth dimension as well as the new world of Relativity, which she embraced.[40] As she declared in a 1940 lecture,

> In abstract art, space-time is the dominant concept. . . . [T]he abstract artist seeks, by using contemporary knowledge, to create new forms to express the new age. Space-time is his medium, and substances of modern science the material with which he works. In its own right, abstract art seeks plastic equivalents for the revolutionary discoveries in mathematics, physics, biochemistry, radio-activity . . .[41]

In works such as *Transversion* (Fig. 8.9), Pereira painted on the back of the newest types of corrugated and textured glass (i.e., "substances of modern science") and then mounted these panes in front of painted panels to incorporate light as energy directly into the work of art. The exploration of light—"light-space," as this approach might be termed—was reaffirmed as appropriate to the "century of Einstein" by sculptor Athena Tacha in a 1967 article on *Sculptured Light*. In her text Tacha addresses both static and kinetic light works by contemporary artists as well as by the pioneers of the 1920s, including Gabo and Moholy-Nagy, whose description of light as "time-spatial energy" she notes. Manifesting the continuity between the 1920s and 1960s in the "century of Einstein," Tacha concludes, "Light-art is intrinsically not just three- but four-dimensional, since time is also an essential quality of it."[42]

FIGURE 8.9   Irene Rice Pereira, *Transversion*, 1946; ceramic fluid and porcelain cement on two panes of corrugated glass on a bottom panel of oil on hardboard. The Phillips Collection, Washington, DC.

Returning to 1920s Berlin, in contrast to van Doesburg's initial advocacy of time in film as the appropriate expression of Relativity Theory, the Russian Lissitzky only espoused time or motion in the mid-1920s. Before that, in the paintings he termed "Prouns" (an acronym for "project for affirmation of the new"), Lissitzky, who had been a pupil of the Suprematist Kazimir Malevich, worked to extend his mentor's investigations of a cosmic spatial fourth dimension in relation to Relativity's new conception of space-time.[43]

Referring repeatedly to Einstein, Lissitzky declared in his 1920 essay "Proun," "Methods which were once employed in a particular branch of art, knowledge, science, philosophy, are now being transferred into other areas. This is happening, for example, to the four coordinates of Minkovsky's world: length, breadth, height, and the fourth one, *time*, are being freely interchanged."[44] Lissitzky preserved Suprematism's freedom from orientation; however, in contrast to Malevich's planes of color floating freely in an absolute white space, Lissitzky in *Proun 30T* (Fig. 8.10) figures a relational space created and curved by the complex forms within it. In many of his Prouns, the artist also

FigURE 8.10   El Lissitzky, *Proun 30T*, 1920; oil on canvas. Kunstmuseum Hannover mit Sammlung Sprengel.

achieved an unprecedented degree of ambiguous spatial shifting through his use of Necker-reversing, axonometrically projected three-dimensional forms.[45]

By the time of his 1924 essay *A.[rt] and Pangeometry*, however, Lissistzky had rejected the possibility of effectively figuring space-time in painting, declaring that "the multi-dimensional spaces existing mathematically cannot be conceived, cannot be represented, and indeed cannot be materialized." Noting that "space and time are different in kind," he concluded that "time [itself] now becomes a factor of prime consideration as a new constituent of plastic F[orm]."[46] Lissitzky referred to the Prouns as a "interchange station between painting and architecture," and by 1923 he had already begun to move beyond painting to incorporate time and motion directly into exhibition spaces, such as his *Proun Room* of 1923 and subsequent designs.[47] Here Lissitzky set his viewer into motion in an environment of geometric shapes painted or mounted on walls, creating a new kind of perceptual experience. Declaring his preference for physical space versus mathematical spaces in *A. and Pangeometry*, Lissitzky touted a new, motion-generated "imaginary" space that would produce a "fundamental change" in the "apparatus of the senses."[48] Yet in

1924 Lissitzky continued to celebrate non-Euclidean geometry's "explod[ing] of the absoluteness of Euclidean space," as he had done in the 1920 "Proun" text, and the curved elements present in many of the Prouns strongly suggest General Relativity's description of the non-Euclidean curvature of the space-time continuum in the vicinity of matter.[49] Lissitzky made this very point in a 1924 letter to De Stijl architect J.J.P. Oud, arguing that the straight line "does not correspond with the universe, where there are only curvatures and no straight lines."[50] Indeed, the non-Euclidean curvature of space-time would become an increasingly important theme in the subsequent figurations of space-time, particular among the Surrealists, whose form language itself was already organic and curvilinear.

"Imagining space-time" effectively designates this fifth approach to Relativity Theory, beginning with Lissitzky and continuing with the Surrealists in the 1940s (and occasionally in subsequent decades). In his 1939 essay "Des tendances les plus récentes de la peinture surréaliste," reprinted in *Le Surréalisme et la peinture* in New York in 1945, Surrealism's founder André Breton discussed the artists Roberto Matta Echaurren, Gordon Onslow Ford, and Oscar Dominguez as specifically concerned with the "four-dimensional universe" of space-time. Of their works such as Matta's 1944 *The Vertigo of Eros* (Fig. 8.11), Breton explained,

> Though, in their forays into the realm of science, the accuracy of their pronouncements remain largely unconfirmed, the important thing is that they all share the same deep yearning to transcend the three-dimensional universe. Although this particular question provided one of the leitmotifs of cubism in its heroic period, there is no doubt that it assumed a greatly heightened significance as a result of Einstein's introduction into physics of the *space-time continuum*. The need for a suggestive presentation of the four-dimensional universe is particularly evident in the work of Matta (landscapes with several horizons) and Onslow Ford.[51]

Although sharing Dalí's organic form language, Matta and his young colleagues rejected the clearly three-dimensional perspectival space within which Dalí's watches deform (Plate 1) in favor of a dimensionally suggestive amorphous space with no definite horizon or clear spatial orientation. The young Surrealists also embraced non-Euclidean geometry wholeheartedly—both for its new association with Einstein and Relativity Theory and for its longstanding function as an iconoclastic sign of the overthrow of traditional absolutes (i.e., Euclid's axioms).[52] By the later 1950s and 1960s, in fact, "space-warp" became a buzzword to suggest curved space-time, its popularity augmented by the Space Age, science fiction films, and the popularization of black holes in the 1980s.[53]

FIGURE 8.11 Roberto Matta-Echaurren (1911–2002), *The Vertigo of Eros*, 1944; oil on canvas, 6'5' × 8'3". The Museum of Modern Art, New York, given anonymously (65.1944). Digital image © The Museum of Modern Art/Licensed by SCALA/Art Resource, New York. © 2006 Artists Rights Society (ARS), New York/ADAGP, Paris.

By the time the Surrealists had come to New York during World War II and were producing such paintings, there had been another round of publicity about Einstein and Relativity in the United States. His initial impact in 1919 and the early 1920s had produced extensive newspaper and periodical coverage, including seventy-seven stories in the *New York Times* during 1921. Einstein visited the United States in winter-spring 1930–31 and 1931–32, lecturing at Cal Tech and generating new publicity with discussion of his latest adjustments to General Relativity.[54] Two events of 1931 graphically register the impact of this visit: the publication of the song "As Time Goes By" and the painter Stuart Davis' speculations on Relativity Theory in his daybooks of the early 1930s. Although "As Time Goes By" is most closely associated with the film *Casablanca* (1942), Hermann Hupfeld wrote the song for the 1931 musical *Everybody's Welcome*. Associating the current era's "apprehension" with "speed and new invention / and things like fourth dimension," the verse then complains, "Yet we get a trifle weary / with Mister Einstein's theory"— surely a response to the new wave of Einstein publicity in 1930–31.[55]

FIGURE 8.12   Stuart Davis, *Landscape with Garage Lights*, 1932; oil on canvas. Memorial Art Gallery of the University of Rochester, Marion Stratton Gould Fund. Art © Estate of Stuart Davis/Licensed by VAGA, New York, NY.

Stuart Davis, America's preeminent painter in the planar style of later Cubism—and a friend of Pereira since the early 1930s—was hardly weary of Einstein's theory in this period. In addition to jazz rhythms, which he saw as a vital new form of expression, Davis took to heart the implications of the newest science for art.[56] In *Landscape with Garage Lights* of 1932 (Fig. 8.12), Davis gave form to his conviction that "through science the whole concept of what reality is has been changed."[57] James Jeans' *The Mysterious Universe* had made that science newly accessible in 1930, and in his daybook of 1932, Davis responded directly to Jeans' diagrammatic explanation of the interrelationship of space and time (Figs. 8.13 and 8.14).

Davis's notebook page includes similar diagrams with the notations "When you draw this—you are drawing this rectan[gular] linear space, potentially," and "you are drawing this angular direction, potentially, when you draw this."[58] Akin to Paul Klee's reconsideration of the basic marks an artist makes, Davis found in Jeans' diagrams a new avenue for the modern painter to configure nature.

Above the daybook drawings in Figure 8.14, Davis noted,

> From any given point the line moves in a two-dimensional space relative to all existing points. . . . Relativity, knowledge of this fact, and the ability to visualize logical correlatives of a given angle allows the artist to *see* the *real* angular value of his drawing as opposed to associative value.[59]

The angular structure underlying works such as *Landscape with Garage Lights* recalls space-time diagrams like Jeans', which register both motion in

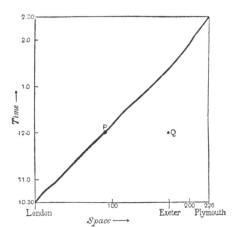

FIGURE 8.13 "Diagram to illustrate the motion of a train in space and time." From James Jeans, *The Mysterious Universe* (New York: Macmillan, 1931), 108.

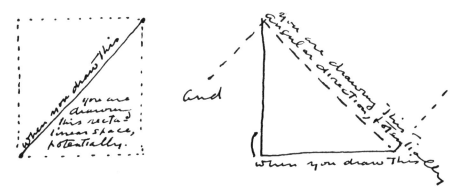

FIGURE 8.14 Stuart Davis, Daybook drawing, 1932. From *Stuart Davis*, ed. Diane Kelder (New York: Praeger, 1971).

space and the passage of time. Indeed, Davis continued in his daybook, "The picture itself could be called a Duration of so many seconds in Time (Einstein has said that space is a fourth dimension). Therefore we must build a picture with four coordinates. . . ."[60] Although Davis would subsequently move beyond his focus on angles, Relativity Theory clearly continued to serve as an inspiration: he adopted the phrase "color-space" (with its suggestion of color-space-time) to describe his space-making system, and in 1947 referred to the act of painting as "a new event in Time and Space."[61] Stuart Davis, not Picasso, was the Cubist painter stimulated by Einstein and Relativity Theory—at the very moment, ironically, when the Picasso-Einstein myth was emerging in the 1940s in the wake of publications such as Giedion's *Space, Time and Architecture*. In a typology of artistic responses to Einstein, Davis represents a sixth category, which might be termed "diagramming relativity," comprised of artists who have drawn directly upon diagrams and illustrations in sources on Relativity.[62]

If Davis was reading James Jeans, Andy Warhol, by contrast, was not engaged specifically with Einstein's physics when he included the scientist in his 1980 silkscreen portfolio based on a series of paintings, *Ten Portraits of Jews of the Twentieth Century* (Plate 2). Instead, for Warhol, Einstein—like Sigmund Freud, Gertrude Stein, and the Marx Brothers, among others—was a popular culture icon, a star like other of Warhol's subjects, from Marilyn Monroe to Elvis Presley.[63] Nonetheless, Warhol's image may well be another example of an artist diagramming Relativity, if in a very general way. In contrast to all of the other, highly colored portraits in the series, the Einstein image is monochromatic—akin to the ubiquitous black-and-white photographs of Einstein and, specifically, to one of Philippe Halsman's photographs of Einstein of May 1947.[64] More importantly, instead of the freely overlaid blocks of color in the other images, the placement of the gray-tone planes and the unusual darker gray bar in the Einstein image strongly suggests the angular shift of frames of reference diagrammed by Minkowski in 1908 (Fig. 8.15).

That diagram of the Lorenz-Einstein transformation was readily available in sources from the 1960s–70s, such as the Dover Publications paperback *The Principle of Relativity*, which also bore a photograph of Einstein on its cover.[65] In addition, the two sides of Einstein's face are rendered differently, effectively splitting his visage into two different frames of reference. Warhol was a keen observer of popular imagery, and, without engaging physics directly, he responded to the photographic record of Einstein (like Höch) and, it seems, augmented his portrait's meaning with a widely reproduced, condensed sign of the significance of Relativity Theory.[66]

Einstein's death in 1955 focused attention once again on the person of the man, whose image was already solidified enough that a fictional Einstein-like

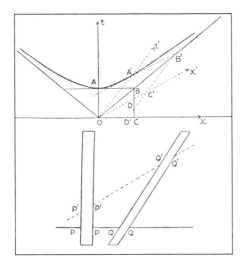

FIGURE 8.15   Hermann Minkowski, fig. 1, lower half [Lorentz-Einstein transformation]. From "Space and Time" (1908), in *The Principle of Relativity* (London, 1923; New York, 1952).

genius could be evoked in the 1951 film *The Day the Earth Stood Still* simply by a blackboard covered with equations.[67] Those equations could be distilled to $E = mc^2$ (or a near approximation, as in the Sidney Harris cartoon, Fig. 8.16) with no loss of meaning, especially when joined to fly-away hair and rumpled clothing. Those same metonymical signs accompany the *Einstein Action Figure*, marketed in 2003. In the Einstein figure's packaging, the scientist is set against a background of star-filled space, with both $E = mc^2$ and the popular sign of the atom with its orbiting electrons floating at his sides. In contrast to the rectilinear grid on which the figure stands, a curvilinear (i.e., warped) line of letters above his head declares that "imagination is more important than knowledge."[68] This sign-oriented approach might be termed "popular iconographical signs of Einstein and Relativity" and added as a seventh element in a typology of artistic responses to the scientist. This is an iconography that may or may not include images of Einstein himself and consists of elements such as the scientist's hair and rumpled clothing, $E = mc^2$, equation-covered blackboards, distorted or impossibly numbered clocks, the phrase "space-time," and curvilinear grids.[69] That the new German Einstein memorial coin does not include an image of Einstein—but rather simply his equation on a spherical mass curving a spatial grid—testifies to the power of this language to communicate.

FIGURE 8.16   Sydney Harris, Untitled cartoon, 1997.
© ScienceCartoonsPlus.com.

CHAPTER EIGHT

This was the Einstein to whom Robert Wilson responded in his collaboration with composer Philip Glass on the 1976 opera *Einstein on the Beach*. According to Wilson, the opera's title was inspired by photographs of Einstein, the sailor.[70] While Glass prepared for the project by reading extensively on the subject, Wilson looked at photographs and talked to people about Einstein: "I just want to know what the man in the street knows because that's what they'll be bringing to the work," he declared at the time. Wilson considered Einstein to be a "fellow dreamer, mystic, and time traveller," and in *Einstein on the Beach*, he took a highly poetic approach to his subject, working to heighten the audience's sense of time over its five-hour expanse.[71] Besides his focus on time, emphasized by the rapid and repetitive mathematical counting in the lyrics, Wilson relied on the familiar iconographical signs of Einstein, dressing all his cast members to look like the scientist. As Wilson has explained,

> Einstein talked about trains, so you see a train . . . in profile, coming across the stage. It is then interrupted by a line, because Einstein said that if you saw a train going across a field it would look like a line. There were all kinds of references [which] I prefer to use . . . in such a poetical sense.

To the train Wilson added Einstein's violin, another image with resonance for anyone aware of the scientist's personal life. And drawing upon the primary aspect of Einstein's impact on modern art, Wilson has declared, "For me, the basis of all architecture, and constructing any work, is time and space."[72]

Wilson had been a student of Sibyl Moholy-Nagy and considers her course in the history of architecture to be "the most important class I had in school."[73] Through Moholy-Nagy's widow, Wilson clearly absorbed Moholy-Nagy's profound commitment to Einstein and the world of space-time. Although Moholy-Nagy died in 1946, his message lived on in his book *Vision in Motion*, which was required reading in art schools in the 1950s and early 1960s. Moreover, that was the moment when kinetic art emerged full blown as an international movement and was regularly discussed as the appropriate art form for the world of Einstein.[74] Moholy-Nagy's 1922–30 *Light Prop/Light-Space Modulator* (Fig. 8.7) was often included in exhibitions as if it were a work of contemporary art, and it regularly figured in chronologies for such shows. Paris was a centerpoint of international activity in kinetic art, including the 1955 exhibition *Le Mouvement* at the Galerie Denise René in Paris. Here, even Marcel Duchamp, enthusiast of the spatial fourth dimension, was recast as a kinetic artist on the basis of his optical experiments with dimensional illusion, such as his *Rotary Demisphere* of 1925 (The Museum of Modern Art, New York).[75]

Robert Wilson was unusual among young American artists in the 1970s in his interest in Einstein and space-time.[76] Between the myth of a Cubism-Relativity connection and kinetic art's dominance as the recognized expression of space-time, Relativity must have seemed used up as a relevant source for younger artists. Robert Smithson, in fact, railed against a fourth dimension associated with Einstein and kinetic art in an unpublished text of 1962, "The Iconography of Desolation."[77] However, Smithson was also in contact with a group of artists around the cooperative Park Place Gallery in New York, where there was discussion not only of the temporal fourth dimension but also of a newly recovered spatial fourth dimension. The ten members of the Park Place group, including Mark di Suvero, Peter Forakis, and Dean Fleming, were the subject of David Bourdon's January 1966 *Art News* article titled "$E=MC^2$ à Go-Go," a title that reflects the dominance of space-time discourse at midcentury.[78] Di Suvero, who saw himself in the tradition of engineer-artists like Calder and the Russian Constructivists, was the group member engaged most specifically with Relativity Theory. "Space-time is the only way you can think since Einstein," he declared to Bourdon. A deep interest in philosophy and science led di Suvero well beyond the popular evocations of space-time to explore the physics of gravity and other forces in the large-scale constructed sculptures (often with moving components) that he has continued to fabricate to the present day.[79]

In his June 1966 *Artforum* essay, "Entropy and the New Monuments," Smithson had referred to Park Place as a "space-time monastic order."[80] However, Park Place artists Forakis and Fleming were energized less by Einstein than by the geometry and philosophy associated with the spatial fourth dimension, which Forakis had discovered in books by P. D. Ouspensky and Claude Bragdon he found in a San Francisco artist's book sale in 1957.[81] Forakis shared Ouspensky's *Tertium Organum* (first published in 1911) with Fleming, and the two friends as well as Smithson also found more contemporary sources on the subject, including the ideas of Buckminster Fuller and the writings of Martin Gardner in *Scientific American* and, for Smithson, in *The Ambidextrous Universe* of 1964. Spatial ambiguity, mirror asymmetry, paradox, and visual illusion were central to the appeal of the spatial fourth dimension for these artists, who sought to undercut "'three-dimensional' logic."[82] In Forakis' *Atlanta Gateway* of 1966 composed of three intersecting 200-foot steel pipes forming two tip-to-tip tetrahedrons with 100-foot bases, the configurations of the sculpture shift radically as a viewer moves around the sculpture (Fig. 8.17). It was Fuller who had promoted the tetrahedron as the building block with which to model "four-dimensional" geometry (with 60-degree versus 90-degree angles), and *Atlanta Gateway*'s unpredictable mutations call to mind the effects of four-dimensional objects in rotation. As Lawrence Alloway wrote in a 1968 article on Forakis, "his work is geometric, but it is a geometry of continuities and double-takes, rather than of stable determinate solids."[83] Although Forakis kept on his studio wall an enlargement of the famous photograph of Einstein sticking out his tongue, Park Place group members were generally more

FIGURE 8.17   Peter Forakis, *Atlanta Gateway*, 1966, steel pipe, Atlanta Gateway Park, Atlanta, GA. Photograph by Angus Winn; courtesy of Peter Forakis.

interested in spatial geometries, including topology. Critics who interpreted their references to the "4D" in terms of Einstein and the temporal fourth dimension alone missed a central aspect of their artistic practice.

The 1960s, however, saw a new phase of work on General Relativity by physicist John Wheeler and others that would recast the study of space-time in far more geometrical terms.[84] This was the Relativity physics that excited painter Tony Robbin beginning in the early 1970s. Once Robbin's attention had been turned to four-dimensional geometry and space, he was remarkably fortunate to be handed a group of early 20th-century books on the subject by a mathematics professor colleague at Trenton State College, where he was teaching.[85] Subsequently, Robbin began a serious study of physics, auditing a course at NYU and working with a tutor, as well as conversing with Wheeler on a lecture trip to the University of Texas. Robbin has chronicled his interaction with mathematicians and physicists, including Paul Steinhardt and Englebert Schucking, in his 1992 book, *Fourfield: Computers, Art, and the Fourth Dimension*.[86]

In the paradoxical structures of four-dimensional space-time, Robbin found an appropriate analog for the complexity of modern experience. For example, in works of the late 1970s, Robbin combined rich textures of color and linear grids in different orientations to create shifting perspectives that refuse to cohere in a single viewpoint.[87] Although Robbin's initial approach to the spatial fourth dimension was largely intuitive, he increasingly engaged four-dimensional geometry. In the early 1980s, he connected with mathematician Thomas Banchoff at Brown University, who was doing pioneering work in four-dimensional computer graphics, and he subsequently studied programming himself. Robbin's painting *Lobofour* of 1982 (Plate 3) combines a patterned background of Necker-reversing four- and eight-sided figures with pairs of projections of sections of the rotating hypercube (both painted on the surface and modeled in wire rods). In addition, the overall collapsing metric of the painting is meant to suggest the curvature of space-time.[88] Robbin's work pioneers an eighth category of responses to Einstein and Relativity, which might be designated "perceptual complexity based on four-dimensional (geometric) space-time." Robbin's continued study of four-dimensional geometry, including its history and its relation to contemporary developments in topology and physics, is the subject of his most recent book, *Shadows of Reality: The Fourth Dimension in Cubism, Relativity, and Modern Thought*.[89]

The first edition of my book *The Fourth Dimension and Non-Euclidean Geometry in Modern Art* appeared in 1983—just a year before the emergence of string theory as well as the field of computer graphics and related ideas of cyberspace, which would all contribute to a major resurgence of interest in higher dimensional space by the end of the 20th century. Following upon the new focus on geometry in work on General Relativity by Wheeler and others, string theory has made dimensionality a lively issue in physics and cosmology and the subject of much science writing for a lay audience.[90] Along with the great popular acclaim Einstein received as the 20th century ended, interest in the scientist and his theories has risen to new levels after a lull during the 1960s, 1970s, and 1980s. Reflecting that renewal, a recent series of works by Pop art pioneer James Rosenquist specifically addresses the moving and stationary frames of reference of Special Relativity. With their distortions and compressions of form, Rosenquist's paintings such as *Mariner—Speed of Light* of 1999 (Fig. 8.18) and *The Stowaway Peers Out at the Speed of Light* of 2000 bring this discussion full circle to the issue of motion-induced contraction explored by Mendelsohn in the *Einstein Tower* (Fig. 8.6). According to Rosenquist, "The paintings are about my imagination as to a new view, or a new look at the speed of light."[91] In *Mariner—Speed of Light*, brilliantly colored forms and their metallic reflections twist, stretch, and swirl, giving visual form to the effects that might be experienced near the speed of light. In paint on canvas Rosenquist produces some of the most visually compelling images of the "vision in motion" that Moholy-Nagy argued

FIGURE 8.18   James Rosenquist, *Mariner—Speed of Light*; oil on canvas, 1999. Private Collection. Art © James Rosenquist/Licensed by VAGA, New York, NY.

was the central characteristic of the new world of Einstein and space-time. Metaphorically, the paintings express Rosenquist's belief in the inevitable difference between his own frame of reference as a painter and that of his viewers and critics, who have not traveled the same lifelong voyage as he.[92]

From Höch to Warhol or from Mendelsohn and Dalí to Rosenquist, Einstein and Relativity Theory have clearly been important stimuli in the 20th century, providing new iconography and encouraging stylistic innovation for a number of artists, architects, and designers. The typology of visual responses herein—images of Einstein, formal distortion-contraction as a sign of Special Relativity, the incorporation of time into art, exploring light and "light-space," imagining space-time, diagramming Relativity, popular iconographical signs of Einstein and Relativity, and perceptual complexity based on four-dimensional (geometric) space-time—is by no means complete, however. Much work remains to be done on this subject, particularly at nodal points where there was a concentration of activity in relation to Einstein's theories, such as Berlin in the 1920s or Paris in the 1950s, with the flowering of kinetic art. Post–World War II Europe and Latin America, on which far less art historical scholarship exists in general, are ripe for the study of art and science and, specifically, the impact of Einstein and Relativity Theory.[93]

Basic exploratory research with an eye to artists' discussions of Einstein and science is the first stage for such a reexamination of 20th-century art. These investigations also need to be undertaken against the backdrop of a more nuanced understanding of the history of early 20th-century science. Duchamp pinpointed the problem accurately when he declared in a 1967 interview, "The public always needs a banner; whether it be Picasso, Einstein, or some other."[94] The sole focus on Picasso and Einstein as fully formed, public icons for much too long, even after the myth of a direct connection between them was debunked, has created a distorted view of the history of modern art and science. Just as Picasso does not embody all of Cubism, Einstein and Relativity do not represent all of early 20th-century science. Most importantly, it is crucial to recognize the continued dominance of the pre-Einsteinian, late Victorian "electromagnetic world view" for laypeople through the 1910s—as well as the loyalty to the ether by both the public and scientists such as Gustave Mie and Oliver Lodge in the later 1910s and early 1920s. Neither the Michelson-Morley experiment nor the publication of Einstein's papers in 1905 spelled the immediate end of ether, as is so often stated.[95] Recently, Peter Galison has issued a similar call for historians of science to "particularize" Einstein and the evolution of his theories, as he recedes further and further behind his words now "splintered into modules, stripped of context, and rendered into slogans."[96]

For cultural historians addressing the engagement of artists with science, the critical interface to be explored is contemporary popular writing on science in magazines and books—e.g., Duncan's *The New Knowledge* (1905), the widely read texts by Eddington and Jeans of the 1920s–30s (kept in print well beyond those decades), or accessible texts written for the public by scientists themselves. Such texts, along with such indexes as the *Reader's Guide to Periodical Literature* or, for the 1960s, the *Whole Earth Catalog*, allow a historian to gauge what scientific ideas were accessible and seen as relevant in the cultural field in which an artist was operating. Gavin Parkinson's excellent book *Surrealism, Art and Modern Physics*, noted above, is solidly grounded in his careful tracing of the popularization of Relativity and quantum physics in interwar France. The same attention to transmission of ideas characterizes two other fine contributions to the study of Einstein and art as well as literature: Kathleen James' work on Mendelsohn, discussed herein, and Holly Henry's *Virginia Woolf and the Discourse of Science*. Finally, Christina Lodder and Martin Hammer's book on Naum Gabo is an exemplar of exacting scholarship on an artist who has long been linked loosely to Relativity Theory but whose range of sources was, in fact, far more varied.[97]

Because the "fourth dimension" was at the center of the Cubism-Relativity myth, this essay has focused primarily on Einstein's and Minkowski's ideas on space and time. Apart from Mendelsohn's focus on energy in his "dynamic functionalism," Einstein's E=mc² equation has been addressed in this essay primarily as a visual sign for Relativity and not in terms of its content. Yet new conceptions of energy and of mass/matter-energy interaction were vital to many artists both in the context of the pre-Einsteinian ether physics as well as in the decades after Relativity's triumph.[98] Artists Athena Tacha and Agnes Denes, for example, have been deeply engaged with Einstein's physics of energy as well as space-time since the later 1960s, with Tacha also reading extensively in quantum physics. In her 1978 essay "Rhythm as Form," Tacha includes "the interchangeability of matter and energy" along with gravity and the "interdependence of space and time" among the concepts of science that she seeks "to render tangible and communicable to others" in her sculpture.[99] Denes has discussed her detailed drawings that diagram or map invisible systems and structures as a response to the dynamic realm of Einstein's "four-dimensional principles of relativity," in which "objects become processes" and "[m]atter is a form of energy."[100]

Finally, artist Matthew Ritchie, who initially came to prominence in the 1990s, is another figure for whom the full range of Einstein's science has stimulated a highly creative imagination. Ritchie's art is grounded in his extensive reading in science and other fields, out of which he creates complex systems of information and narrative, as in the 2003 exhibition *Proposition Player*. Yet Ritchie is primarily a painter, and his central expressive vehicle is an organic language of line and color drawn and painted on walls and floors and even suspended in space. Ritchie's swirling imagery regularly evokes comments about its energy, and curator Lynn Herbert's phrase "energy-filled continuum" is particularly apt for an artist who is adding a unique new phase to the tradition of 20th-century artists responding to Einstein.[101] Now augmented by more recent issues in physics, including the eleven-dimensional universes of string theorists like David Gross and the membrane theory of Lisa Randall (see their essays herein), as well as time travel, dark energy, and dark matter, the "romance of many dimensions" of artists with Einstein and Relativity Theory is clearly continuing into the 21st century.

9

RENDERING TIME

Caroline A. Jones

ONE CAN MEASURE THE CONTINUED relevance of Albert Einstein for artists through a careful analysis of popular and pictorial understandings of relativity—and this crucial task represents the scholarly achievement of Linda Henderson. An alternative approach is to examine the consistent preoccupations with a spatialized time that infuse the culture from which Einstein himself emerged. This allows us to gauge his impact precisely by comparing the perspectival shifts evident in the visual arts *before* Einstein to those *after* the wide popularization of relativity. Rather than examining texts and historical documents, this brief intervention will focus on works of art, using formal analysis (among other tools) to capture the shifting modes by which painters induce in viewers a particular relationship to space and time. As exemplars, the work of Claude Monet will be seen to exhibit the profound tensions in apperceptions of time and space during the long 19th century, while Matthew Ritchie's projects will emblematize the Einsteinian "relativistic" perspectives that proliferate in contemporary art today.

Roiling all thinking subjects of industrializing urban modernity, obsessions about the stability of spatio-temporal perceptions can be seen to have contributed to the necessary 19th-century conditions for the eventual uptake of Einstein's ideas. Without investing in ideas of "influence" on Einstein or his popularizers, I offer a conceptual juxtaposition that folds telling moments of pre-Einsteinian territory against equally salient ones within the Einsteinian terrain of contemporary art. This juxtaposition specifically brackets early 20th-century artists' grapplings with Einstein's new theories. The purview of other scholars in this volume (and of most writers on Einstein's impact), these direct encounters, misprisions, and creative misunderstandings are purposefully folded out of view, precisely to lay out as starkly as possible the contrast posed by my comparative analysis. First, I will set up what I take to be Monet's modernism and locate it in a dialectic between flux and *petite sensation* that itself aimed to resolve conflicts in earlier spatio-temporal visualizations. Second, I will parse the contemporary work of explicit Einsteinian Matthew Ritchie. The primary comparison set up by this argument pivots between the epoch before Einstein and our own. Einstein is the unspoken fulcrum. This is not to say that physicists were the only ones to grapple with the meaning of time and space in modernity (Freud, Marx, Virginia Woolf, Pablo Picasso—the century unfurled scores of comparable titans), but it *is* to claim that Einstein's theories were part of the mix, crucial for 21st-century modes of representing space and time in the known universe.

To abstract these claims: Monet kept faith in a transferable point-of-view, a positivist phenomenology in which sensations that had been generated over time, and registered in the artist's body, could become universal through the work of art. Yet there was a fundamental instability in his work, a spatio-temporal ambivalence if you will, that the work of Einstein would help later artists to crystallize and critique. At the other end of the 20th century, Ritchie—with his own unique vision of Einstein's universe—is the beneficiary of such questions. Ritchie has no illusions about the artist's capacity to transfer specific "sensations." He paints universes of dizzying possibilities where the spatialization of time is a shifting, multiply perspectival affair and where Einstein's science is central to attempts at comprehending a relativistic cosmos.

A Brief Introduction to Western Pictorial Temporality

There is nothing exhaustive about the two positions I have apostrophized as Ritchie and Monet, nor were their particular pictorial solutions unique. On one account, to look at the history of art is to see an endless succession of attempts to spatialize humans' experience of time. Yet what I want to remark upon in the long history of time's artistic visualization is not its variety, but rather its ruthless *standardization* in the modern period—through the development and institutionalization of what was called, appropriately enough, "perspective." I will not belabor this history here, but suffice it to say that art historians have struggled long and hard to dismantle the teleological view of Western vanishing-point perspective as somehow inevitable because it was "correct."[1]

The longer prehistory of polymorphously spatialized time is important to understand, in part because those modes were definitively flattened and simplified—first by Renaissance perspective, then by French academic realism, then by a triumphant modernist formalism. What I want to offer here is a glimpse of the ways in which Einsteinians such as Matthew Ritchie have moved beyond debates over perspective (both the traditional Renaissance kind and its "avant-garde" sequel, a flattened modernist abstraction) to find a relativistic path allowing a reinvention of temporal complexity.

In the next chapter, Ritchie adds his voice to ongoing critiques of claims that the photographically frozen temporal instant somehow replaced what painting had always done. He reminds us (and we always do need reminding) that the syncretic art of pictorial representation, in his words, "is a succession of discrete actions, the performance of which, inside the artist's mental equivalent of Hilbert space, generates more and more local complexity that then acquires the appearance of order." What I want to focus on here are the particularly distinct ways in which this "order" has been defined by human bodies rendering time.

Roughly speaking, Western artists' attempts to render (and thus spatialize) time pictorially can be seen to pursue two general strategies since the Classical age: phenomenological—for example, they induce the physical experience of time through its re-creation in the viewer by means of "reading time"; and rhetorical—for example, they suggest temporal shifts in a pictorial narrative through signification, by means of the presentation of particular *signs* for the passage of time. Neither exclusive nor definitive, these approaches could be mixed, and the balance between them is what will prove so suggestive in examining the pre- and post-Einsteinian spatialization of time.

Western European artists inherited various systems of visualizing time in two spatial dimensions that essentially utilized reading or viewing time to evoke different moments in a narrative, or distinct registers of experience (divine, secular, and so forth). Early Medieval codices, for example, attempted to maintain what we might call "papyrus-time" (imported from Egypt to the Greeks and thence to classical literature and early Christian texts)—the slow tempo of an unfurling scroll transferred representationally onto the pages of the new medium of the book.[2] It is at this point that scroll traditions of left-to-right, top-to-bottom reading systems were brought from Greek and Latin practices into European painting, introducing an orderly temporality that had once flowed from the flexible scroll's sequences of texts and pictures passing through the hands, and under the eyes, and that was now transferred to the flat, bounded picture on the page of a book. The pictorial conventions of the scroll could be only awkwardly mapped onto the new format of the single page, with "sections" of a scroll simply visually transferred to the top, middle, and bottom of each page.

The vagaries of time in such works were themselves harnessed to the primary goal of medieval eschatology—to remind the viewer and reader that secular time, with its seeming linearity and immutable progression, was but a confusing mask that occluded the eternal and its well-springs of immortality.[3]

Critiques of secular time were mounted through *simultaneous* (i.e., on the same page) depictions of events narratively constructed (i.e., in the accompanying parables) as having taken place at different times. In this sense, the book afforded possibilities that the scroll, with its literal linearity, had restricted. The more randomly accessed (and composed) pages of the codex *destabilized classical representations of time*. Codices' more open structure forced the medieval artist to visualize pictorial time; the spatialized order of the page could now inscribe eternity against the secular through the manipulation of two newly separable systems: reading-time and narrative or iconic temporal signifiers. The mapping of these two systems allowed a new, highly complex sacred geometry of the page.

The integrated calibration of spiritual and secular time continued to be a consuming pictorial problem for humanists in the Renaissance. Visualizing mortality in the plane of the page or the panel, and placing it conceptually against eternal duration, were goals that met a new challenge, as the pictorial field (in illuminated manuscripts no less than portable panel paintings) became singular. Was "reading time" lost in the fictive instantaneity of linear perspective? Order was now to be contained within a single, unified geometrical-spatial system, presented to the view of a single eye (imagined to be that of the owner of the increasingly portable panel picture). Only the snuffed candle or the eternally blooming lily could speak of the end or endlessness of time.

As Heidegger would observe of this epoch, once Renaissance artists and theorists acknowledged that painting could have a "perspective," the possibility arose that there might be others—introducing relativism into the unified world picture at the very moment of its origin.[4] Thus, even in one of the most powerfully unified pictorial systems the West had ever conceived, representations of space and time contained the seeds for their own destabilization, much as the unproved postulates at the base of Euclid's compelling system of geometry became the fissures through which relativistic space-time could emerge.

Academic pedagogy would suppress such instabilities in the spatio-temporal frame of art. The royal or princely institutions called "academies" were highly successful in establishing Renaissance perspectivalism in Italy, France, England, and eventually Germany and the New World. It is no accident that the artistic innovations around temporality that make Monet so interesting to art historians today were forged in conscious opposition to the French *Académie des Beaux Arts*. It could be said of the academic system that it existed to transmit temporal stasis: repeatable formulae for producing spatially stable images of a given temporality. The great "machines" of history painting constituted the most valorized achievement for any academic painter, machines by which the rush of increasingly industrial time could truly be tamed into (ideological as well as pictorial) stasis. Yet at the shaky origins of these academic establishments were artists whose work was not yet formulaic, as one can see in the spatio-temporal oscillations of a truly gifted artist of the French Enlightenment, Nicolas Poussin. Working more than 200 years prior to Monet, he proposed a dynamic "neo-classicism" that would allow France to take up the mantle of a classical past that had actually occurred elsewhere (Rome, where he apprenticed himself for many years), even as that past could be continually reinvented for the present. Poussin's painting was itself a response to vacillating post-Renaissance schemes for depicting space and time, and its function was to still the fluctuations of the moment to achieve the time-lessness of eternal art. Pictorially, the movement of the depicted body (rhetorically spatialized time) and the movement of the eyes across the image (phenomenologically spatialized time) were increasingly stilled in favor of static tableaux that could be mastered by a steady gaze—a development apostrophized by Poussin's two versions of a motif he titled *Et in Arcadia ego*. Dated 1620 and 1638, the two compositions both address the ultimate stilling of time through death ("and even in Arcadia am I"), yet they go about this quite differently.[5] Each image arranges three shepherds and a shepherdess in a bosky landscape, but in the earlier painting they tumble toward a stone tomb that bears the "et in arcadia ego" inscription, while the later work shows them calmly arranged on either side of that same tomb. The earlier canvas wants both narrative and "reading time," the later work, eternity. The first depicts a left-to-right progression of awareness among the figures, ending with a skull that literally terminates the story. In the second, the painted figures are positioned like symmetrical bookends, closing movement the way parentheses close a digression, to keep it from breaking out into the text. The icon of the skull is part of the first version's rhetorical strategies for spatializing time; in the second version, Poussin eschews such iconography for a static, single-point perspective that stills the reader's own body (a physical rather than conceptual allusion to death). The closed composition that freezes phenomenological reading time works powerfully to evoke the eternity that is his theme.

One could argue that time and its spatialization are what these paintings are actually *about*. For our purposes, what is interesting is Poussin's shift from a narrative flow, which is located in the viewer's reading body, to a punctiform, static eternity, which imagines that body can be transcended through a masterful, universal gaze. This dialectic—which I will designate as flow versus point—is what Monet would visually restage as the central problematic of modern existence. It would be taken up philosophically by Bergson and scientifically by Mach, Einstein, and others.

Whereas these different systems of order were brought in by Poussin to tame what Ritchie calls "local complexity," what I want to emphasize here is how a variety of temporal devices were still available to the young painter. Increasingly, artists standardized and simplified their modes of rendering time to reinforce the putative instantaneity of vision itself. One need not take sides or produce a Whiggish history to argue that visual "instantaneity" is itself an artifact of positivism, buttressed by the genre policing of Gotthard Ephraim Lessing on the theoretical front (for example) and a thousand academic "machines" of history painting in the field of praxis. The erasure of the situated, positioned body and its dynamically relative point of view would be precisely where Einstein (among others) would take a stand. Monet stood productively in-between.

Le Petit Sensation and Embodied Time

Around the 1850s and 1860s, discourses of positivism became the common coin of French modernity, and art became a primary vehicle for capturing the almost infinite subtlety of *le petit sensation*. This little sensory signal was seen as coming from the world to register itself with great precision in the artist's body, after which it would be filtered through the artist's temperament—it was a stimulus from which more complex perceptions of space and time would be built. The "little sensation" or "fresh sensation" sought by the Impressionists replaced earlier notions of the painting as a reflection of *disegno interno* (the ideal design as it existed in the artists' mind). The internal ideal, much like the authority of the ancients, was dismantled in favor of a questioning, testing, and infinitely reciprocal relation to empirical sense-data streaming in on the body from the modern world.[6]

Influencing these 19th-century ambitions to deploy subjective vision strictly in service of data from the world was doubtless the physiological psychology of Victor Cousin, delivered through the centralized French educational system to every bourgeois schoolboy in the country.[7] Trained to attend to his attending, this newly self-conscious subject formulated his sensations from a Cartesian and then a Comtean confidence in the observing self. The complex interplay between "subjective" impressions of "objective" sense data furthered a literary and artistic modernism in which the artist's goal was to summon an embodied sensory apparatus to produce what was called, appropriately enough, the *cuisine* on the picture's flat surface.

Inheriting, adapting, and disputing the developing notion of *le petit sensation* was philosopher Henri Bergson, who had more impact on French modernism in the 19th century (and well into the 20th) than perhaps any other intellectual at the time. Bergson was in touch with the scientific and mathematical ideas of his age, and although he would later meet Einstein himself (in order to disagree elegantly with his ideas), he knew both Georg Riemann's and Nikolai Lobachevsky's non-Euclidean geometries well before. The grouping of non-Euclidean geometries as a class (by Riemann himself in a 1854 speech, published in France in 1867), ensured that both positively and negatively curved space would be seen as *plausible alternatives* to a strictly Newtonian universe—that is, not as competing and incommensurable descriptions of the real, but as relativistic possibilities within it.

Bergson's central philosophical contribution lay in his insight that time could be subjected to the same critique. Although he opposed relativity with a poorly founded argument against Einstein's view of simultaneity, his philosophies were equally opposed to the instrumental (and industrial) segmentation of time, its separation from lived experience. For Bergson, experience was durational; he gave that claim spiritual as well as political dimensions. In this, I will argue, he was in perfect synchrony with Claude Monet (whose work was important for Bergson's ideas). Against the traditional "clockwork" universe of René Descartes, with its mechanistic atomized movements aggregating to a regularized (and regularizable) structure of time and space, Bergson (in ideas recently recirculated by Gilles Deleuze) emphasized flux and *durée*. What makes Bergson useful (to Deleuze and to European artists of his own time) was the way he used arguments about space-time to theorize subjectivity itself.

In the years Bergson was completing his doctorate (received in 1889) on the topic of *Time and Free Will*, the nature of the observing subject had already been a pressing issue for the French painters who called themselves the Impressionists. These artists were attempting to work out how the subject's enduring personality is not a fixed thing prior to perception, but is produced through shifting sensory encounters. Fueling Bergson's own thinking were these theories of the self as a consciousness emerging through flux and change, cemented only by memory into coherence. Significantly, Bergson and his interpreters were explicitly opposed to the interpretation of time in spatial terms (whether the scansion of a second hand or the plotting of a work day on a calendar grid). Such segmenting modes were, for Bergson, an *in*appropriate rationalization of an inherently intuitive life process. Thus, paradoxically, it was Impressionist painters' obsessions with a spatialized sensation (*le petit sensation* that materialized, through the temperament, as a dab of paint on canvas) that drove Bergson's critique, even as the philosopher's identification of the problem fed the visible anxieties of an artist such as Monet.

Segmentation of the little sensation would be, for Bergson, industrialization's mark on the embodied consciousness of the artist. But of course segmentation was what the good Cousinian schoolboy was trained to do, and artists contemporaneous with Bergson did it extremely well. To be modern, an artist such as Claude Monet had to render each retinal sensation as a concrete, positivist fact. (As Paul Cézanne would later opine: "Monet is only an eye. But good Lord, what an eye!" [*Monet ce n'est qu'un oeil. . . . Mais, bon Dieu, quel oeil!*])[8] Artists could hardly dodge the problem of spatialization, inherent as it was in the two-dimensional challenge of making a painting. But Bergson made them understand the cost of freezing that space in a single universal *punctum* of experience.

Thus there is nothing so simple as a causal force linking Bergson's philosophies with contemporaneous painting such as Monet's, but one can still be struck by their shared obsessions with embodied time[9]—obsessions that I want to emphasize here. Coming at the end of the 19th century and the beginning of the 20th when Impressionism had fallen out of fashion, Bergson's theories explicitly rejected some of the earlier painters' atomizing approaches.

Mathematician Edouard Le Roy (Bergson's chief interpreter) put it this way in 1912: "Our aim is to modify the habits of imagination . . . to break through the mechanical imagery in which we have allowed ourselves to be caught; and it is by awakening other imagery and other habits that we can succeed in so doing."[10] The tension between the atomized moment codified by *le petit sensation* and an integrated experience of flux and *durée* was already palpable by the late 19th century. I will argue that we can see that tension in the paintings of Monet, fueling Bergson's antimechanistic campaign and preparing the way for debates over Einstein's relativity in the decades to come.

The paintings of Monet constituted precisely the kind of modern art Bergson and his followers had to worry about. Such worries would not arise from their subject matter, whether steaming trains or placid poppy fields, but their *complete renegotiation of the pictorial conventions for rendering and spatializing time.* Those conventions, as I argued above, had been gradually standardized since classical times; I want to suggest that they were pictorial equivalents to the mathematical universe of Euclid and Newton that even the French Revolution had not shaken. Yet his own experiences of modernity pushed Monet to take the spatio-temporal frame of perception through far more dramatic changes than painting had suffered in the several centuries before him. His intentions were first signaled in the paintings of modern transport—trains steaming into the 1877 *Gare Saint-Lazare*, for example (Fig. 9.1). They would be confirmed by the grand Impressionist series he completed in the late 1880s and 1890s.

Monet's views of train stations celebrated the new tectonics of steam—those unprecedented industrial atmospherics, belching from strange iron horses, that flattened the picture plane and joined the railroad itself in "annihilating space and time" (to borrow a phrase from one contemporaneous writer).[11] Whereas the older repertoire of spatialization devices had used reading conventions to

FIGURE 9.1   Claude Monet, *The Gare Saint-Lazare: Arrival of a Train*, 1877; oil on canvas (31⅝ × 38⅝ in.). Harvard University Art Museums, Fogg Art Museum, Bequest from the Collection of Maurice Wertheim, Class of 1906, 1951.53. Photo: David Mathews © President and Fellows of Harvard College.

simulate the passage of time (toward death, as in Poussin's first "Arcadia" painting) or used perspective and symmetry to evoke the stillness of eternity (Poussin's second version), Monet and his contemporaries began to flatten perspective and avoid narrative, the better to capture a purified present composed of infinitely simultaneous sensations. It may seem that this infinite simultaneity resembles, phenomenologically speaking, the mastering gaze from the Enlightenment, and both are similarly suspended from the entropic rush of embodied time. Yet Monet's fictive simultaneity had a very different historic role to play. The new spatio-temporal system crafted by the Impressionists was obsessed with the momentary; it sacrificed the fullness of time and change for the apparent flatness and paradoxical permanence of the fleeting slow impression, forever fixed by paint.

Just how was a self to be constituted from such perceptions, these little sensations wrested from the *flux*, drawn up out of Bergson's flowing *durée*? Is the reality of such an "impressionistic" world to be trusted? (Bergson would say no.) Do the rapid accelerations characteristic of modern life disprove the old certainties of a clockwork universe and its unmovable prime mover, or do they instantiate it? The base unit of modern artistic phenomenology—*le petit sensation*—was to be the tool for hammering away at such questions, dismantling the barrage of sense data both crowding and constituting the modern self. In the practice of Claude Monet, the little sensation was made analogous to the unit of the brushstroke itself.

This can best be seen in Monet's late 19th-century paintings of the Rouen cathedral (Plates 4, 5, 6). Painted as a series and titled for the time of day that their "impressions" were first noted, they were constructed to be compositionally identical. Depicting a single great French gothic facade, they have sometimes been seen as a step back from the earlier restlessness of those modern train stations. In fact, by stripping the nominal modernity of the subject matter away, these canvases enforce the fact that what there is to experience as modern in the painting *resides in the mode of painting itself*. Clearly, the modernity of the motif was no longer necessary. Twenty years after the railway stations, the cathedrals seemed to be their antipodes—paintings of architecture whose stones function as frozen precipitates of eternal, nominally Christian time. But seen in the context of Monet's work as a whole, they can still summon an entirely modern anxiety—not only because the cathedrals fissure into that series of temporally localized views, but also because they offer an almost unbearable tension between *le petit sensation* and the durational self, instantiating deep intellectual, political, and moral conflicts over perception and time.

These conflicts were already alluded to in Monet's earlier paintings, in which space is similarly flattened (or effaced through clouds of steam), and time compressed to an instantaneous moment. But this shockingly modern denial of conventional pictorial reading regimes and narrative devices also participated in contemporary obsessions with time and space. Even as one form of standardization (perspectival realism) was being challenged, another (temporal convention) was at hand. Driven by the speed of the railroad, there was a civic premium on *de*-relativizing time in industrial nations and in confirming their imperial hegemonies, as Galison has reminded us.[12] Industrializing Europe (and its New World) suddenly needed regulated time zones. When Monet was painting those 1877 pictures of signalmen and iron horses, there was still no such temporal regulation in the United States; train travelers could reset their watches to some two hundred local time zones on the transcontinental trip from Washington to San Francisco. France itself was only fully regulated by 1912, when President Raymond Poincaré celebrated with a world time conference in Paris. This temporal regulation culminated at 10:00 in the morning on July 1, 1913, when the Eiffel Tower beamed the first global time signal out into the world.

But what Monet was learning as he codified *le petit sensation* as the tool of Impressionism was that clock time had little to do with perceptual time. In this pre-Einsteinian epoch, it was hard to conceive of the potential elasticity of the clock, its capacity to share the distortions of other types of measuring devices within particular frames of reference, *its links to convention and perception*. But one could easily sense the fluidity of time in embodied perception as against the newly omnipresent mechanical clock (how time flies! Yet also, how it can drag . . .). The spatialization of time had always been one of painting's quintessential problems. Now, time would also be its subject. Even as Monet registered the tick of regularized time in his paintings' titles (*Rouen Cathedral, Noon*), his pictorial choice was also to leverage that clockwork universe by using a phenomenological approach to fix the highly relative human *experience* of time and space in an age that was busy standardizing its mechanical measurement.

There is, to be sure, a clear aspiration to scientific precision in Monet's recording of each turbulent sensory cue. This is no Luddism of the paintbrush. One has only to compare Monet's railroad paintings with their famous antecedent by Turner (the misty *Rain, Steam and Speed* from 1844) to get the point. For all his debt to and interest in Turner's technological sublime, Monet wants to record the precise relation between the stimulus of machines and the sensing perceiver. He has no interest in actual sublimity, in being "overwhelmed." Even in its subject matter, Monet's painting of the *Gare Saint-Lazare* foregrounds the system of signals, engineers, and engines that installed themselves in the heart of industrializing Paris, increasingly regulating contemporary life.

It is Monet's discrete and fattened brushstroke that links the systematic paintings of railroads with the later cathedral series. The brushstroke begins to be an almost palpable equivalent to the punctuated sensory input from the trainyard, even as it complicates all regulatory pretensions. It offers a contrapuntal insistence on the time of the body, against the mechanisms of industrial time.[13] In this way, Monet perfects an experiential scientism that spatializes time by radically compressing it, flattening conventional perspective. But his scientism also imagines that this emblematic unit of the brushstroke will read the same for all viewers—their bodies will replicate the scientifically observant condition of Monet's body, when that eye first clocked its *petit sensation*. There is a single eye, with one (universal) perspective. This positivism will stand in marked contrast to the expanded poly-perspectivalism characteristic of art after Einstein is taken fully on board.

The evident scientism of Monet is pertinent to my argument, reminding us that the artist's obsession with time and perception was neither nostalgic nor uninformed. Famously, he made a portrait of his wife Camille on her deathbed, pushing himself to use the mechanism of his eye (and its positivist protocols) to screen emotion from perception, even as he captured a moment from the flux of shifting colors and sagging lines as her face passed into death. Atomization, that segmentation of the eye, has important work to do in this context—it literally removes Camille's passing from an *occurrence over time* to a moment's impression, while also, of course, fetishizing that moment as a passage *out* of time, into an eternity to which the mechanized eye of Monet will not allude. To recapitulate the dialectic: on the one hand, a *flux* theory of meaning emerging in *durée*; on the other, a positivist *atomization* and abstraction of the perceptual act. The one would give Bergson some of his best ideas; the other, his nightmares.[14]

How does Monet's pictorial dialectic maintain itself at all, given the irreducible atomization of that brushstroke? The answer lies between hue and percipient. For all its sensory segmentation, Monet's practice does not rest with the pure particularity that *le petit sensation* implied. Although each perceptual moment is fragmented into its constituent stimuli, these units are fixed on a canvas such that they must be blended in the eye to make the art— *the flux, denied by the "atom" of le petit sensation, is reconstituted through the act of perceiving the painting.* In this way, long-term conventions of "reading time" are brought back in by Monet's intensely phenomenological strategy. No longer assigning the eye its conventional trajectory of left to right or top to bottom, movement now takes place in a larger unit, the body. Movement toward or away from the canvas—a noticeably modern viewing practice—becomes a child's game of fragmenting and annealing the image. This movement of the body produces alternate frames of reference, literally different *perspectives* on the image that are also the very signal for the competing views of time and space at the heart of Monet's project.

The tension I am describing here emanates from within Monet's practices: a tight oscillation between flux and *durée* (the painter and viewer's embodied reading of the Impressionist painting) versus atom and instant (the dab of paint that crystallizes the momentary *petit sensation*). This tension, first suggested in the railway paintings from the 1870s, came to an apogee in the more systematic series (not only the cathedrals, but grainstacks and lily pads) that Monet began painting in the 1890s. The issues mobilized by these later cycles had real significance given the industrialization of time that was proceeding apace in the world as a whole. The resistance to that mechanization was propelling Bergson to publish the book that came out of his doctoral thesis, *Matter and Memory* (1896), just as it contributed to Monet's move from city steam to country rhythms. Flux and theories of *durée* could be understood as reasserting *private* time (recall the reflective time of the Cousinian schoolboy); where the atomized instant Bergson argued against was the synchronized, coordinated, measurable public time of Poincaré's 19th-century train stations (and Monet's dab of paint). It was Einstein who would take coordinated time and make of it an observer-based relativity, but Monet and his interlocutors were still trapped in a dialectical relationship between two seemingly irresolvable positions. Flux was the ebb and flow of individual consciousness, versus the atomized instant visible in the scientific click of a second hand on a crisply demarcated public clock. Two notions of time—and two modes of access to spatialized temporality in painting—pulled against one another. Only after Einstein could they begin to be understood as coexistent frames of reference, inherently plural perspectives available to be "seen" in the world, or in the work of art.

In his choice of icons for this struggle, Monet is clearly torn. He is attracted to the rigor and excitement of the scientifically measurable instant, but the subjects he wants to paint can be seen as powerful metaphors for *resistance* to this industrialization of our circadian rhythms. Leaving railroad time behind by the 1890s, Monet paints gothic cathedrals and bends in the river, shimmering mountains of hay and voluptuous lilypads, emphasizing seasonal and diurnal rhythms that seem to confirm the eternal verities of life, land, and religion.

But look again at the images of those cathedrals. "Timeless" archetypes of faith, yet measured in punctiform two-hour increments. It is the artist's own obsessive rendering of these motifs that turns them into types of clocks, archaic things that become modern in spite of themselves, sundials synchronized with the march of an atomizing episteme. The shadow scrapes across the sculpted portal, and we go from noon to two o'clock on a sunny afternoon, experiencing the ticking of Monet's brushstrokes as unwavering measurements of the otherwise unremarkable passage of a village day.[15]

Monet's paintings are clocks. Let me clarify this polemical metaphor with a question: are they the synchronized clocks that knit late 19th-century Europe together, and finally regulated the sprawl of the United States? Precisely not. They are perceptual clocks, circadian timepieces that locate a bodily difference between the fresh vision of morning and the sagging eyelids of a hard-working afternoon. But even if they register time's passage as passive sundials rather than complex machines, and even if their subjects are objects of artistry that recall the very dawn of French history—marking out sites miraculously preserved from the flood of mass culture, department stores, and commodities rising in Paris—they are still clocks. However they perform themselves as experiential, internal, and primordial, they are still clocks, forming a remarkable series as regular as the Parisian *horloge mère*.

The cathedrals in particular announce themselves as clocks of perception, invariant in their composition and further regulated by the dabs of paint that break the flux of seeing into units, *les petit sensations*. The cathedrals are endlessly repeatable icons of timelessness that helplessly register the passage of time, forced to count the passing hours in a mercilessly illuminated, temporally dissected pictorial regime. When you see such paintings strung out one after the other along the wall in a retrospective exhibition, their underlying modernity is palpable. They get more and more regular, with Monet's motifs increasingly invariant across shifting perceptual conditions. Suddenly, their endless repetition can be read as itself a figure of industrial production—scores of iterations of a single motif.

The spatialization of phenomenological time emerges forcefully in one small and consistent feature of Monet's cathedral pictures (see detail, Fig. 9.2). The rose window contained in the triangle above the cathedral door becomes a kind of oculus in the painting, a static, staring center that is made oddly parallel with the surface of the canvas rather than following the shallow perspective Monet sets out in the representation as a whole. As if exaggerating the monocularity of Western perspectival painting into Cyclopean dimensions,

FIGURE 9.2 Claude Monet, *Rouen Cathedral, Façade,* 1894 [detail]; oil on canvas (39⅝ × 26 in.). Museum of Fine Arts, Boston, Juliana Cheney Edwards Collection. Photograph © Museum of Fine Arts, Boston.

the single eye of the cathedral's west-facing window looks out at us, waiting to be blinded. It bids us a baleful farewell from the caboose of a receding freight train vanishing into the past.

Monet remains pinned in the stare of his own painting—he can neither wholly embrace the atomization of his eye, nor resist the increasing demand to register that very process as embodied and durational. How productive and fascinating is his oscillation! It captures at the level of perception and the body the deepest struggle of modern painting—to be of one's time (echoing Baudelaire's injunction) yet survive for the ages. It is a definitively pre-Einsteinian gaze, locked into a Manichaean struggle between the imagined fixity of eternal *durée* and the volatile experience of *le petit sensation*. The instant is annealed in a field of onrushing perceptions, and that interaction becomes a metaphor for modern consciousness itself.

The Flatness that Is All-Dimensional: Temporal Spatialization around 2000

Modernism's obsessions with time and space continued, but when that large episteme was itself confronted by postmodernism, such concerns took new forms. Painting ceased to be the primary medium for such investigations, and Einstein came centrally into the artistic imagination with Philip Glass's 1976 opera, *Einstein on the Beach.* It is telling that the most powerful contemporary artist to deal directly with the cultural meaning of Einstein is a composer whose medium is sound in time. For Glass, who developed this opera with theater designer Robert Wilson, Einstein was a figure for the tragic humanist in a postmodern and possibly post-humanist age.

Glass says, variously: "As a child, Einstein had been one of my heroes. . . . The emphatic, if catastrophic, beginnings of the nuclear age [were] the most widely-discussed issue of the day." Elsewhere he refers to the physicist as a "witness" and describes *Einstein on the Beach* as "an opera about a great mathematician who loved music." The figure of a mustachioed, white-haired actor is often on stage in the five-hour opera; you hear him playing violin over the chanting of numbers and the singing of mesmerizing solfege syllables. Glass is concerned not merely with Einstein the myth, but with the new sense of time that he inaugurated: dense, textural, coeval with space, susceptible to gravity and event-horizons. But amazingly, working out of modernism as he is, Glass still hybridizes Einstein's measurement-driven observational physics with the language of flux and *durée* so familiar from Bergson's day. Criticizing the sense of time driving most of Western music, Glass critiques it in typically Bergsonian terms: "As if you were to take a length of time and slice it the way you slice a loaf of bread." In place of that atomization, Glass wanted to construct musical structure through rhythmic cycles and repetitions, like a spiraling orbit in space time from which we never quite return to where we were.[16]

Glass and Wilson's extraordinary opera may or may not have put the war between flux and atomization to rest, but it clearly energized new possibilities for a 20th century marked by the myth and science of Einstein. In many ways, their postmodernism is also the beginning of our thoroughly Einsteinian age. Some of this moves "beyond" Einstein, inaugurating the age after Einstein's death, the age after Q.E.D. (quantum electrodynamics) took relativity in directions the physicist himself could not endorse. But in other ways it is the final phase of comprehensive uptake for Einstein's ideas. Around the time the artist Matthew Ritchie emerged, the pervasive implementation of Einstein's central theories was attested by every satellite-tracking cellphone on earth, and Phillip Glass's opening onto postmodernism (via Einstein) could be taken for granted.

Ritchie's Einsteinian universe extends beyond the artistic landscape inaugurated by Glass (still a romantic Minimalist, after all) and expands even the multiple perspectives suggested by Wilson. If those artists were remanent modernists, Ritchie is resolutely postmodern. His obsessions with time, space, and the body register an entirely different episteme.[17] Ritchie's highly personal periodic tables, voodoo cosmogenies, computer avatars, dark matter maps, and tarot card patterns mix the stuff of hard science with older or more vernacular modes for dealing with the universe—revealing that pretensions to scientism are not what they were in Monet's day. This is not to say that Ritchie has no interest in scientific views—he is as informed as any layperson on the subject of strings and quantum energies. For Ritchie, it is the miraculous flatness of space-time itself that the painter aspires to inhabit. The precise geometry of the universe is analogized to the pictorial condition, resulting in a complex flatness nothing like the negative, ascetic restraints on painting that the 20th century once imposed.

Those earlier ascetic restraints had grown out of pictorial decisions by painters such as Monet—recall the staring oculus that flattened the canvas of the cathedral painting to the imaginary depth of a playing-card. But in Ritchie's thoroughly Einsteinian universe, flatness is merely a dimension, not a negation of space. I am speaking metaphorically, forcefully connecting pictorial discourses of flatness with physicists' views, in which "flatness" is merely one possible outcome for cosmological space-time, dependent on the average mass-energy density of the universe. Rather than displaying the geometries of Lobachevsky or Riemann, where space is curved negatively (and parallel lines expand infinitely away from each other) or positively (and parallel lines eventually cross), Ritchie loves the fact that current astronomical observation seems to confirm an entirely eerie equipoise—or very close to it. Einstein himself mistrusted the Cosmological Constant he theorized almost a century ago, but it was the only thing that made sense of the equilibrium observed in our universe, the only thing that could allow us to exist.[18]

As Ritchie knows, there is nothing "classical" about this equilibrated flatness, since its apparent functionality requires that what we perceive is balanced by far vaster quantities of mass and energy that we cannot perceive or measure—dark matter (the mass that must be out there to make observational astronomy consistent) and dark energy (the latter constituting Einstein's Cosmological Constant, accelerating the expansion of the universe). In other words, the calm "stasis" of the universe (and, by my metaphorical extension, Ritchie's poly-perspectival paintings) relies for its existence on the parallel existence of enormous mass and energy that cannot be perceived (and can be detected only at enormous scales and energies). Perception must multiply perspectives (for the Einsteinian), yet some perspectives are not available to perception at standard human pictorial scale. Confirming the almost Futurist stylistics of speed in Ritchie's paintings, contemporary astrophysicists now surmise that the universe is both flat and accelerating in its expansion (a theory awaiting confirmation from NASA's WMAP project).[19] But if science cannot yet fully record (or perceive) events at these scales, Ritchie can still imagine them. The conceptual topography this artist wants to invoke is at a scale of almost inconceivable dimensions, confounding classical physics in the same way that a cosmic "lens" formed by the gravitation of galaxies confounds the palm-sized lenticular crystal of Newton's optics. Like the WMAP's "'baby picture' of the universe," Ritchie's works are "only images," but in such images are countless perspectives and cosmological claims.[20]

The secular miracle that animates science (and Ritchie) is the remarkably limited range of elements and energies in the world—limits that allow us to become conscious and aware of them *as* limits. Einsteinian in its pragmatic embrace of the given, Ritchie's work is far from limited in its referentiality. Physical limits (mass, temperature, gravitation) restrict the world but not what it can signify. Ritchie's similarly restricted colors (proprietary rather than "primary") could be compared to the astonishingly restricted protein chains that make up organic life—they could arrange themselves in trillions of ways, but in fact they exist in relatively few and harmonious combinations.

The trembling brushstrokes that registered Monet's eye and hand were part of the ticking of a perceptual clock. Ritchie's works substitute for this tick, tick, tick of *les petit sensations* a drift, drip, deluge of data—cascades of information from different reference frames that necessarily take many forms and entail radically different body practices. Yet although his pictures court virtuality, they hardly dissemble their awareness of the viewers' bodies; on the contrary, they count on those bodies to move through the physicalized armatures made out of drawings, or past the palimpsest of informational structures deposited on the pictorial surface. Ritchie, like other artists of his generation, is acutely aware of site, often adjusting works to incorporate specific elements of the room's architecture (sockets and soffets, to say nothing of stairwells, ceiling tiles, and swimming pools). See in particular Ritchie's *Games of Chance and Skill* (Fig. 9.3), an anamorphic image incorporating elements from seven different representational systems, extending in stretched formation on the sides and ceiling of a long corridor next to a swimming pool at MIT's Zesiger Center.

No longer remotely monocular, Ritchie's implied viewer moves within an obviously shifting Einsteinian frame of reference. This mobile viewer is imagined to possess scanning eyes that are cradled in a skull, poised on a neck, mounted on a torso, and mobilized from below.

Time enters into the viewing of these artworks as it did for Monet's—through the annealing of a view. But there is no "snap" into focus, where the cathedral reveals itself as an icon or the haystack settles into a landscape. The viewer traversing (and indeed surrounded by) Ritchie's forty-foot "painting" has no possibility of composing a single view. Viewers move in to read the small units of color, pull back to grasp the vertiginous baroque ensemble, slip across to feel the strangely virtual pictorial world colliding with peripheral vision, and come back to test the current of these often spiraling compositions. Beyond any influence that could be attributed to his understanding of Einstein, Ritchie is drawing on the popularity of gaming and simple mouse-moves—mental disciplines that have trained the average viewer in modes of navigating the virtual. Here, however, the whole meat machine gets into the act.[21] Duration and depth is invoked by viewers' own movement, mobilizing the frozen tableau that "pictures" (from Poussin to Monet) once seemed fated to produce.

FIGURE 9.3 Matthew Ritchie, *Games of Chance and Skill*, 2002 installed at MIT Zesiger Center, Cambridge, Massachusetts. Courtesy Andrea Rosen Gallery, New York.

Ritchie's paintings intensify and layer information so that it can be read at various levels—from the close-up of comic books (with their scrolling sequences evoking papyrus time) to an imagined aerial view of the Big Bang.[22] Crucial to what we might call the "expansion mod" of these pictorially virtual programs is their scalable quality—these paintings' capacity to refer to microbes or parsecs, depending on the frame of reference or mood of the viewer.[23] Twisting, folding, playing with extensive surfaces that we intuitively understand are the source of any possible interiority—this is the experience of the subject that Ritchie's works summon and produce. No coincidence either that this concept of folding-from-the-world is also affecting the look and feel of architecture, design, and some sciences in our Einsteinian present.

Monet's pre-Einsteinian space was the flat and stable surface of a grid, a grid made up of the constituent brushstrokes that reiterated the rectangle of the canvas in small. But as we saw, the Euclidean, machinic modernism of the grid was complicated by the flux of viewing, which always threatened to dissolve pictorial order into a hail of incomprehensible sensations. Ritchie takes this grid and torques it, bearing an Einsteinian debt to the spiral (understanding that "Einsteinian" is the larger cultural fold that brings with it Hermann Minkowski's theories and non-Euclidean mathematics in general). Again, the metaphoric evocation here is to the spiral of space-time—a shift from Heraclitus's temporal river (linear and simple in its topology, however complex are the waters that rush in), and from the Roman-to-modernist spatial grid (regularization across the legions being controlled).[24] The Einsteinian spiral found in the art of Ritchie's generation is intended to critique the grid. The spiral is still systematic, still plottable. But it is inherently disorienting, producing a different kind of viewing subject. Proprioceptively, we never know quite where we are on its curve. If modernist art historian Rosalind Krauss wrote the definitive analysis of grid systems in modernist art,[25] we now need to understand the subject-viewer that is produced in an encounter with the Einsteinian spiral. In the current moment, the spiral is rarely a mathematically precise form (although the natural order of fractals and Fibonacci numbers can occur). More often, spirals in contemporary art are interrupted, crisscrossed, and difficult to parse; they may be figures produced by moving algorithms rather than stable progressions. Gone is the tidy grid, which kept things predictable for so long—whether you were a Roman general confident in the invariant layout of the next garrison town, or Clement Greenberg finding "hallucinated uniformity" in an Abstract Expressionist canvas.[26]

Pictorial underdetermination in Ritchie's paintings has its analogy in the dark energy now speculated to drive the universe, or the dark matter that huddles beyond our instruments' grasp. It remains paradoxical that such brilliantly colorful canvases should summon metaphors of darkness, but this is precisely the way fluorescence functions: to summon energy from imperceptible wavelengths and recalibrate it to radiate in the visible—bringing dark energy into view. The unknowable "look" of the universe in the first three minutes becomes the backdrop for Ritchie's poly-perspectival passion plays. His works function best if we can apprentice ourselves to his conceptual topography, learn its runes, and totter through the allusive scenography with its map in hand. Then we can watch how Ophiel (stand-in for Monera, the bacterial kingdom) draws energy from the Dead (dark matter? At any rate, one of Ritchie's narrative cycles), eventually blossoming on The Family Farm (the cycle that immediately follows it) to bring us into our Holocene timeslice.

The place of embodied time in pictorial experience has been my leitmotif, and Ritchie's work has helped bring it to the present. It is a present that I have argued could only be possible in the multiplied frames of reference Einstein bequeathed us. Staring at Ritchie's abstractions induces reverie (the gaze), but moving around and among them generates a higher informational uptake (scanning, the glance, the unconscious operation of pattern recognition). Suddenly the conscious and unconscious movements of our body—at the level of breath, glance, and stride—are pricked by what the subliminal scanning operation has put together at one of many referential frames. As we look at *The*

Eighth Sea (Plate 7), for example, a "punctum" occurs through our awareness of the depicted bodies and abject fragments that are smuggled into the Pollockian sprawl of lines and the Picabia-like facets of color. The subconscious fires a missile into the neocortex, unravelling the picture in our interpretation even as the picture's depicted body unravels before our eyes.

The blue coils (recurring motifs that have summoned the doomed Trojan priest Laocoön in reviewers' minds) at once cradle and devour this body, nicely complementing the sanguine tones of dissolving flesh. In Ritchie's cosmology, the ropy blue strands stand for Lucifer, light-bringer. And/or the cerebellum. And/or Protista (the kingdom of few-celled eukaryotic organisms). And/or infinity. The eyeballs-become-whirlpools are further confirmation of the scalability of this universe—we could be viral phages looking for a host, or godly creatures watching cycles of entropy on a cosmic scale.

If the Enlightenment bequeathed to Monet a commitment to the pictorial as productive of a single moment, apostrophized in the atom of sensation (albeit tortured by the unresolved experience of *durée*), a fully cultural Einstein allowed Ritchie the freedom of spiralling space-time and multiple-perspective frames of reference. Philip Glass was in the middle of this trajectory. His own anxieties about spatialization (resisting a "loaf of bread" materiality) and his commitment to abstraction left him straddling the modern-postmodern divide. For contemporary artists, strategies for spatializing time are no longer locked into a single frame of modernist flatness. They have been freed by culturally assimilated concepts of relativity, allowed to occupy an expansive, narrativized flatness that spirals out, producing multiple viewpoints in space and time. Our present, then, is resolutely Einsteinian, whether we are scientists listening to the birth-cry of a newly mapped universe and snapping a few baby pictures (the sound and image of space stretching at the origin of time), or art viewers parsing Ritchie's complex palimpsestuous artworks (time layered in multiple narratives and pictorial alternatives of cosmic origin and ludic games).

Eyeball to eyeball, we leave Ritchie in an Einsteinian face-off with Monet, the dizzying ocular whirlpools of a new millennium up against the cyclopean rose window of a 19th-century modernist past. The modernist remains forever trapped in his oscillation between flux and *le petit sensation*, his painting ticking with perceptual units in a brilliant, regimented, somehow tragic performance. The Einsteinian swims and plays in alternate worlds that are only imaginable through relativity, in an expanded universe made possible by Einstein's miraculous year. Monet's single oculus, identified with the church and pinned to the edifice of French culture, stares back at Ritchie's deliriously multiplying optical whirlpools. Being human, we have no choice about which time to live in, but being Einsteinian, we know which one we'd choose.

10

INTO THE BLEED: EINSTEIN AND 21ST-CENTURY ART

Matthew Ritchie

"THE PAINTER, THE POET, the philosopher and the natural scientist each try to make their cosmos and its construction the pivot of their emotional life."[1]

I once heard that Einstein said he got all his best ideas from looking at the sea. An odd myth, since he lived in Switzerland before the publication of the four papers that would change the world, and he was a notoriously bad sailor. But the sea as a metaphor for his work is essential. It is impossible now to imagine the magical sense of scale that existed when it still took weeks, not hours, to travel between countries, when long, slow boat journeys were how you saw the world. But in Einstein's time and still today, the sea is the only physical encounter we have that represents infinity. To look at the night sky is be an observer. When you are in the sea, you are a participant, a swimmer in all the seas of the world. At a sufficient distance from land, there is no distance, no time, no color, no direction that is not a property of the sea. The sea is time and space, thing and thought, environment and element. Alone in the vastness of the sea is where reason yields most easily to an ecstatic awareness of the world-signal, where logic takes its rest on the threshold of the unknowable, infinitely curving paths of the space-time continuum, the darkness quietly lapping in with the tide, into the caldera of the skull, closing the gap between the world and the word. This space is where artists live.

In a larger sense, visual art had always been waiting for Einstein—for as long as the existence of humanity itself, perhaps even longer. After all, it has been an inherent contradiction in artistic practice since cave painting to present an illusion of coherence that soberly reflects the prevailing cultural model while actually working inside a fractured continuum that reverses all traditional notions of causality. Some collective cave paintings in Brazil have contributions dating across hundreds of years. And studies begun with monkeys in the 1970s have consistently shown that behaviors (such as the production of paintings) that were previously understood as essentially human are byproducts of mental processes basic to all primates. And in a way, the kind of inductive, "refinement of everyday thinking" that Einstein liked, the nonverbal perceptual tools that Einstein often spoke of, that he used to see what no one else could see, have always been the tools of the artist. The world of the "picture" has always been a world where light is understood as a fixed and external value separate from the accumulated image, where an unknown agency has provided the momentum, and where about 95 percent of the information is missing—a space we call negative space, and oh—paper is also geometrically flat. As well, artists have always believed our knowledge of nature to be fundamentally limited—since every artist knows that as soon as we grasp one part of the picture, another part slips through our fingers.

But despite these extraordinarily obvious affinities, a portfolio of subjective criteria that embodied questions of experienced and local time for five hundred years, no artist seriously placed any of these issues at the center of their practice. Relying instead on a simplification of Euclidean geometry as a sufficient philosophical endorsement of the artificial construct called perspective, and accepting Newton's optical and mechanical space as a useful if invisible justification for narrative conventions that had originated in a partitioned tribal history that sought its political and theological capital in the representation of the metamorphic miracle, artists, like most people, got along to get along.

After all, Newton himself contributed to a visual culture that relied on the acceptance of certain unreliable models as gospel. He fancied himself a mystic, even an artist of sorts, prone to an obsessive diagramming of the lost temple of Solomon, in a recessive attempt to match geometry to meaning, seeking a parallel model of the hidden universe that would eventually consume as much of his life as Calculus and Universal Gravitation.

Only as logical positivism itself was collapsing in the late 19th century under the concentric pressures radiating from central Europe and perspectival space was revealed as a pictorial gallows did artists scramble for the exits. Nothing made the problem clearer than the appearance of photography. The recording of a single moment, frozen in time, exposed the ancient truth: painting is a succession of discrete actions, the performance of which, inside the artist's mental equivalent of Hilbert space, generates more and more local complexity that then acquires the appearance of order.

Painting was not, nor had it ever been, real. From 1905 to 1917, special relativity, quantum theory, atomic motion, mass energy equivalence, and general relativity contributed to a global psychological submission to scale, an elision between conscious and unconscious perception and a mechanized exchange of time and space that would ultimately transform every social and intellectual subject and progressively eradicate any idea of the "naturally" perceived world being the center of the universe from the practice of art. The exposure of paintings' limitations of locality by photography led first to the defensive fiction of the picture plane, an early form of "brane" that, more than any other mechanism, sustained the transition of modernism from an investigation of the impossibility of truly representing time to an acceptance of the inevitably subjective representation of any image. This high modern model of the artist was exemplified by Willem de Kooning—last paladin of the questions raised by Cubism. Incorporating and then abandoning subjectivity and the relational plane was essential to modernism's ability to reckon with the changes wrought in our understanding of physical reality by Einstein and Niels Bohr. As Linda Dalrymple Henderson has shown in her chapter in this volume, Einstein was so bound to these iconic discoveries of modern physics that his gesturing image, typically scrawling equations on a blackboard, casts a long shadow across this early modern period, even if his work was often only superficially understood or used as a general stand-in for advanced concepts only distantly related to relativity. During the 1950s, for example, a highly charged discussion developed among the abstract expressionists as to whether painting itself was, or took place, in another dimension, a discussion impossible without Einstein and yet as far from his work as the contemporary representations of the fourth dimension that appeared in various efforts allegedly to represent certain kinds of mathematical and physical space. A flattened and collapsed pseudo-geometric space became a shorthand for science itself, creating a popular iconography that was subsumed into the visual language of popular science, a strange cocktail that mixed Piet Mondrian, Salvador Dali, and cartoons.

But placing this new subjective "local" self at the center of the universe was not to last either. Ironically, it was the revival of interest in Marcel Duchamp's work by artists such as John Cage, Robert Rauschenberg, and Jasper Johns (ironic because Duchamp had no real connection to Einstein or his theories) that triggered a second declension of modernism, one manifested through a revived interest in time, chance, and process in the 1960s and 1970s—and thus intertwined with Einstein's legacy. At the same time, there was a strong interest in combining cognitive and behavioral science with the emerging field of information theory to examine certain unsolved questions about perception. Coincidentally or not, this was also the period of the Aspect experiments, designed to review certain philosophical questions stemming from quantum mechanics initially raised by Einstein. By the late modern period as it transitioned into the early postmodern, science, especially science epitomized by the presentational methods favored by Einstein, Marvin Minsky, and Richard Feynman, had become an invitation to indulge in conceptual and creative gamesmanship that found itself inexorably drawn to the limit cases of modern science.

The early 20th-century model of Kurt Schwitter's *Gesamtkunstwerk*, intended as a representational totality that would both extend and mock Wagnerian monumentality, returned as the physical constraints of phenomenological and empirical art bore down on most artists. A more nuanced and temporally fluid approach to the history of information provoked a sustained effort to produce codices and world models through the generation of a map, or continuum, where all positions are potentially occupiable, equally possible in or out of sequence.

As George Steiner put it: "The original *Merzbau* was proposed as a continuum, defiant of any circumscription in time and place. A gigantic metaphor whose components are brought into proximity to affirm their disparity."[2] Artists as diverse as Joseph Beuys, Arakawa, Robert Smithson, Mel Bochner, Barry le Va, Oyvind Fahlstrom, Hannah Wilke, Sigmar Polke, and Alfred Jensen pursued diverse strategies that emphasized the unpredictable effects of accumulated information without context. They were never scientists; Beuys' diagrams, Jensen's and Polke's paintings are useless as scientific maps, as are any other artist's cosmological diagrams. They thrived instead on contradiction and refusal, and resist both conceptual and aesthetic categorization, capitalizing on the confusion that underlay what was now two generations of mass cultural acceptance of Einstein's theories and his benignly magisterial image as the man who in some way "understood" the universe. Joseph Beuys, whose strategy of personal magnetism and glyphic pronouncements accompanied by an information system served up in blocks, countered this iconic image with an artistic equivalent, an incomprehensible universal mythography delivered through lectures (in that same heavy European accent of authority) that served both to amplify and to refute the idea of a dialectical genius. Robert Smithson, his American equivalent; dove deep into the choppy waters of thermodynamics and chiral theology, surfacing with his masterpiece "Spiral Jetty" in the 1970s just as the many worlds theory re-emerged through the causal loopholes in the Copenhagen interpretation. The observer and the observed had changed places. Ideas had become art and artists sought to sculpt reality itself.

Ultimately, this elision of content and form led to a series of collapses and pseudo-revivals of various styles of art-making under the aegis of the "postmodern" that slowly gave way to the primary underlying theme of much contemporary work, information itself. By the 1990s pure information was understood as both presence and commodity in the material world. But a time bomb was waiting for everyone. The revival of a cosmological science built on the unanswered questions in Einstein's work and rebuilt on information flooding in through a shift in information processing took place through an extraordinary new mechanism that has perhaps been described best by Krzysztof Pomian: "We practice every day and at an enormous scale, a kind of cognition that is, despite its being extrasensory, nevertheless a physical fact."[3]

This new form of agency, a vast instrumentality of sensors, telescopes, computers, and detectors that burrow through all distinctions and categories, often in service to experiments that begin by investigating the very questions raised by the Einstein-Podolsky-Rosen paradox, has revealed that underneath us, inside us, behind the inner veil of creation is a universe that is constantly generating its own frame of reference and mapping its own parameters, a continuum that generates our physical world even as it generates our field of consciousness. A further point has been made: the continuum does not contain information; it is information. Whether in the form of qubits, mass/energy, stories, places, essential forms or rules, art that seeks to describe reality in any way, no matter how obscure or personalized, must always describe relationships between competing, opposed, or cooperating forces, penned in by the illusory membrane of perception itself, while the various positions within the structure are all open for business simultaneously in an ecology of information that draws much of its positive but argumentative character from Einstein's own attempts to reconcile gravity and quantum mechanics and from his discomfort with the ghostly underpinnings of his craft.

This new cultural context has bled through into the practice of contemporary art in surprising ways. In direct contrast to the work of the early 20th century, contemporary artists are no longer trying to force an accelerated understanding of an idea by reducing it to a cartoon, the method that ultimately devolved into an illustrational cliché that illustrated neither the theory nor the art. Their visual languages draw a parallel between the real world and the invented, accepting that any metaphor must both include and distort reality in its frame of reference. All their methods stress the individual point of view—they are artists after all, understanding that the observer is never neutral, is always a participant. The camera, the sculpture, or the installation become for them the equivalent of the photonic detector, with the same questions of truth, decoherence, and choice lurking in the background.

It is not enough simply to reference science; that's easy. The harder job is to construct a genuine, if personal, investigation of time and space. Artists as diverse as Doug Aitken, Ricci Albenda, Carsten Holler, Hiroshi Sugimoto, and myself all operate inside overall conceptual models that I think can fairly be linked to competing influences that come in large part from Einstein, influences that have radically redefined our conceptions of order, space, and time and that converge in contemporary art.

CHAPTER TEN

Hiroshi Sugimoto's work from the 1980s and 1990s proposed three kinds of time as an answer to the postmodern confusion then prevalent (Figs. 10.1 to 10.6). A series of sea views shot all around the world emphasizes the essentially unified character of the sea, no matter where or when it is photographed. Although they have different titles, names we give them, nothing really distinguishes the Sea of Japan from the Caribbean Sea or the North Sea. A simultaneity of place is implied.

In Sugimoto's movie house photographs, a single exposed frame captures the light emitted by an entire film, hours of time are collapsed and the idea of narrative, or emotionally experienced time is reduced to a purely physical phenomena.

In his images of museum dioramas, a re-invented model of a static event is treated as if it was a real, temporally limited scene and is supplied with the properties of ambient motion through extended exposure. A repeatable moment is given the illusion of uniqueness.

Carsten Höller's work *Neon Circle* from 2001 (Plate 8) directly attempts to simulate the experience of an accelerated observer's relative appreciation of space-time through the fixed medium of light. Two bands of neon bulbs rapidly accelerate around the observer, giving a surprisingly effective illusion of movement.

Doug Aitken's *Interiors* (Plate 9) uses multiple projections of heavily edited DVDs to build an overlapping environmental symphony of vision. Aitken

FIGURE 10.1 *Perminian Period*, 1992. Courtesy of Hiroshi Sugimoto, Sonnabend Gallery.

FIGURE 10.2 *Devonian Period*, 1992. Courtesy of Hiroshi Sugimoto, Sonnabend Gallery.

FIGURE 10.3 *Lake Superior*, Cascade River, 1995. Courtesy of Hiroshi Sugimoto, Sonnabend Gallery.

FIGURE 10.4   *Caribbean Sea, Jamaica*, 1980. Courtesy of Hiroshi Sugimoto, Sonnabend Gallery.

synchronizes and desynchronizes four disparate narratives that collectively form a soundtrack to a quadratic movie whose whole is greater than the sum of its parts. Like most artists trying to force an approximation of the apparent temporal paradoxes of relativity and create loose analogies to entanglement and other exotica, he is probably constitutionally disposed to challenge the scientific consensus that the effect of quantum activity and large-scale gravity are too small to notice. Artists exist only on a local scale.

My own interest in generating a visual representation of the space-time continuum catalyzed at MIT with a commission for the most Newtonian location of all, the gym. The representation of information in motion became a stimulus for considering the relationships that bind time, space, and information into a coherent space. This in turn led straight to the labyrinth of unresolved questions and eerie theories that spin out from Einstein's legacy. Entanglement and quantum teleportation, dark matter and negative pressure—bring it on; this kind of talk from physicists sends a tingle through the part of the imagination that still thrills to the idea of voodoo science.

In a later work, *Proposition Player* (Plates 10–12), information is consistently used as a generative substrate and as a site of resistance. Einstein's famous lines, "it's hard to sneak a look at God's cards" and "God does not play

Figure 10.5 *U. A. Walker, New York*, 1978. Courtesy of Hiroshi Sugimoto, Sonnabend Gallery.

dice," can be revisited through these pieces: a hand of cards that contains all the physical processes of the universe mapped onto the deck and a craps table where visitors can throw dice to build digital projections that construct atoms from fundamental particles.

The large sculpture, *The Fine Constant* (Plates 13–15), can be understood as a kind of "brane" penetrated by linear elements, which could be seen as representing gravity, and is surrounded by a vast wall drawing titled "The Hierarchy Problem," referencing the formidable imbalance we find in the four universal forces. Around the room and over the wall drawing, paintings describe the timeline of the universe like snapshots of epochs, as if the universe could be seen and understood as a single, continuously evolving object. For me, Einstein's theories and their influence on contemporary cosmology triggered an interest not only in the large scale, but in the broken or even reversed narrative, to occlude conventional readings of story, time, and place.

The convergence of interests became clearest of all to me through Einstein's interest in David Hume, the philosopher he read most avidly just before publishing the special theory of relativity. In his famous analyses of

FIGURE 10.6   *Paramount, Oakland,* 1992. Courtesy of Hiroshi Sugimoto, Sonnabend Gallery.

causality and induction, Hume argued that there is no logical justification for believing that any two events that occur together are connected by cause and effect—or for making any inference from past to future, anticipating some of the conclusions of the general theory by three hundred years. Of equal interest are his lesser known contributions to the history of Scottish law supporting the unique practice of allowing three alternative verdicts in a criminal trial: Guilty, Not Guilty, and Not Proven, a position entirely consistent with Hume's understanding of reality, equally consistent with Erwin Schrödinger's equation, and generally useful as a metaphor for the operation of the scientific method. In the terms used by Anton Zeilinger, the verdict of Not Proven can be understood as the superposition of the states of 1 and 0 (Guilty and Not Guilty), or as a statement that includes more information than its outcome.

Strangely, an analog to this superposition is one of the key states in contemporary art. For an artist, the gaps in the general theory, the Not Proven verdicts, are, to be honest, as intriguing as the parts that work. Einstein's interest in the provable and his consequent struggles with the idea of a statistically determinate universe create one of the most compelling narratives of 20th-century intellectual life. The idea that the universe is a collective form;

a single dynamic entity inside which pockets of activity, like human beings, act as variable focal points, and where the hidden and the revealed information is not only embedded *in* but is a condition *of* the surface, is new and startling. And if everything is information, then information can be understood as having a direct physical effect on the universe. In effect, it *is* the universe; so perhaps emblems that represent pure information, like paintings, are more powerful than we could ever imagine.

The space opened up by Einstein replaced Newton's scaffold with an edgeless sea, a perceptual and permanent bleed that changes thought through perception even as it changes perception through thought. It makes us think about thinking while we're looking at the something that is being thought about. By insisting, even if reluctantly, on the verdict Not Proven, Einstein left an open form for his legacy, a dialectical pragmatism that deals simultaneously with local and universal issues—a final luxury infinite in scope, like all masterpieces.

11

EINSTEIN AND MUSIC

Leon Botstein

FEW ASPECTS OF EINSTEIN'S LIFE and personality are cited with such regularity, in nearly identical phrases and commentary, as his devotion to and love of music. Every biographical account tells the same story. His mother, a pianist of reasonable amateur proficiency, wanted her son to play the violin. It is speculated that she wished to have a partner in the family for *Hausmusik*. Musical culture in the home, which meant chamber music, usually with piano, was a highly prized symbol of successful middle-class acculturation in late 19th-century, German-speaking Europe. It signaled *Bildung*, a sign of status and achievement particularly prized by assimilated, urban German Jews. Of all the cultural practices of gentile Europe, music was the arena most attractive to post-Emancipation Jewry, for in its instrumental forms it did not demand a sacrifice of or retreat from Jewish identity.[1] Einstein is said to have started the violin at five, but he liked neither his teachers nor practicing. He rebelled against several teachers and the regimen of exercises with considerable violence.[2]

However, according to his own recollection, accepted by his biographers, at age thirteen (or thereabouts) he discovered the violin sonatas of Mozart. Einstein fell in love "with their artistic content and unique gracefulness" and learned to play them, thereby improving his skills. This was accomplished—again according to his own account—without "ever practicing systematically." "[L]ove is a better teacher than a sense of duty," Einstein concluded with respect to violin playing.[3] Before accepting this as a truism or an indication of Einstein's skills, readers (particularly aspiring violinists) would do well to recall the dictum attributed to Jascha Heifetz: that without at least three hours of practicing a day, one can't be truly good; if one requires more than six hours, one has no talent. Einstein's account of his own evolution as a violinist does not encourage an optimistic conclusion regarding his basic proficiency.

Nonetheless, from the moment of his adolescent epiphany upon encountering Mozart's violin sonatas, music assumed a pivotal and permanent role in Einstein's life. He played with regularity, using the engagement and concentration (including the unique perceptual distortions of temporal experience) induced by playing an instrument to "gather his thoughts" before finding his way through scientific problems.[4] But even beyond time spent playing alone in this Sherlock Holmesian manner, Einstein was an enthusiast for chamber music and playing with others, often with an audience of friends. Chamber music was a regular part of Einstein's social life in Bern, Zurich, and Berlin, where he played with Max Planck and his son, among others.[5] Later in his life in his Princeton years, owing to his age, he gave up the violin but enjoyed improvising on the piano.[6]

What kind of musician was Einstein? How good was he on his beloved instrument? How literate, in the sense of reading music and knowing theory, history, and repertoire, was he? Biographers cite an 1896 cantonal examiner in Aarau who deemed Einstein's playing of a slow movement of one of Beethoven's violin sonatas as remarkable and revealing of "great insight."[7] This oft-quoted remark, however, circumvents any direct observation regarding his technique. Playing an adagio from a Beethoven violin sonata (probably from the first five sonatas, published before 1801, and not from op. 96, op. 47, or even those from op. 30) with notable musicality tells us just one thing: it is clear that Einstein communicated a sense of feeling, if not musical form and meaning, whether or not he played in time or in tune. What struck the inspector was that Einstein's music making was neither impersonal or imitative (in terms of interpretation), nor a dutiful mechanical exercise. He displayed a deep love of the music, a quality that was and remains in short supply. Music possessed an unusual meaning for this student. He conveyed this sense of meaning convincingly, some audible technical shortcomings notwithstanding.

The only reliable witness to Einstein's violin playing, one undoubtedly but not overly influenced by the aura surrounding the "person of the century" (to whom he owed a personal debt), was Boris Schwartz (1906–1983). Schwartz was himself a fine professional player and an eminent scholar, particularly on violin playing. He studied in Berlin with Carl Flesch in the early 1920s, returned to Berlin in the early 1930s to pursue musicology, and in 1939 performed under Toscanini as a member of the NBC Symphony Orchestra. While in Berlin, he began playing for Einstein even as a youngster and met with him with some frequency.[8] Einstein admired young Schwartz and was instrumental in rescuing him and his family after 1933 by bringing them to the United States.

Schwartz's account leaves one in little doubt that Einstein possessed a respectable amateur technique adequate to a very limited repertoire, but played with intensity and understanding. He was not incompetent, but neither was he capable of playing demanding parts that required, for example, dexterity, speed, and control of the bow or the fingerboard beyond third position. His repertoire was therefore limited largely to easy Baroque pieces, including the music of Corelli (among his favorites), the Bach Concerto for Two Violins (not an easy piece), the piano and violin music of early Beethoven, and music by Mozart and Haydn. It is likely, as was the case in many comparable amateur households, that not all of this repertoire, or even all of the movements in a particular work, were performed. There is no evidence that Einstein took part in a regular quartet or attempted anything beyond the range of the Beethoven op. 18 quartets and the early trios.

There is no mention of the Bach solo sonatas and partitas, a striking absence since these six works form the core of every violinist's ambition, particularly one who revered Bach as much as Einstein did. But they require a sophisticated variety of left- and right-hand techniques, which is why there are so many well-crafted exercises of the type against which Einstein rebelled. They are designed to enable students to learn to play not only Bach, but Mozart and Beethoven as well. There is no easy music, but in fairness most amateurs of Einstein's generation could negotiate only a limited range of repertoire because the strictly technical facility deemed minimally socially acceptable (including the tolerance of spotty intonation) was modest but inadequate for most chamber music written after 1814. Later Beethoven and Brahms appear to have been beyond Einstein's reach, as was the entire early and mid-Romantic repertoire for piano and violin, as well as the 19th-century music for trio and quartet. Given Einstein's account of how he learned to play, the hypothesis that his violin playing was of sufficient but quite limited capacity seems reasonable. That within his circumscribed skills the capacity to communicate real feeling and intensity came through lent a considerable aura to Einstein's music making. The appreciation of Einstein's refined and intense sensibility grew only after he became famous.

Toward the end of Einstein's life, the young Juilliard Quartet visited him in Princeton, and he joined them in playing second violin in Mozart's great G-minor Quintet (a work, incidentally, that Richard Strauss considered the finest piece of music ever written).[9] Their praise and delight notwithstanding, the quartet was forced to take the tempi very slowly to accommodate the great man. Nonetheless, the quartet was impressed by Einstein's level of coordination and intonation. Perhaps their expectations were low. But it was clear to them that Einstein had a discerning ear for pitch and rhythm and was highly musical.

A proper analogy can be made with tennis. If we claimed, hypothetically, that Einstein played tennis and loved it, we would discover that he could return a soft serve and sustain a slow-moving, high-arc rally, but probably had a weak serve and backhand, and no net game. He was a rank amateur who looked like one, but took the game seriously nonetheless. He could hit a ball back and sustain a game, but he was no competitive contender, not even in the most provincial of country clubs. Later in life, fine amateurs and distinguished professionals were delighted to play with him because he was devoted to the game, loved to play, and was, after all, Einstein. It is important not to construe Einstein's musical skills as analogous to Vladimir Nabokov's accomplishments as a butterfly expert, Ernest Bloch's work as a photographer, Arnold Schoenberg's skills as a painter, or even Paul Klee's violin playing. Except for his being Einstein, his tastes, habits of playing, and skill set were commonplace among educated, German (particularly German Jewish), middle-class adults at the turn of the century.

Nonetheless, there is no reason to doubt the emotional and intellectual centrality and significance of Einstein's attachment to music. Beyond science, neither art nor literature competed for his attention. Einstein is said to have remarked that "music has no effect on research work, but both are born of the same source and complement each other through the satisfaction they bestow."[10] Einstein's classic papers from 1905 and his lifelong criteria for what constituted a convincing scientific theory and explanation revealed consistent aesthetic qualities and criteria. These included simplicity, elegance, coherence, clarity, and formal transparency. These attributes were precisely those that defined his musical tastes from his adolescence on.

Furthermore, a parallelism has been argued between the role of so-called creativity and intuition in the conduct of science and in the arts, particularly music. For John S. Rigden, Einstein's solution to the inadequacies of the notion of ether and the reconciliation between theory and the constancy of the speed of light in the special theory of relativity can be compared to the achievement of Mozart.[11] On the one hand, both achievements reflected the terms of argument and issues of the day, and thus possess a recognizable historical context. On the other hand, both demonstrated an intuitive leap and originality of conception that was transformative and, in retrospect, entirely distinctive and out of the range of reigning conventions.

What is striking in Einstein's case, if one follows Rigden's line of reasoning, is that at age thirteen, Einstein might just as well have developed an enthusiasm for Wagner or music from the Romantic era, such as Schumann, Chopin, or Mendelssohn. Indeed, his passion for the Mozart of the early 1880s, which preceded the great Mozart revival of *fin-de-siècle* Europe,[12] was ahead of its time. Most comparable thirteen-year-olds might have started their love of music with Mozart, but they would have shifted allegiances to later composers. Einstein never did. This may have had decisive consequences for the history of science and modernity.

To return to Einstein's own habits of music making. The concentration and concomitant daydreaming implicit in Einstein's violin playing, as his son observed, may have permitted him a structured nonlinguistic and temporally self-contained space in which to think intuitively, particularly when faced with an impasse.[13] The player of music reconfigures time against and apart from the clock. This recalibrated sense of time occurs in the listener as well, whose sense of musical time, however, remains different from the performer to whom he or she listens. Music renders the notion of absolute time problematic because the experience of time is so radically transformed by music without uniformity or predictability. Music can alter one's sense of space as well. Sound, when organized and sustained over time, often forces a confrontation with the visualization of space and its empirical observation through sight.

Einstein's method of working involved, as is usually the case, an interplay of solitary thinking and intense exchanges with others (with Michele Besso and the Olympia Academy and later with Planck and Bohr). This suggests a parallel to musical activity. Some allege that music was a refuge for Einstein, "the window into the place where Einstein sealed all his emotions in order to avoid dealing with interpersonal relationships."[14] This claim, however, is unconvincing. Einstein used music making most often for the opposite reason, as part of a quite ordinary function, as an activity that rendered otherwise awkward socializing palatable. It seems likely, as well, that music making was a dimension of Einstein's legendary, if not notorious, attraction to women.[15]

Music making in 19th-century, urban, domestic life developed a long and well-documented connection to courtship and seduction.[16] Einstein's attachment to music was, in this regard, entirely unexceptional for his generation. It fits very neatly into the uses of *Hausmusik* as a mode of middle-class socializing, often with a sexual undercurrent. Einstein's music making with women has a familiar ring to it, reminiscent of Leo Tolstoy's short story "The Kreutzer Sonata" and that once-familiar logo of the Lanvin perfume My Sin, in which a man holding a violin passionately kisses his female accompanist apparently in mid–music making.[17] Rather than a refuge, playing the violin for Einstein was all about rendering social interaction easier, even though the time he may have spent playing alone could very well have provided important self-contained moments of intuitive reflection.

For the historian of science, there is the plausible claim that, in contradistinction to a work by Mozart, the breakthroughs represented by Einstein's work might ultimately have been realized by others, though perhaps later and in a less elegant and remarkable fashion. No music historian would argue the historical inevitability of any composer or any work of music. The confirmation of the existence of atoms, the photoelectric effect, the corrective to the wave theory of light, and special and general relativity possess, however, as Einstein would have vigorously argued, an objective reality. These discoveries might have been made by others, given the intensity, circa 1905, of the engagement with the scientific problems of the day. There is, by contrast, no Mozart sonata or Beethoven symphony out there, hidden and waiting to be assembled or discovered.

For this reason, Einstein himself, while linking the conduct of science with aesthetic inspiration, also differentiated between the two enterprises. Fantasy and invention were crucial in making art in a way that was not entirely comparable to the conduct of science. Yet might there be in the arts a question of proof or matters of truth-value, much less of utility (construed as a matter of ethics) as understood by engineering and applied science? At first one is inclined to say no. However, as Einstein's aesthetic prejudices suggest, it is possible to understand his firm musical tastes as asserting that, for him and those who shared his aesthetic views, some universal system of true aesthetic combinations might have existed. Art in the cosmos might possess a governing logic that, if understood, could furnish criteria of acceptance and rejection of an artwork's validity or claim to beauty. In this way, indeed, a Mozart work would somehow be immanent and implied, if not present, within the universe, much like a valid claim of science.

Given Einstein's deep attachment to music as central to his life, the question turns to what kind of music Einstein liked as an adult and with what aesthetic criteria he judged music. Ernst Mach may have doubted the existence of atoms as a result of a larger philosophical outlook concerning human consciousness and external reality. In a parallel manner, Einstein's particular conception of music as an adult may hold clues to an underlying epistemological outlook. As is well known, Einstein revered Bach, but held Mozart in the highest esteem. He is even alleged to have believed precisely that which has just been suggested: that Mozart's capacity to form melodies could be described as a process of selection and organization out of the universe of possibilities.[18] Described in this way, Mozart's music could be deemed as true and valid. His genius could be understood as analogous to a scientist's correct insights and intuitions about nature and God's creation.

Einstein's attachment to Beethoven was consistently limited to the more Mozartian aspects of early Beethoven.[19] Beethoven's more dramatic and Romantic characteristics in the middle and late periods seemed to him artificial and strained. Einstein's aversion to the heroic Beethoven is itself reminiscent of Jacob Burckhardt's aesthetic logic, articulated in his influential 1860s classic *Die Kultur der Renaissance in Italien* [*The Civilization of the Renaissance in Italy*]. The work draws a contrast between Raphael and Michelangelo, in which Raphael is held up as representing a near-perfect normative and harmonious artistry, and Michelangelo is presented as a towering but disturbing genius of individualism and theatricality of gesture.[20]

Einstein's resistance to middle and late Beethoven make his comparable lack of affection for Handel understandable. The dramatic (as opposed to the formal) imagination in Handel led Beethoven to regard him as the greatest of his Baroque predecessors. Interestingly, Einstein's view of Beethoven was akin to many early 19th-century music critics, as late as those from the 1840s, who considered op. 18 to be the pinnacle of Beethoven's music. The works written after op. 59, particularly the late compositions and including the Ninth Symphony, were regarded as reflecting the composer's increasing willfulness, isolation, and even madness. Einstein found no inner core of sentiment in Brahms; Richard Strauss was too decorative and facile, a view held widely by many of Einstein's contemporaries; and Wagner was altogether not to his liking.[21] Predictably enough, Einstein loved the Italian Baroque composers.

Einstein's views and tastes explain why, despite his prominence as a modern revolutionary during his Berlin years (a notion that became overwhelmingly popular in the 1920s), he seems to have evinced no interest whatsoever in revolutionary developments in contemporary music.[22] He showed no preference for any living composer. He agreed to be the honorary president of the Ernest Bloch Society in the late 1930s only to demonstrate solidarity with a Swiss American contemporary in need whose Jewish identity formed the center of his self-definition.[23] One could argue, however, that Einstein could have liked Bloch's music because it was remarkably conservative, despite the fact that Bloch was Roger Sessions' teacher, and because Bloch and Einstein once played chamber music together.

Though Berlin in the 1920s was the center of new musical developments, Einstein seems to have been oblivious to its modernist musical life. The music of Busoni, Schoenberg, and their pupils passed by without his comment. He seems never to have heard of the composer Franz Schreker and took no interest in new approaches to pitch organization of the era, whether those invented by Alois Hába, Josef Matthias Hauer, or Schoenberg himself.[24] None of the polemics on the aesthetics of music by Busoni, Berg, and Hans Pfitzner that dominated the era and the Berlin musical scene caught his interest. These were topics covered by newspaper journalism.

In Berlin, Einstein did meet famous performers of the historic repertoire, including the violinist Adolf Busch and the pianist Artur Schnabel. His interest in Schnabel, who wrote twelve-tone music, was in his role as the authoritative interpreter of the Classical era, particularly the music of Mozart, Beethoven, and Schubert.[25] Schnabel's advocacy certainly corresponded to Einstein's own tastes.

In the Berlin of Einstein's years, Schnabel and his violinist partner Carl Flesch helped spearhead a novel attitude toward the interpretation of the repertoire of the past, particularly the Classical era, that asserted a valid, historically based, anti-Romantic reading of the texts. To discover the right way to render a select canon of masterpieces in performance became a lifelong effort that would later define the ambitions of Rudolf Serkin and his circle in post–War War II United States.[26] In Berlin, Modernism in composition thrived alongside a radical reversal of the free and highly personalized performance practices and readings of the Classical musical repertoire characteristic of the later 19th century, notably the generation of performers after Liszt, Hans von Bulow, and Anton Rubinstein. A revolution in interpretation and canon formation located in a historicist and antisentimental conceit regarding an "objective" style took shape in Berlin alongside early 20th-century innovations in new music. Insofar as Einstein's years in Berlin brought him into contact with music, it was on these grounds.

For a scientific revolutionary, then, Einstein was profoundly conservative in his aesthetic tastes vis-à-vis the serious music of his own day. There is no indication that he had a taste for light and popular music. The easy explanation is, of course, that he liked only the music he could play, but the more plausible answer lies in a nascent aesthetic absolutism. Einstein's musical preferences were akin to his resistance to the idea that the statistical nature of quantum mechanics as outlined by Bohr and Heisenberg might be the last word. Coherence, symmetry, and order were necessary attributes of musical beauty. The elaboration of thematic materials, rhythmic patterns, and above all, the integration of small-scale and large-scale structures were understood as being governed by clear and distinct laws. These attributes were necessary for musical beauty and meaning. For Einstein the claims of aesthetics were at once normative and rational, despite the role that invention and imagination played in the creation of music. The absence of contradiction lay in the nearly infinite possibilities for beauty that were present within the universe.

This aspect of Einstein's musical tastes puts him in league with conservative music theorists of his day, as will be further noted, particularly Heinrich Schenker and to a lesser extent Pfitzner. For Schenker, tonality was natural and normative; it determined, in turn, structure and form.[27] For Pfitzner, genuine musical inspiration led to form and order of a certain character.[28] It could not be achieved mechanically, abstractly, or outside of the boundaries of tonality. For Einstein, the rationality inherent in true beauty had a stronger link to ethics, as in the 18th-century tradition of aesthetic philosophy exemplified by Shaftesbury, Moses Mendelssohn, and Schiller. For Pfitzner, genuine musical inspiration was associated with the irrational emotional capacities tied to post-Herder notions of history and race. Yet, unlike either Schenker or Pfitzner, Einstein did not include as exemplary Brahms or any other mid- or late 19th-century composer.

The comparison between Einstein and Schenker is even more apt if one takes into account the hierarchical conceits within the conservative wing of the musical culture during Einstein's Berlin years. These included the privileging of instrumental music over music with words or instrumental music with literary programs. The music that Einstein favored was exclusively instrumental music. He favored music organized into formal genres without so-called extra-musical significance. Opera, song, and even the innovations of quite accessible *Zeitoper* from the 1920s and early 1930s (as audible in *The Threepenny Opera* and *The Rise and Fall of the City of Mahagonny* of Kurt Weill) did not fascinate him. Music, if divine in its order, was abstract in its elements and content. Einstein's aesthetics are reminiscent of the highly influential tradition of thought initiated by Eduard Hanslick in his famous 1854 essay "Vom Musikalisch-Schönen [The Beautiful in Music]."[29] Music was a self-referential system of sounding forms in motion that artificially reconstructed time.[30]

One might be inclined to speculate that this tacit assumption concerning music's essence helped suggest the notion of the absence of priority of any frame of reference in special relativity and, at the same time, the notion of equivalents that was crucial to the formulation of general relativity. The experience of music can, one might argue, suggest a non-Newtonian connection between space and time. Because Einstein was an adherent (without a conscious awareness of the origins of his beliefs) of a fixed notion of value derived from a theory of absolute music, his understanding of music may have helped shape his original, yet stubborn and self-confident mode of inquiry. His search for overarching patterns, which began in the years preceding 1905, may have contained, in this sense, a musical impulse.

Einstein's aesthetic conservatism was, however, not entirely reactionary for the 1920s. His contemporary Schoenberg was, after all, notorious in the Berlin of the 1920s as a radical conservative. Schoenberg's leap into a system and method of twelve-tone composition was put forward as a means of restoring an 18th-century aesthetic akin to Mozart's. Schoenberg's revolution was a decidedly anti-Romantic and anti-Wagnerian effort. He sought to give modern, contemporary expression to an 18th-century tradition of musical aesthetics. Formal musical logic was the outgrowth of constituent cells of musical materials whose meaning and significance were not contingent on any manner of illustration, representation, or tone painting through music. Schoenberg's rationalism was not distant from Einstein's. The key difference was that in place of tonal logic, Schoenberg posited an invented, rational system of organizing and combining pitches.[31] The conservatism inherent in Schoenberg's brand of Modernism can also be found in widespread neoclassicism among French, Russian, and German composers of the 1920s, notably in the work of Stravinsky, Poulenc, and Hindemith. Yet Einstein paid no attention to them, either.

Einstein's recalcitrant lack of interest in or even sympathy with these contemporary evocations of contemporary historicism and rationalism in musical logic remains striking. One equally renowned contemporary, the Viennese philosopher Ludwig Wittgenstein, does provide a parallel case of studied disregard for contemporary music in the face of a deep and abiding love of and obsession with music. Like Einstein, Wittgenstein was an avid music lover with limited amateur skills. Their musical education did differ. Einstein, who came from a bourgeois home, received the limited instruction characteristic of a German Jewish home with modest resources. Wittgenstein's home was highly assimilated, converted, and extremely wealthy. His mother and siblings were excellent pianists and legendary patrons (Wittgenstein's brother became a professional pianist). Wittgenstein's own amateur playing, limited to the clarinet, was marginal, far more so than Einstein's playing of the violin. Yet Wittgenstein's training in music theory and literature seems to have been more extensive than Einstein's. Wittgenstein shared Einstein's disregard for and incomprehension of musical Modernism.[32] He did however, admire Brahms and a few neo-Brahmsian composers such as the blind Viennese composer Josef Labor, whom the Wittgenstein family supported. Wittgenstein, like Einstein, suspected that, beginning with Beethoven, music declined from the ideal achievement represented by the Viennese Classicism of Mozart and Haydn, let alone the formalist genius of Bach. Wittgenstein's views of contemporary developments in music and his negative attitude to Wagner were similar to Einstein's. Wittgenstein's complaint that he could not understand certain philosophical arguments because they were couched in "language on holiday"—language gone berserk, outside its normal framework—applied to his view of the way Wagner mishandled musical materials.[33]

Both Wittgenstein and Einstein had interesting notions about visualization, mental images, and pictorial logic, particularly in relationship to music. In Wittgenstein's case, his reflections on music reveal much about his idea of mental pictures, conception of family resemblances, and notion about games in language. Wittgenstein's views on how music and language compare are an ideal starting point. Interestingly, both he and Einstein were committed to the metaphysical transcendence and priority of music as an art form.

The two have another point of common contact—Ernst Mach—though Einstein's debt to him is more profound. Mach had an influence on aesthetics and literature, certainly within Viennese circles. Consider Robert Musil's dissertation on Mach.[34] Mach's theories about sensory perception intersected with debates on the aesthetics of music. Mach held the antirealist position that musical understanding derived from the functional relation of the sensations. This functional, nonrealist attitude to the perception of hearing is related to skepticism about the objective reality and priority of tonality and the overtone series. Significantly, Mach also placed a tremendous premium on the economic use of language in the description of sensations.[35] This concern for the language of explanation is mirrored in the writing styles of both Einstein and Wittgenstein.[36]

In the cases of Einstein and Wittgenstein, therefore, one is faced with a contradiction between the popular reputation of these figures as intellectual revolutionaries who contributed to the novel, intellectual arsenal of modernity and their aesthetic judgments. What in Einstein's case resolves this apparent contradiction is the popular misunderstanding of his scientific breakthroughs, particularly the abuse and wrongheaded understanding of relativity.[37] As has been frequently argued, Einstein retained a nearly theological conviction in a concept of order and simplicity, and a Spinoza-like attachment to the rational and logical character of creation. Relativity theory was a breakthrough in understanding the nature of reality. It did not undermine, in some Nietzschian sense, the notion of truth. Rather, relativity theory undermined a form of common sense and an inadequate epistemology of space and time without justifying a nihilistic sensibility. The defeat of the Newtonian concepts of absolute time and space by eliminating the priority of any frame of reference did not destabilize the project of understanding reality and the rational progress of science.

The confusion of relativity theory with philosophical relativism was both a popular and philosophical development that misappropriated Einstein's work. It is precisely this misappropriation, however, that became profoundly influential in the development of the arts during Einstein's lifetime. This fact contains a particular dimension of irony vis-à-vis the relationship of the history of physics to music. For it was the work of Helmholtz and Mach that helped popularize the notion that the claims to objectivity on behalf of the Western system of musical organization, particularly tonality, were false. Helmholtz's work in acoustics made it quite clear that the Western systems of pitch organization and definition had no priority in nature and were culturally determined.[38] Mach's work on the psychology of perception made it clear that issues of form and meaning were the result of learned processes of perception; aesthetic qualities were not inherent in the external sounding objects of sonic reality.

Just as Einstein's 1905 paper on Brownian motion established the reality of the atom (despite Mach's skepticism), so too did his work on relativity establish an understanding of time and space that superseded Newton but redeemed the project of understanding the principles and laws in the universe. This fact was quickly overlooked. Einstein's conservative musical tastes might have restrained contemporary post–World War I misunderstandings among artists. That Einstein believed the music of Bach and Mozart did possess truth-value as evocative of the character and structure of the universe, notably its aesthetic character and contours, should have provided a warning. The superiority of their music rested not in subjective judgment or taste but in a deep parallel to nature.

Because Einstein's attachment to music was well known in his lifetime, musicians and theorists from the early 1920s on, with a certain amount of self-aggrandizing, borrowed the idea of relativity to posit theories about music. Two examples reveal the divergent consequences between the misappropriation of relativity theory and its informed use. The aesthetic conclusions are, predictably, diametrically opposed.

In 1922, the American composer and fantast Dane Rudyar wrote an essay called "The Relativity of Our Musical Conceptions."[39] Though his argument was based on a complete misunderstanding of relativity, he used its authority to declare the equivalence of all rules regarding the nature of music and musical form, thereby demolishing the priority of the traditions of European music making. Not only that, but Rudyar denied the essential claims of European musical theory, which, in his view, wrongly insisted on the primacy of the note as opposed to sound. In a haze of rhetoric, Rudyar (an interesting composer in his own right who abandoned conventional composition and later became a California guru) projected that once the absolute priority of the European system of music had been destroyed, a new order (or rather lack thereof) would emerge from the chaos. Rudyar anticipated a unified field theory of sound that, in stark contrast to Einstein's goal, had no organizing logic. Form became an obsolete notion.

The second example is a more sophisticated version of the relationship between Einstein's theory of relativity and music. It was put forth in the 1950s by the Russian-born Jewish composer and music director of the Temple Emanu-El in New York City, Lazare Saminsky. Also a physicist, he wrote an essay about the metaphysics of music.[40] He concluded that tonality could be compared to gravity and that musical sound "[bore] a telling similarity to a nucleus of physical energy and its field."[41] Tonality functioned as an analogue to an equivalence and, by allowing an imperative, generated normative criteria for aesthetic judgments that transcended historical style.[42] Saminsky's interest in the notion of invariance in Einstein's work led him to argue that music must assume discernible form. He claimed that music ultimately takes part in a formal coherence and symmetry that underlies the entire universe. Saminsky's words for this coherence suggested a pattern of "Ring, Cycle, and Return."[43]

Rudyar's and Saminsky's Einstein-inspired theories of music should be compared with Einstein's own claims about music, which are extensive. The most interesting is a 1921 interview in the Berlin journal *Menschen*. In answer to the question "What do the artistic and scientific experiences have in common?" Einstein wrote that "as free beings admiring, learning, and observing . . . we enter the realm of Art and Science. If what is seen and experienced is portrayed in the language of logic, we are engaged in science. If it is communicated through forms whose connections are not accessible to the conscious mind but are recognized intuitively as meaningful, then we are engaged in art."[44]

Form is the operative and normative concept. For Einstein, both science and art were depersonalized. They begin when our quotidian experiences and concerns and volition are set aside, and we observe the world as free individuals. Einstein's claim about the intuitive meaningfulness of art implies the notion that art is communicated through form. In this sense, Saminsky's adoption of Einsteinian notions was far closer to the physicist's own claims than was Rudyar's. This comes as little surprise since Saminsky actually understood Einstein's work.

Einstein's adult relationship to music cannot be defined as a consequence of scientific work that led him to embrace Modernist art and aesthetics. Rather, the clue to the significance of Einstein's relationship to music rests in the opposite. The aesthetic choices the very young Einstein made were ones from which he did not deviate. During Einstein's youth, he could have engaged with the three basic developments in German musical culture that were taking shape, but he did not. These three developments would overlap with his greatest scientific achievements.

First was the advent of Modernism in the form of Richard Strauss. Strauss essentially stripped music of its transcendental and metaphysical priority. In the *Symphonia Domestica* (1902–03), for example, and the late tone poems, Strauss used the techniques of music, both formal and representational, against the metaphysical conceits about music as a mode of critique. The followers of both Brahms and Wagner made grand claims about the power and significance of music. On the one hand, the Brahmsian aesthetic of absolute music was heralded as a nonrepresentational but meaningful nonlinguistic signifying system. The Wagnerian aesthetic, on the other, was understood as a profoundly psychological, representational, and poetic medium in substance and experience. As far as Strauss was concerned, music was just a form of life, expressive of ordinary experience. Music in his hands (in contrast to Mahler) became a radical instrument of an antiphilosophical realism. Music should represent a baby crying, a card game, or an orgasm. By using a virtuosic skill in tone painting, Strauss challenged the claims of instrumental and absolute music to a higher, nobler experience, much to the outrage of many music lovers, Einstein among them. At the same time, Strauss negated through extreme realism the Wagnerian myth and the claims of art to the status of religion and metaphysical philosophical doctrine.[45]

Second was the debate Ferruccio Busoni began in a 1907 essay called "Entwurf einer neuen Ästhetik [Sketch of a New Aesthetic of Music]."[46] Busoni claimed that the rules of chromatic, late-Wagnerian composition, which were primarily concerned with harmony, represented only one set of musical rules. He argued that the whole concept of tone should be expanded to include microtonalities, that a more experimental and universal system of music making lay beyond the Western system. He conceived of a sounding universe and called for the creation of new, mechanical musical instruments adequate to modernity. Busoni's position came under immediate attack by a conservative faction led by Hans Pfitzner, who responded in 1917 by writing a pamphlet called "Futuristengefahr [The Danger of Futurism]."[47] Whereas Busoni argued that the system of making music should be expanded into new areas of making sound and based on new rules, Pfitzner argued that music had a normative natural basis in tonality.[48]

Third and comparable to the Busoni-Pfitzner argument was the aforementioned debate regarding the objective basis of Western pitch definitions and the resultant system of harmony. Two important harmony textbooks exemplify the debate. In 1906 Schenker claimed in his text *Harmonielehre* that tonality based on the overtone series represented an objective, natural system.[49] Schenker suggested that music is an abstract, functional system with metaphysical priority, natural roots (like the innate grammar of language) and an organic logic of self-relation. Great music is that which is most economical and simple, and that finally refers, despite all its variation, to some very fundamental invariances.

The other textbook, also called *Harmonielehre*, was written by Arnold Schoenberg in 1911.[50] It argued for what he called the emancipation of the dissonance or the notion that there is no normative, tonal, gravitational pull. In each formal context, any pitch eventually creates its own frame of reference. In this way, the distinction between consonance and dissonance becomes contextual, if not relative, and certainly not objective. Schoenberg's later formulation of the so-called twelve-tone system of writing took this notion further. It assumes that all tones are of equivalent value. Their interrelated functionality is generated by the series, the tone row, on which each piece is based. In 1911 Schoenberg took issue with Schenker, claiming that there was no priority to, for example, the relationships between the tonic, dominant, or subdominant except within the context of a work or style, and therefore no objective reason to view dissonance as either negative or stable in its definition. The Schenker-Schoenberg debate about music took place during the years Einstein was in Bern, Zurich, and finally Berlin.

One can conclude that Einstein's love for a certain genre of music, for a particular and clear logic and formal expectation in music, may have been very useful to him, albeit indirectly. His economy of explanation, his use of aesthetic criteria to judge scientific explanation (e.g., his discomfort with the later results of quantum theory), and his distrust of statistical resolutions ("[God] does not throw dice"[51])—all these characteristics betray a deeply held aesthetic prejudice located in childhood sensibilities profoundly connected to music. One could argue that later Einstein's concern for some deep, unified theory that would align discoveries about quantum mechanics with his own work has this musical, aesthetic origin. Music informed not only his scientific achievements but also his unwavering attachment to a certain ideal of truth that in turn limited his appreciation to a specific era of music and sustained his insularity from Modernism and other contemporary musical developments.

Einstein's conception of music was indeed the inverse of the Romantic Wagnerian theory. In the Wagnerian conception of music, the artist becomes God. Through Einstein's normative acceptance of what we call the Classical style in music, he imagined the opposite—that God becomes an artist. The artistry of a God-created world that the scientist reveals becomes analogous to a kind of Mozartian creation. Einstein's search for scientific explanation has an aesthetic premise that inverts the Wagnerian formula and explains the deep anti-Romanticism and anti-Modernism of Einstein's musical tastes. This allegiance set him apart from his contemporaries, whose enthusiasm for late Romanticism in music may have prevented them from thinking differently, in opposition to convention, along the lines Einstein did.

Science since Einstein has proven him largely right. In the context of postmodernism and the late 20th-century retreat from Modernism and atonality, one might be tempted to say—tongue in cheek—that he may yet be proven right with respect to music, at least in the West.

Acknowledgments

This is a revised version of the speech delivered January 22, 2005, at the conference "Einstein for the 21st Century" in Berlin, sponsored by the Einstein Forum. I would like to thank Matthias Kross, Matthew Deady, Yehuda Elkana, Jane Smith, David Botstein, Gerald Holton, and Yaron Ezrahi for their assistance.

SEEING THE UNSEEN

E. L. Doctorow

WHEN I WAS A STUDENT at the Bronx High School of Science in New York City, our principal, Dr. Morris Meister, had an image for scientific endeavor and the enlightenment it brings: "Think of science as a powerful searchlight continuously widening its beam and bringing more of the universe into the light," he said. "But as the beam of light expands, so does the circumference of darkness."

That image would certainly have appealed to Albert Einstein, whose lifelong effort to find the few laws that would explain all physical phenomena ran into immense difficulties as the revolutionary light of his theory of relativity discerned a widening darkness.

Of course, to a public celebrating its own mystification, that hardly mattered. The incomprehensibility of his space-time physics, and the fulfillment of an early prophecy of the theory of relativity when Sir Arthur Eddington's experiments confirmed the bending of starlight as it passed by the sun, was enough for Einstein to be exalted as the iconic genius of the twentieth century.

This was a role he could never seriously accept; he would come to enjoy its perks and use it as he grew older on behalf of his various political and social causes, but his fame was an irrelevancy at best and did not accord with the reality of a life lived most of the time in a state of intellectual perplexity. To be a genius to someone else was not to be a genius to oneself. Acts of mind always come without a rating.

Einstein would say by way of calming his worldwide admirers, "In science . . . the work of the individual is so bound up with that of his scientific predecessors and contemporaries that it appears almost as an impersonal product of his generation."

Could this statement have been something more than an expression of modesty on his part?

Einstein came of age in a culture that was in hot pursuit of physical laws. In Europe some of his scientific elders—Albert Michelson and Edward Morley, Hermann Helmholtz, Heinrich Hertz, and Ernst Mach, to name a few—determined that electromagnetic waves move through space at the speed of light; their work called into question the concepts of absolute motion and absolute rest, everything in the universe moving only in relation to something else. So the science leading up to Einstein's breakthrough was in a sense premonitory—it gave him the tools with which to think.

If we look outside the scientific enterprise of his time to the culture in general, we discover that this same turn-of-the-century period in which Einstein conceived his theory of relativity put him in the national German-speaking Jewish company of such contemporaries as Sigmund Freud, Franz Kafka, the revolutionary atonalist composer Arnold Schoenberg, the critic Walter Benjamin, the great anthropologist Franz Boas, and the philosopher of symbolic forms Ernst Cassirer. They joined the still-living precedent generation of Friedrich Nietzsche, who had proclaimed that God is dead, and Gustav Mahler, whose freewheeling First Symphony was written while Einstein was still a child. Mahler's First, a big kitchen sink of a symphony, with its openness to idea, its structural relaxations, its excesses of voice and extravagance of mood, all coming after the unified and majestic sonorities of Brahms, for example, was in effect a kind of news broadcast: "This just in: the nineteenth-century world is coming apart."

Frederic V. Grunfeld's book *Prophets Without Honor* is the definitive account of this cultural florescence of German-speaking Jews. A multibiographical study of some of the artists and intellectuals of the period, it finds as their common characteristic not only an intense work ethic but also a passion that would drive them to take on the deepest and most intransigent questions. As Freud would plumb the unconscious in his effort to "understand the origin and nature of human behavior," so Einstein would set off on his lifelong quest for a unified field theory that would encompass all physical phenomena.

Of course, outside Germany some world-shattering things were going on as well: in Paris, Braque's and Picasso's cubist paintings and Stravinsky's *The Rite of Spring*, which brought on a riot at its premiere; in Bologna, Marconi's experiments with radio waves; at Kitty Hawk, the Wright brothers' first flight. So Einstein came of age at a moment not only in German culture but in world history—those early years of the twentieth century—that if I were a transcendentalist I might consider as manifesting the activity of some sort of stirred-up world oversoul.

The English poet and essayist Matthew Arnold speaks about such historic moments of creative arousal in literature in his 1865 essay "The Function of Criticism at the Present Time." "The grand work of literary genius," says Arnold, "is a work of synthesis and exposition . . . its gift lies in the faculty of being happily inspired by a certain intellectual and spiritual atmosphere, by a certain order of ideas, when it finds itself in them; of dealing divinely with these ideas. . . . But it must have the atmosphere, it must find itself amidst the order of ideas, in order to work freely; and these it is not so easy to command. This is why great creative epochs in literature are so rare; this is why there is so much that is unsatisfactory in the productions of many men of real genius; because for the creation of a master-work of literature two powers must concur, the power of the man and the power of the moment, and the man is not enough without the moment."

Arnold's thesis puts me in mind of the debate among historians of science as to whether science at its most glorious (for example, the work of Copernicus, Galileo, Darwin, or Einstein) is a revolution or whether it emerges incrementally as evolution. Perhaps it is both evolutionary and revolutionary. Perhaps there *is* an evolving communal intellect, and its role is periodically to be stunned and possibly outraged by the revolutionary ideas that it had not realized it was itself fomenting.

Thus, to speak of the power of the moment does not gainsay the power of the man. Opinions vary as to when, if ever, the theory of relativity might have been articulated if Einstein had not lived. Some scholars have said it would have taken generations. The eminent English astrophysicist Sir Martin Rees believes that it would have been conceived by now, but not by just one theorist working alone.

So what are we to make of Einstein's own reference to the communal context of creativity, whereby the scientific work of an individual "appears almost as an impersonal product of his generation"? As always, he was being totally honest. Yet we must ask to whom the work appears as an impersonal product—certainly not to the world that applauds it and names its producer a genius. Rather it appears impersonal to the producer himself, the revelation of such work coming to his mind always as a deliverance, at a moment in his thought when his personality, his psyche, is released from itself in the transcendent freedom of a revelation.

The creative act doesn't fulfill the ego but rather changes its nature. You are less than the person you usually are.

CHAPTER TWELVE

Einstein's theory of relativity was an arduous work of self-expression no less than the work of a great writer or painter. It was not accomplished without enormous mental struggle. It was created not merely from an intellectual capacity but also from an internal demand of his character that must have defined itself in his nightmares as Atlas holding up the sky with his shoulders. It was a matter of urgency to figure things out lest the universe be so irrational that it would come down around his and everyone else's head. The term *obsession* is woefully insufficient to describe a mind so cosmologically burdened.

We have to assume also that there was the occasion of lightning clarity when that formula $E = mc^2$ wrote itself in his brain, the moment of creative crisis, the eureka moment. And here a writer can only scrub about in his own field to find a writer's equivalent moment, as described by a giant of his profession, Henry James.

In his essay "The Art of Fiction," James speaks of the "immense sensibility . . . that takes to itself the faintest hints of life . . . and converts the very pulses of the air into revelations." He celebrates the novelist's intuitive faculty "to guess the unseen from the seen," but the word *guess* may be inadequate, for it is a power, I think, generated by the very discipline to which the writer is committed. The discipline itself is empowering, so that a sentence spun from the imagination confers on the writer a degree of perception or acuity or heightened awareness that a sentence composed with the strictest attention to fact does not.

Every author from the writers of the ancient sacred texts to James himself has relied on that empowering paradox. It involves the working of our linguistic minds on the world of things-in-themselves. We ascribe meaning to the unmeant, and the sentences form with such synaptic speed that the act of writing, when it is going well, seems no more than the dutiful secretarial response to a silent dictation.

This feeling, I suggest, may be the same as the scientist's in his eureka moment, when what he has discovered by seeing past the seen to the unseen has the character of appearing as "an impersonal product of his generation."

And there must be something common to the creative act, whatever its discipline, in James's assertion that from one evocative fragment of conversation overheard by the writer an entire novel can be written, that from the slightest bit of material a whole novelistic world is created. We may represent this as the Little Bang of the writer's or scientist's inspiration, thinking analogously of the Big Bang, that prime-moving happenstance when the universe blew out into its dimensions, exploding in one silent flash into the volume and chronology of space-time.

If the analogy seems grandiose, I remind myself that the writers of the ancient texts, the sacred texts of our religions, attributed that Little Bang of their own written cosmologies not to the impersonal product of their generation but to God. The God of the universe was the author of what they wrote, so awed were they by the mystery of their own creative process.

But whether the creative mind feels it is dutifully transcribing a silent dictation, or that its work appears almost as an impersonal product of a generation, or that it is serving as a medium for the voice of God, what is always involved is a release from personality, liberation, an unshackling from the self.

That self was wildly manifest in Einstein's youth, when he seems to have renounced both his German citizenship and his Jewish faith; it was manifest in his adulthood during the course of two difficult marriages and an affinity for extramarital wandering. His biographers tell us how, in his student days as an assimilated Jewish boy in a German gymnasium, one of his teachers held up a rusty nail and, looking directly at Albert, said such spikes were driven through Christ's hands and feet. That brought home to the boy the social isolation he was born to, a position he came to relish because looking in from the outside, he saw clearly the pretensions and lies and dogmas upon which the society fed. He would come to distrust every form of authority. He was from the beginning, as he himself said, "a free spirit."

It was in childhood that Einstein's difference as a quiet, unflinchingly observant Jewish kid allowed him to hone the skepticism that as an adult he applied to intellectual postulates that had been in place for centuries. His society's resentment grew as Einstein's mind grew, exponentially. By the 1930s, a winner of the Nobel Prize, he was at the top of Hitler's enemy list. He was designated for assassination, and even when he was out of the country, in Belgium, authorities insisted that he have bodyguards. Einstein's biographers agree that he was always philosophical, always calm in the face of personal danger. As his fame grew, he had necessarily to apply his mind to social, political, and religious issues. He brought to these nonscientific issues the same clarity of thinking that was evident in the only definitions of time and space that he could allow himself: time, "something you measure with a clock," and space, "something you measure with a ruler." God he called *Das Alte,* or "the Old One," identifying the only attribute of God he could be sure of—old in nominal existence solely. He applied that same beautiful and scrupulously pragmatic clarity of thought to the famous ethical conundrum most forcefully postulated by Immanuel Kant: How can there be an ethical system without an ultimate authority, without the categorical imperative of an *ought*—in short, without God?

Here is how Einstein cut through that problem: "Ethical axioms are found and tested not very differently from the axioms of science. Truth is what stands the test of experience," he said. "For pure logic, all axioms are arbitrary, including the axioms of ethics. But they are by no means arbitrary from a psychological and genetic point of view. They are derived from our inborn tendencies to avoid pain and annihilation, and from the accumulated emotional reaction of individuals to the behavior of their neighbors. It is the privilege of man's moral genius . . . to advance ethical axioms which are so comprehensive and so well founded that men will accept them as grounded in the vast mass of their individual emotional experiences."

There is one more point to be made in the futile project of trying to plumb the creative mind of this genius: throughout his life he found excuses, almost apologies, for his prodigious accomplishment. "Sometimes I ask myself," he once said, "how it came about that I happened to be the one to discover the theory of relativity. The reason is, I think, that the normal adult never stops to think about space and time. Whatever thinking he may do about these things he will already have done as a small child. I, on the other hand, was so slow to develop that I only began thinking about space and time when I was already grown up. Naturally, I then went more deeply into the problem than an ordinary child."

Einstein had a sense of humor; a sly diffidence was one of his stocks-in-trade when dealing with the press, and this was a sweetly funny thing to say—except that in this case I think he was quite serious. For hidden in this remark is an acceptance of himself as an eternal child. This prodigy of thought was eternally a child prodigy. And if that would seem to diminish the man, remember that it was a child who cried out that the emperor had no clothes. All his life Einstein would point to this or that ruling thought and reveal its nakedness, until finally it was the prevailing universe that had no clothes.

Dare we think that a mind of this immensity—independent, self-directed with such a penetrating clarity of thought, and driven with a rampant curiosity—must have had, too, a protective naïveté about the nature of itself? There was a confidence in reality that must have protected him from the philosophical despair of Ludwig Wittgenstein, another genius born to the power of the moment, just ten years after Einstein, and the most influential European philosopher of his generation.

Wittgenstein revolutionized philosophy by dismissing everyone from Plato to Hegel as purveyors of metaphysical nonsense. All philosophy could do was to understand thought. He was a philosopher of language who used linguistic analysis to distinguish those propositions that were meaningful from those that had no justifiable connection to the existing world. "The meaning is the use," he said. Wittgenstein's philosophy, a technique more than a teaching, was almost directly attributable to the appropriation by science of the great cosmological questions that had traditionally been the province of philosophy. Certainly Einstein's discoveries were the salients of this scientific encroachment. Yet Wittgenstein believed that science, even at its most successful, by its nature could go only so far. He articulated the most desolate intellectual pronouncement of the twentieth century: "If all possible scientific questions are answered," said Wittgenstein, "our problem is still not touched at all."

What did he mean? He meant that even if Einstein, or we, find the final few laws to account for all phenomena, the unfathomable is still there. He meant all science hits a wall.

Wittgenstein's is the steely gaze of the inconsolable and ultimately irretrievable spirit directed into the abyss of its own consciousness. His is the philosophical despair of a mind in the appalled contemplation of itself. Such a despair was not in the nature of Einstein's beautifully childlike contemplations.

Einstein was directed outward, his face pressed into the sky. The universe had always been there, as it was, regardless of how it was conceived by humanity, and so the great enterprise was to understand it as it was in the true laws by which it operated. It was a matter for wonder and mental industry. The crackling vastness of black holes and monumental conflagrations, the ineffable something rather than nothing, such an indifference to life as to make us think that if God is involved in its creation he is so fearsome as to be beyond any human entreaty for our solace or comfort or the redemption that would come of our being brought into his secret—this consideration did not seem to be part of Einstein's cosmology.

Einstein's life spanned the terrors of the twentieth century—two world wars, the worldwide Great Depression, fascism, communism, the Holocaust, the threat of nuclear war—and he was never less than steadfast and rational in his attention to the history of his time. He lived as he thought, in the thrill of the engagement. He was a scientist, a secular humanist, a democratic socialist, a Zionist, a pacifist, an antinuclear activist, and never, so far as I know, did he succumb to a despair of human life. So finally, even if in his Einsteinian pragmatism God could only be accurately described as the Old One, surely there was a faith in that image, perhaps an agnostic's faith, that made it presumptuous for any human being to come to any conclusion about the goodness or incomprehensible amorality of God's universe or the souls it contained until we at least learned the laws that governed it.

For Albert Einstein a unified field theory needn't be the end. It can just as well be the beginning.

PART

III

SCIENCE AND WORLD

THE ASSASSIN OF RELATIVITY

Peter L. Galison

SEEMINGLY PARALLEL

After all that has been said about Einstein, today a rather different story, perhaps one not usually heard in the canon of Einsteiniana. Friedrich Adler fought and followed his father, Victor Adler, head of the Austrian Workers Party. When his father pushed him to study something practical like engineering, he chose the abstractions of physics and epistemology. When his father argued for more professorial ambition, "Fritz" plunged into leftist politics (a dynamic that may not have surprised Victor Adler—he was, outside politics, a psychiatrist, rather well known). In fact, Sigmund Freud dreamed of Victor Adler, as he reported in his epochal *Interpretation of Dreams*, where the Dream Dr. Adler warned the dreaming Freud off excessive materialism in psychological explanation. During his theory years, Adler and Einstein had known each other in Zurich. They had the same thesis advisor, Alfred Kleiner, and both Adler and Einstein battled furiously with Kleiner over everything from physics to their careers. In 1908 both Einstein and Adler applied for a university job that had just come open, Adler as a *Privatdozent* at the University of Zurich, Einstein as a clerk from the Bern Patent Office. Friedrich Adler wrote his father on 19 June 1908 that his competition

> is a man named Einstein, who studied at the same time as I did. We even heard a few lectures together. Our development is seemingly parallel: He married a student at about the same [time] as I, and has children. But no one supported him, and for a time he half starved. As a student he was treated contemptuously by the professors, the library was often closed to him, etc. He had no understanding of how to get on with the important people. . . . Finally, he found a

position in the patent office in Bern and throughout the period he has been continuing his theoretical work in spite of all distractions. Today he is in the school of Boltzmann, and one of the most distinguished and recognized. And this school, not that of Mach, is the mode today."[1]

Ludwig Boltzmann's statistical mechanics, not Ernst Mach's sense-data positivism, may have been the mode, but it was definitely not the fashion Adler wore. His loyalties were to Mach in the realm of physics, as well as on epistemological and political grounds. But Einstein's epistemic-physics classification with Boltzmann did not sway Adler for a moment from backing Einstein for the position he himself was pursuing. To the educational authorities in 1908, Fritz wrote: "If it is possible to obtain a man like Einstein for our university, it would be absurd to appoint me. I must quite frankly say that my ability as a research physicist does not bear even the slightest comparison to Einstein's. Such an opportunity to obtain a man who can benefit us so much by raising the general level of the university should not be lost because of political sympathies."[2]

The Einsteins and the Adlers took up residence in same Zurich building, 12 Moussonstrasse. Both Adler and Einstein had married Slavs, both had young children. From that building, Adler wrote his father: "We are on very good terms with Einstein, who lives above us, and, as it happens, we're closer to them than any of the other academics. They run a . . . Bohemian household—like our own." Their young children played together, and the two physicists retreated to the attic for discussion. "The more I talk to Einstein," Friedrich Adler added, "the more I realize that my favorable opinion of him was justified. . . . We find ourselves in agreement on questions which the majority of other physicists would not even understand."[3] Such questions no doubt included a shared (Machian) disdain for absolute space or absolute time.

Adler's high estimation of Einstein was reciprocated. By the end of 1910, most everyone knew that Einstein was to be offered a professorship at the German University of Prague: "One can now read in the professional journals about my appointment at Prague. But I have not yet received any communication. If it really comes, then I wish that [Friedrich] Adler would become my successor."[4] Adler had written extensively on Mach's phenomenological approach to physics and had written some more technical papers (that Einstein respected) on thermodynamics, bashing Ostwald's attempts to reduce all physics to energy. But Adler's succession to Einstein was not to be. With some reluctance, weighted down by financial concerns, Friedrich Adler abandoned epistemology and physics for party work in Zurich and, after 1911, back in Vienna—Victor's unsuccessful struggles to persuade him to continue his work in theory notwithstanding.[5] Einstein and Adler remained in contact; in fact, in 1911, Friedrich brought Einstein to his parents' residence in Vienna where they spent a memorable evening in discussion.

In the summer of 1914, Leon Trotsky, then in exile in Vienna, visited the Socialist Workers Party to meet with the Adlers and recorded this:

> In the secretariat I came across Friedrich Adler or "Doctor Fritz" as he was called in top party circles. . . . Thin, of quite a good height, slightly stooping, with a fine brow over which his curly shining hair fell down and with the imprint of a perpetual thoughtfulness on his face, Fritz always stood as one apart amongst the quite numerous Vienna party intellectuals so ready for wisecracks and cheap anecdotes. He had spent a year and a half in Zurich as assistant professor of physics and as editor of the local party newspaper *Volksrecht* [*People's Right*].[6]

Trotsky continued: the young Adler "with his inimitable revolutionary temperament" disdained this old system run by his father. But Fritz Adler's real criticism, according to Trotsky, "his distrust and his hate," were directed at the Habsburg regime. During their last meeting (on 3 August 1914), Friedrich Adler showed Trotsky the government's appeal to the population: hunt the suspicious foreigners. Trotsky said: "It was with concentrated abhorrence that he [young Adler] spoke of the rising orgy of chauvinism. But his outward restraint merely masked his profound moral shock."[7] Then Victor Adler stepped into the secretariat, suggesting that Trotsky immediately accompany him to the chief of the political police (a striking measure of Adler's authority) to find out what the government intended to do with Russian exiles. In the car, Trotsky gestured to the festive reaction the city had to the beginning of war. Victor responded: "all the unbalanced and all the insane are coming out on the streets: this is their day. . . . War throws open the field to every Instinct and every type of lunacy. . . ." *Irrenarzt* Adler, according to Trotsky, viewed politics, especially (as he often put it) *Austrian* politics, as a form of psychopathology. Adler apparently had enough clout with the government to get an immediate audience with political police chief Heyer, who told the leftist duo that leaving town would be an excellent idea. Tomorrow? Trotsky asked. No, today, Heyer responded. Now. Three hours and ten minutes later, the Adlers had Trotsky and family on a train for Switzerland.[8]

Sane and Insane Killing

Furious as he was at the slaughter of war, Einstein in late 1915 was immersed in the hardest work of his life, modifying and extending the equations of general relativity. In the third week of November 1915, he realized that the theory correctly described the precession of the perihelion of Mercury. Overcome with emotion, he told a collaborator he had the feeling that something actually snapped within him. Friedrich Adler's passionate opposition to the war was directed elsewhere. At lunch, on 21 October 1916, Einstein's "parallel-life" friend took his Browning pistol, walked up to Count Stürgkh, Prime Minister of Austria (who was dining at the prestigious Meissl & Schadn Hotel), and put three bullets through the politician's head. Immediately arrested, Adler was matter-of-fact (*sachlich*) in his responses to the authorities' questions. In May 1917, the Viennese authorities hauled Adler before an Exceptional Court (*Ausnahmegericht*) in Vienna. A *cause célèbre*, the courtroom was packed, the windows opened, and hundreds hung on every word. In his defense, Adler adduced a wide range of offenses by Stürgkh, fastening on two.

Under the former leader's brutal bureaucratic rule, no one, Adler argued, knew who really ruled Austria. Go to a minister and you would be told that it wasn't his department, you had to see Stürgkh. Go to Stürgkh and you would be sent to the military. Stürgkh, Adler contended, "was not a man of the politics of *fortwursteln* [Austrian slang: get-along, go-along]. He was a man who with clear intent and conscience wanted to transform Austria into an absolutist state, and who with a clear course and a steady hand steered for that goal. . . . He was an opponent whom one had to notice, and with whom one had to battle relentlessly. . . . "[9] Second, it was Stürgkh who had dissolved Parliament, Stürgkh who had wrecked the progressive road toward democracy that most civilized countries were already following, Stürgkh who stood as a "dead point" (*toter Punkt*) that had to be bypassed if Austria was to progress.

To save his son's life, Victor Adler pleaded for the court to find his son insane. He invoked Friedrich's past history in which his drive to work had surpassed all reason; he gestured to the mental illness found on so many branches of the family tree. And Victor hired a superb defense team to demonstrate the incompetence of his son, as the court deliberated whether he should hang. Aiding the defense was a testimonial from the Physical Society of Zurich, introduced at the trial and duly recorded in the transcript, along with another by Professor Heinrich Zangger. Adler's lawyer, Dr. Gustav Harpner, insisted on the "spontaneous" nature of these glowing reports, both of which spoke to Adler's scientific, philosophical, and personal qualities. Harpner: "I don't need to say that I did not ask for this testimonial. It was a fully spontaneous intervention [*Kundgebungen*]," and that he hadn't known until that very day either of the scientific society or of Professor Heinrich Zangger, who delivered testimony for the defense from Zurich's juridico-medical institute. "Spontaneous"? Not quite. Barely forty-eight hours after receiving Friedrich Adler's communication, Albert Einstein had intervened. "Dear Zangger," Einstein had written,

> The day before yesterday I got a scientific letter from my old friend Fritz Adler. You surely know what the man has perpetrated. I know you never thought particularly highly of him because of his scientific orientation. But as a person, he is an exceptional fellow and extraordinarily selfless, which has gotten him into this stew, as a matter of fact. My compassion for him has grown so strong that I really would like to do something for him.

Behind Zangger's "spontaneity" stood Einstein.[10]

In April 1917, Einstein also wrote Adler directly (in prison):

> I hope you have received my package. I now have an unusual request: When your affair is deliberated in court, I would like to be summoned there as a witness; you should apply for this. Do not think this senseless; for not only the circumstances directly connected to the event are clarified through witness accounts, but statements are also brought forward that shed light on the character of the perpetrator.—How much I would like to discuss the relativity problem with you![11]

As we will see in a moment, that discussion was in train.

But politics weighed heavily on Einstein too.

The war, the physics of simultaneity, politics—all were fused even among Einstein's friends, including the scientist-philosopher-industrialist Walther Rathenau, former head of the war provisions board and author of *Zur Kritik der Zeit*. Modifying Einstein's *Gedanken-experimentalische* description of the coordination of clocks, Rathenau dropped a letter to Einstein just before Adler's trial: "Your illustration of the two flashes of lightning and the train really gripped me here (Incidentally, I turn it into two dynamite explosions and a czar train). What startles the czar twice is only *a single* matter for the assassin."[12]

Einstein so feared Adler's execution that he set aside some of the most productive work of his life, on a generally relativistic cosmology, to write the Emperor:

> His Majesty!
>
> Under the pressure of an unavoidable duty, I take the liberty of communicating a plea to your majesty.
>
> The political murder, of which Fritz Adler is guilty, shook the well-being of every rightly sensitive person in the deepest way. With not a single word will I prettify this gruesome act. With regard to the psychological situation of the perpetrator, however, it seems to me to have to do with a tragic accident rather than a crime. Few can have known Herrn Adler so well as me.

I have known Herrn Adler since we were studying physics together 20 years ago in Zurich. He was for several years still my closest colleague as an assistant professor in [physics] at Zurich University; we also were, at that time, housemates. In those years I took Adler to be a man of the purest character, of an almost unparalleled selflessness. Few people have shown themselves to me, to be so unqualifiedly reliable and honest as he; few have, in such measure, overcome their own desires and devoted their strength to other than personal things.

Under these circumstances, I feel an unavoidable duty, if I herewith submit to your majesty, from the bottom of my heart, a plea for you to invoke the law of clemency, in the event that Adler is sentenced to death. A valuable life could be so preserved.[13]

On 18 and 19 May 1917, Stürgkh's assassin broke his public silence: hauled before the *Ausnahmegericht*, as Einstein feared, Adler faced hanging. Debate raged. Victor Adler's hope for an insanity ruling had its best chance through the psychiatric report. But the commission found nothing *Wahnhaftes* in the motivation of Adler's act.[14] Indeed, as they saw it, "Adler's act, from the standpoint of morality, must be considered as in some ways in contradiction to his character as it has been demonstrated up to the present. . . . One must say that his life has been, up to now, idealistic and all his efforts have been in that direction. He is however a fanatic . . . far from the normal Type." The fanatic and the mentally ill might be branches from the same trunk, the commission allowed; they conceded too that Adler was hereditarily a "psychopathological person" who suffered depressions and shallow manic states. But the experts ruled that "the act of which Adler is accused did not take place under [such psychopathology] instead [it] must be understood as the response of Adler's personality to given outer circumstances."[15] These experts concluded: Adler may be psychopathological but he is responsible.

Against his own (survival) interest but in favor of his (political) ambition, Adler insisted he was fully responsible for his act as an expression of political conviction. "Already before the attack, I knew full well," he told the court, "that people would at first say that . . . committed this act as an incompetent [*Unzurechnungsfähiger*]. I was prepared—I *had* to be prepared that the press, above all, would be under the influence of the regime. . . ."[16]

Battling for Friedrich Adler's sanity—and fate—were two oddly paired sides. On the side of insanity sat Victor Adler, who, as father and psychiatrist, was desperately trying to make the case for his son's hereditarily induced loss of equilibrium, joined by the vicious right-wing press seeking to strip the assassination of political import. Meanwhile, in February of 1917, Trotsky recalled in writing Victor Adler's biting words about the psychopathology of Austrian politics:

> How far [Victor] was at that moment from the thought that his own son would carry out a political assassination. . . . I mention this here because after Fritz Adler's act of assassination the Austrian yellow press . . . attempted to present the self-sacrificing revolutionary as unbalanced and even insane, from the standpoint of their own base "sanity" that is. But the judicial medicine of the Habsburgs was forced to capitulate before the courageous tenacity of the terrorist. [With] what cold contempt [Fritz] would have treated the retorts of the eunuchs of social-patriotism . . . if their voices had reached him in prison.[17]

This prosanity alliance of the State, the medical-psychiatric panel, Trotsky, and Friedrich Adler himself left bad odds for the defense. But before continuing with the chronology, we need to step back because, from prison, awaiting trial, Adler attacked.

Einstein's final synoptic paper on the general theory of relativity was finished in March 1916, and Einstein himself was deeply affected by politics. During the war, Einstein had faced off against most of his professorial colleagues. Back in October 1914, dozens of luminaries—including the biologist Ernst Haeckel and the discoverer of X-rays, Wilhelm Roentgen—had signed a "Manifesto to the Civilized World," berating critics of the German military and defending German militarism in general and war actions in particular. Einstein with two colleagues responded in 1915 with a minority manifesto of their own, lobbying for a League of Europeans and an end to the catastrophic war.[18] To the French author Romain Rolland, Einstein lamented the idiocy of militant nationalism that was tearing apart three centuries of cultural achievement. (Einstein once remarked that he felt roughly the same emotional bond to the nation as he did to his life insurance company.) Even scholars, he sadly remarked, had thrown their lot in with the insanity of nationalism.[19] Einstein's wartime interventions were frequent, and both public and private. When he heard of the assassination, he acted.

Einstein wrote to Katja, Adler's wife, on 20 February 1917, four months after the shooting, three months before the trial: "Your misfortune and that of your husband, whom I have always admired, touched my heart deeply like little else that I have witnessed in this hard world. He is one of the most splendid and purest men I have ever known. I cannot judge his deed, since I cannot assess the motive behind it; I do not believe him capable of rash acts; he is much too conscientious for that. If there is something you think I can do for him or for you, think of me and write me."[20]

About the same time (mid-February), Adler wrote Katja and his parents from prison about Einstein: "I am suddenly a physicist again! I have made a discovery that one would not be overestimating if one described it as the greatest possible in the current state of physics. I awoke Saturday with the solution to a small problem which has bothered me for more than a year, namely the Foucault pendulum experiment. In the course of working through the consequences of my little insight I have arrived at generalization of the Newtonian principles—an elementary law which up until now has simply been overlooked." It would, Adler believed, give clarity to all questions of mechanics and thermodynamics. A few days later, he wrote to say his trial had been postponed until after Easter 1917—a good thing since he was very satisfactorily plunged deep into physics. "It has unexpectedly come to pass that I have found a decisive criterion in relativity theory that rules against Mach and Einstein and for Hendrik Lorentz. It puts everything in a new light. A big part of it is ready to print." Or as he put it in his letter to his parents: "What Mach searched for, I have found. . . . Like all great discoveries" it was simple and full of consequences.[21]

Friedrich Adler's discovery opened for him a new campaign. For Victor Adler, too, it launched a second front on which he could fight to have his son declared insane: the insanity of opposing Einstein. Adler the elder shot copies of his son's new manuscript to Einstein, Philipp Frank, and others, asking if the inner workings of the physics constituted evidence that could be used to show the derangement of his son. That put the physicists in a bit of a quandary, as Frank recalled a bit later: "The experts, especially the physicists, were placed in a very difficult situation. Adler's father and family desired that this work should be made the basis for the opinion that Adler was mentally deranged. But this would necessarily be highly insulting to the author, since he believed he had accomplished an excellent scientific achievement. Moreover, speaking objectively, there was nothing in any way abnormal about it except that his arguments were wrong."[22] Meanwhile, Friedrich Adler had every intention of maintaining his sanity against his father, the courts, the right-wing screeds—anyone challenging his epistemo-physical work. He was willing to die for his sanity.

Back to physics. With the return address "Vienna VIII, 1 Alser St." (which was the prison), on 9 March 1917, Adler wrote to his relativistic twin, Einstein: "Now that I have plenty of free time at my disposal, I have taken up my studies on the foundations of physics again, which I had abandoned seven years ago. I intended to assemble in a book my previously published and unpublished papers on Mach. . . . The work had already progressed quite far when I started the chapter on relativity. Then something quite unexpected happened to me."[23]

Adler's philosophical work, *Ernst Mach's Conquest of Mechanical Materialism*, written in prison (and published in 1918), had two goals.[24] First, it was a rousing defense of many of Mach's doctrines. But it also reconstructed Mach, removing him from dogma. Like Mach, Adler wanted to free physics of absolutes in space, time, motion, and mass. It would be the counterpart, Adler argued, of Marx's historicization of the basic categories of an all-too-human society. Like Marx, Mach had grounded his views on experience (materialism) and made them irreducibly historical (equivalent, said Adler, to "dialectical").[25] In fact, for Adler, Mach's historical materialism of the inner core of science was precisely the needed complement to Marx. To Einstein, Adler wrote that he'd taken Mach's view since 1903 and that he'd advocated the importance of the relativity of motion in his 1907 courses back in Zurich. He recalled for Einstein the arguments they'd had in the attic rooms of Moussonstrasse 12: "I was very irritated, you see, that in taking the centrifugal effects into account, you rejected relativity for rotations. . . . So much the greater was my joy when your general theory of relativity came out." Adler believed Einstein had accepted Mach's position "entirely, including centrifugal phenomena." As Adler told Einstein in early February 1917, that joy was short-lived:

> I was in ecstacy when suddenly, 4 weeks ago, a turning point came in my considerations which reveals the whole problem differently from how I had seen it previously. I found, first in a more recent discussion of Foucault's pendulum experiment, and then generally, a criterion that you and Mach do not take into consideration, or at least, not sufficiently, which sheds new light on it all. I believe I have found where the error in the assumptions not only by Mach but also by you lies.[26]

Adler's new system, he continued, "does not involve a return to 'absolutes' but a criterion of a relativistic nature for preferred reference systems." Here, he hoped was the true spirit of Machism rescued from Mach's and Einstein's over-extended development of relativism.

There are three particularly interesting points about the technical criticism. First, Adler went after what he considered to be Einstein's (and Mach's) anthropocentrism. Second, on good Machist-positivist grounds, he told Einstein that his own way did not involve the ether. No ether as intuitive (Alfred Kleiner), as necessary physical hypothesis (Lorentz), or even as a pragmatically useful fiction (Henri Poincaré). Nor was Adler interested, on Machian anti-atomic grounds, in the detailed structure or dynamics of the electron. Lorentz, Max Abraham, Alfred Bucherer, and Poincaré were above all intrigued by the possibility of spelling out the electron's structure. That said, Adler's work *does* involve a privileged frame—but not because he believed the earth (for example) is special. No, he argued that the privileged frame emerges from a rigorous application of the relativity principle itself. None of Adler's arguments moved Einstein, but Einstein did engage Adler in detail—with as much care as he lavished on any other critic. "Healthwise I am fine," Adler assured Einstein, "and, since I have the opportunity rarely offered in life being able to work in peace, I am completely happy, even though the time still available to me can be counted only by the week."[27]

The trial proceeded—and ended—dramatically. At the final judgment, in May 1917, the court pronounced the defendant guilty and responsible. The punishment—according to section 136 of the sentencing protocols [*Strafprozeßordnung*]—was death. Four days after the verdict, on 23 May 1917, an interview with Einstein appeared on page 2 of the morning edition of the *Vossische Zeitung*: "Friedrich Adler as Physicist: A Discussion with Albert Einstein." Undeterred either by the sentence or by the broader wartime repression that might have affected him, Einstein declared Friedrich Adler to be a clear thinker, someone who acted objectively—and selflessly—in his life as well as his studies. Einstein related how, years before, Adler had sacrificed his own excellent chance to land the University of Zurich professorship—even though it was an existential necessity for Adler that he get it. Adler had removed himself in favor of someone else because he considered the other candidate far superior from the point of view of pure science. Einstein did not mention that he, Einstein, had been that competitor.[28]

Adler's work in physics, Einstein continued, consisted in the further development of the Machian point of view. According to Adler, it was Mach who had "brought it to consciousness" that mechanics had not been essentially altered since the time of Newton. There had been formal adumbrations to be sure, but it was in 1888 that Mach himself disentangled "classical mechanics" (handed down from the 17th century) from an "ideal mechanics" (that would be predicated only on the relation of objects to one another.) Heinrich Hertz had developed this point of view further—and then, through General Relativity (according to Adler), Albert Einstein had brought this "ideal program" to a "more beautiful and more unified" form that would overlap in everyday physics with the classical form. It was Adler's goal (so Einstein related) to describe the past of a body's behavior in terms of classical physics, and to use this "ideal physics" to tie a body to the rest of the world. The difference between classical mechanics and ideal mechanics consists in this: in ideal mechanics there is only the direct description of a body's behavior relative to other bodies: no speed, no acceleration, no inertia, and no force as classically conceived. Adler: "With Hertz and Einstein the real [*wirkliche*] path of the body is the 'straightest path'; in classical mechanics however, the real path is the superposition of two straightest paths" (that is, the straightest paths of velocity and of acceleration). Despite Einstein's respectful description, he was clear—Adler's project was not Einstein's. In particular, Adler was determined to locate a "preferred reference frame" grounded in the center of gravity—for example, the center of the earth or the center of the earth-moon system, when dealing with terrestrial phenomena. In common with relativity theory, Adler recognized multiple equivalent and equally justified frames of reference. But Adler's point of view was necessarily narrower than general relativity, grounded as it was in the center of gravity of ponderable bodies. Einstein reported that Adler considered this description as simpler and more natural.[29] Obviously Einstein did not agree.

Adler's execution was stayed, and he was transferred to the fortress at Stein-on-the-Danube, where he was to serve hard time in solitary confinement. On 6 July 1918, he once again wrote Einstein: "Dear Friend, It has been almost one year since I wrote you, but I have been thinking of you incessantly, for I was occupied the whole time with relativity theory. . . . I know that you are so convinced of the correctness of your foundations that you do not expect anything from further discussions of it. And yet I would like to burden you with the perusal of my book, for I imagine now having really caught Ariadne's thread, leading to a compelling derivation of the necessity for a preferred reference system from your transformation equations."[30] Simple diet, good air, quiet solitude—death row agreed with him: "In virtually all respects, I am doing better here than in Vienna. . . . In short, in this topsyturvy world we now live in, it is in actual fact considerably nicer *intra muros* than *extra*."[31]

Adler lambasted Einstein for his relativity of space and time. Einstein fired back a detailed response on 4 August 1918 that argued this way: At the root of all that Adler had done, according to Einstein, lay two crucial assumptions: (a) no Lorentz deformation of moving rigid bodies, and (b) no influence of motion on the running rate of clocks. Assumption (a) as Einstein noted, leads to Max Abraham's motion of an electron (and therefore was in trouble with electron deflection experiments). Assumption (a) also led inevitably to a direct contradiction with Michelson's experiment if it is valid that for a preferred reference frame K in which light travels same speed always in a vacuum. Such considerations led Einstein to lay out for Adler precisely the constraints that Einstein understood to hem in any putative theory of the electrodynamics of moving bodies. So the ether was neither in motion nor at rest. Einstein insisted that "It was these facts which compelled the formulation of the special theory of relativity." As far as Einstein could see, Adler had not even tried to address these constraints.

The struggle continued. One of Adler's strong claims in *Ortszeit* was that "The error of Einstein's considerations is hidden in the circumstance that Einstein silently has smuggled in the assumption that clocks in all systems have been coordinated [*gerichtet*]."[32] Einstein bridled: "It is self-evident that I treat all times of the system as equivalent since I set out from the postulate of relativity. I cannot be reproached for not making any use of your 'zonal time'." Einstein insisted that Adler's "symmetry system K" was, under another name, nothing other than Lorentz's ether—and so Adler had inherited, willy-nilly, all of Lorentz's problems. "Your sentence 'The error of Einstein's . . .' is completely incomprehensible to me. If a system's clocks are not set rel[ative] to one another (in some way or other), their data are then incoherent and cannot serve for a time definition, not even in a 'symmetry system'." "My [Einstein's] view" is that one can have a "system time" but not a "universal time." This latter can be had only through instantaneous signaling or motion-independent clocks. "The existence of either of these things must be questioned in principle, though, *a priori*."[33]

Adler went after the Einsteinian clock-coordination in other ways. Identifying his own work with the strict examination of measurement that Mach had pursued elsewhere, Adler argued that inequalities in the running rate of clocks could be due either to their mechanism or to the partition of their dials.[34] Einstein's sharp tongue, otherwise held in check in correspondence with Adler, let loose: "the dial is a bit odd, why not the handle and the polish as well? We imagine my standard clocks as having been produced identically somewhere and sometime by a clockmaker who enjoys a world monopoly and then having been brought to the different locations and into the states of motion of the various locations."[35]

Adler tried again a few weeks later (20 September 1918), claiming that the Lorentz theory was preferable in certain respects; that was nonsense, Einstein replied: "A decision between Lorentz and Einstein is impossible . . . since *factually* Lorentz's theory agrees entirely with the special th[eory] of rel[ativity]; it is just a more specialized (exclusively electromagnetic) theory."[36]

The up-the-river theorist then turned to clocks and the "twin paradox," as it came to be called: "you can see that I did not overlook your 'standard clock,' rather that it is the actual point of departure of my reflections. I have been making a serious effort for a year now to understand it, but an *insurmountable logical contradiction* in your assertion remains for me": one twin rotates with the earth, comes back, the other remains still; each looks the same to the other, yet the clocks are supposed to be different. Einstein stands his ground: "Only if systems K and K' were both *justified* systems in the sense of the special principle of relativity would there be a contradiction with the (special) principle of relativity; they are not both justified, however, *because K' is not a Galilean (acceleration-free) reference system*. Only when general relativity is taken as a basis are both frames of reference equivalent. In this case, the difference in the rates of the clocks is explained by the combined effects of the influence of the velocity and *the gravitational potential*. Nowhere is a contradiction evident."[37] Then Adler's piece de resistance:

At the heart of Adler's argument in his book manuscript lay the view that, when Einstein had taken special relativity to encompass *all* inertial reference systems as "equivalent," he had generalized too far. Instead, Adler allowed two reference systems, S and S', to be equivalent if they had equal and opposite velocities with respect to a "symmetry system" (*Symmetralsystem*) K. The infinite set of *pairs* of systems—(Sm, Sm'), (Sn, Sn'), . . . —were equivalent, by Adler's lights, but, emphatically, when the systems had different velocities, Sm was *not* equivalent to Sn'. He then introduced three key definitions. *Ortszeit* (local time) was the time that S or S' should use; it was equal to the quantity Lorentz had introduced some years before, $t' = t - vx/c^2$. This quantity had been taken up by Poincaré and given a physical definition, but for Lorentz it was purely a mathematical device that would simplify the equations. Later, in 1904, Lorentz modified the result, so it read $t' = k (t - vx/c^2)$, where $k = 1/\sqrt{(1 - v^2/c^2)}$. This allowed Lorentz to make the equations of electrodynamics look exactly the same in arbitrary inertial reference systems as they did in the system that was at rest with respect to the ether. Adler wanted to use only the older form of Lorentz's local time, and it is this that he designated "*Ortszeit*."

Next, Adler defined "Zonal Time" (*Zonenzeit*) following, quite explicitly, the everyday notion of a similar name that had proven of such use in scheduling trains and other practical matters. Indeed, Adler's text referred frequently to the details of *real* clock coordination techniques—he discussed master clocks (*Normaluhren*); cable links between sites; and the distinction in particular cities, such as Geneva, between "train time" *(Bahnzeit)* and "city time" *(Stadtzeit)*. Just as Middle European Time specified a common time for the region of the earth's surface from 0° longitude (through Greenwich, England) to 15° east, so Adler (his comparison) used "Zonal Time" to indicate a conventionally synchronized set of clocks based on a single local time in a particular domain.[38] Finally, "System Time" (*Systemzeit*), according to Adler, was what you had if the Zonal Time extended throughout the *entirety* of the domain in question—for example, if you conventionally made the whole world use the time of one particular place, that would be a "System Time." It was Adler's view that Einstein had wrongly applied his light-exchange method of clock coordination to the *entirety* of the domain of *all* inertial reference systems, that is, (Adler said), Einstein had promoted clock synchronization by light exchange to a "higher" *system* concept valid everywhere.[39]

This over-extension was Einstein's great mistake, according to Adler. The famous physicist had "silently" smuggled in the assumption that his light-coordination method could rightly be applied to *all* inertial frames of reference. On the contrary, the prisoner wrote, there is one system, the privileged "symmetric" system K, where this Einsteinian time coordinating "correction" need not take place.[40] We will return to Adler's best argument for K in a moment.

In many respects, Adler understood the special theory of relativity in ways that Einstein found lucid. The good prison air left Adler not the slightest bit vague, for example, about several points on which he found Einstein's relativity theory to have substantively advanced physics. These were not trivial points, either—each had become a point of contention by numerous critics: Adler wrote lucidly in *Ortszeit, Systemzeit, Zonenzeit* about Einstein's new kinematics, how Einstein had studied the properties of rulers and clocks *before* undertaking questions about the forces that moved particles around—and many, even among the world's best physicists, had not truly understood this about Einstein's work. Adler grasped too that Einstein had *derived* the Lorentz transformation, not *assumed* it as had both Lorentz and Poincaré. And Adler noted very clearly that Einstein knew how to add one velocity to another relativistically—Adler, and practically no one else, saw immediately the importance of Einstein's velocity addition formula. Then Adler wheeled around and attacked: Einstein's equations, he insisted, hid a covert preferred reference frame.

Figure 13.1 illustrates Adler's anti-Einstein machine. His Figure 2 shows the machine before anything is set in motion. Two identical meter sticks are used, one (below the horizontal line) with conducting leads attached from each end of the stick to a galvanometer labeled G. The stick above the horizontal line also has two conducting wires attached to its ends, but these go to a battery, *E*. If the upper (battery-bearing) stick moves right to left while the lower stick is stationary then, says Adler, according to Einstein, the upper stick will contract (Adler's Fig. 4): no contact is made between electrical leads, so the galvanometer does not deflect. Conversely, if one takes the point of view of the battery-bearing stick, then the galvanometer stick contracts (Adler's Fig. 3). Again, no deflection of the galvanometer. But, says Adler triumphantly, if we put ourselves in the "symmetry frame" in which the two sticks are moving with equal and opposite velocity, then the conductors touch: the galvanometer deflects. Since the galvanometer cannot both deflect and not deflect, he concludes that the nonsymmetric frames do not reflect reality—which is contained only in the symmetry frame (Adler's Fig. 5).

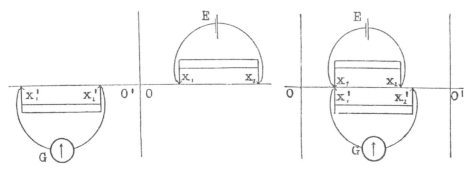

Fig. 2. Die Systeme vor Bewegungsbeginn.

Fig. 4. Die Koinzidenz von x_1 und x'_1 gesehen im System S'_n zur Systemzeit t'_1.

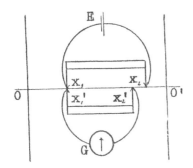

Fig. 3. Die Koinzidenz von x_1 und x'_1 gesehen im System S_n zur Systemzeit t_1.

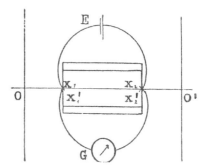

Fig. 5. Die Koinzidenz von x_1 und x'_1 gesehen zur Zeit τ_1 des Symmetralsystems.

FIGURE 13.1   Adler's anti-Einstein machine. Friedrich Adler, *Ortszeit, Systemzeit, Zonenzeit und das ausgezeichnete Bezugssystem der Elektrodynamik. Eine Untersuchung über die Lorentzsche und die Einsteinsche Kinematik* (Vienna: Wiener Volksbuchhandlung, 1920), 128f.

Adler: relativity is smashed. "The galvanometer will say that the two relativity theorists [*both* Einstein and Lorentz] are wrong, that neither of the two slabs has contracted." Bang: Einstein's relativity theory is dead.[41]

This time Adler's target did not drop. "[Y]our bias for absolute time or for the instantaneous signal is exposing itself," Einstein replied. "*Under no circumstance* will a current flow if the contacts made are so brief that their duration is small against the propagation time of light or of an electrical wave between the two contact points."[42] Neglecting the necessity of keeping contact long enough for light to make the crossing is, in principle, at the heart of the problem. But making the contacts longer so this condition was fulfilled in the "symmetry frame" (Adler's Fig. 5) turns out to ensure that the galvanometer deflects in all the moving frames. Einstein liked the example so much that he wrote a new and related one for Adler, illustrating again how the relativity of simultaneity resolved the new paradox, and considered including Adler's galvanometer example in the dialogue between an anti-relativity critic *(Kritikus)* and a relativist that he was writing for *Die Naturwisschaften*.[43] Bullet deflected, its lead recycled as movable type pro causa:

> Kritikus: I see, you have defused my objection, but I must tell you, your argument leaves me more convicted than really convinced. . . .

If anyone was more convicted than convinced, it was Friedrich Adler in late October 1918. By then, the Imperial armies were on the run. Einstein wrote to Adler on 20 October wondering "who between us will be the first to manage to come and see the other. Who can know?" Within days, the Social Democrats were hovering about the levers of power; the Emperor was urgently pleading for assistance; and desperately, the old Council of Ministers offered amnesty for all political prisoners. Friedrich Adler emerged from the Stein fortress on 1 November 1918 and was immediately greeted as a hero of the working class, not only in Austria. That said, it took extraordinarily little time after his release for Adler to clash with his father's moderate positions. It was their last conflict: Victor died on 10 November 1918.

Turbulent times. Both Friedrich Adler and Einstein pushed for a socialist, internationalist position, but one that would eschew the antidemocratic drive of the Bolsheviks. Einstein went to one of the soldier-worker councils to plead for democracy, and on 13 November 1918, pressed for a National Assembly:

> [A]ll true democrats must stand guard lest the old class tyranny of the right be replaced by a new class tyranny of the left. Do not be lured by feelings of vengeance to the fateful view that violence must be fought with violence, that a dictatorship of the proletariat is temporarily needed in order to hammer the concept of freedom into the heads of our fellow countrymen. Force breeds only bitterness, hatred and reaction.[44]

Adler too was similarly vocal in his challenge to the dictatorial left. Despite such increasing doctrinal tensions within the left, in January 1919, Lenin and Trotsky offered Adler the position of Honorary Secretary to the Third International, honorary commander-in-chief of the Red Army, and honorary chairman of the Central Soviet.[45] He refused. To Lenin's and Trotsky's dismay, Adler—at this point far more famous than Einstein—did worse: he joined the center-left Second International as it met in Bern (February 1919), only to dispute and split with *them* because they purged the Bolshevists. In short order he had created the political version of his "privileged frame of reference" in which he could follow no laws but his own.[46] Vladimir Lenin returned the compliment—he denounced Adler in December 1918:

The bourgeoisie are compelled to be hypocritical and to describe as "popular government," democracy in general. . . . The Scheidemanns and Kautskys, the Austerlitzes and Renners (and now, to our regret, with the help of Friedrich Adler) fall in line with this falsehood and hypocrisy. But Marxists, Communists, expose this hypocrisy, [as they aim] to replace this dictatorship with *the dictatorship of the proletariat.*[47]

Trotsky concurred, adding in April 1922 a postscript to his earlier, laudatory *Novy Mir* profile of the Adlers; now he lamented that Fritz Adler's personal courage was not matched by his "strength of thought." A fallen hero, Trotsky lamented, for "now Adler acts as a leader of the 2nd International serving the cause against which he had attempted to stand up if only by staking his life. . . ."[48] Political satirists, but also Trotsky, dubbed Adler's position the "Second and a Half International." When the Moscow Trial of 1–9 March 1931, aimed its sights at "saboteurs" supposedly encouraged by a scheming Labour and Socialist International (expanded from the Second International), Adler and his allies blasted Stalin's falsehoods, showing, with the cover photograph of their book, *The Witchcraft Trial in Moscow,* that one of the accused, Abromovich, could not possibly have been in Moscow in August 1928 because he was in Brussels at the LSI meeting.

Then Adler turned personal. "What I have so far set out could have been written by any other socialist . . . ," Adler wrote, though he never counted himself among the many. "I am compelled to speak *personally* for my attitude to the problems raised by the Moscow Trial is not so simple as that of those who reject individual terror 'on principle.'" He recalled his words to the Special Tribunal in which he asserted that in fact every act of violence was justified against the rulers of Austria—trials like his were a travesty of justice, awakening in him "a feeling of offended honour, a feeling of shame at being an Austrian." Adler added: "Every act of violence was justified against the rulers" of Austria, but now the terrorist should stay his hand against the Soviet Union.[49]

To justify this distinction he continued, "I always held firmly to the belief that questions of individual terror must be decided under a dual aspect: whether they correspond to a *natural consciousness of justice on the part of the people* and whether under the given circumstances they are a *suitable method* in the proletarian struggle for emancipation." Now—November 1936—it was not a time for such acts. The "Stalin experiment" had succeeded in certain economic achievements, but that did not mean for Adler that we should "allow ourselves to be forced to play the part of dumb curs or mendacious whitewashers. In this we differ from the puppets of the Communist Parties." Adler opposed individual or revolutionary acts of terror against the Soviet government. But he absolutely would not "give up the rights of criticism, a criticism which [is] indispensable, not to the detriment but in favour of a peaceful and evolutionary development in the Soviet Union towards . . . the rights and liberties of the people." The Soviets, Adler declaimed, would be the fortress against Hitler's Germany and Mussolini's Italy.[50]

Unlike Trotsky or Lenin—or the increasingly restive Second International, for that matter—Einstein already, by 1930, sketched the Adlers in an elegiac light. By then, of course, he had left behind the fame of ordinary physics. From the moment the pacifist physicist Arthur Eddington had announced the bending of starlight in November 1919, Einstein was the subject of poems and polemics, architecture and arch enemies. In 1930, Emma Adler (Fritz's mother) wrote to Einstein, "I write to you not as the world famous Einstein . . . but to the Einstein who years ago spent an unforgettable evening with us in the Blümengasse," asking if he would contribute to a memory book of Victor. Einstein did, recalling their day together in 1911.[51] In the older Adler, Einstein saw a man whom every political camp trusted, a philanthropic spirit that grasped human frailties, a vestige of an older, tolerant Vienna no longer imaginable in postwar desperation, a priest-like figure of a forgotten patriarchal past. Were such a man alive, Einstein ruminated, perhaps the powers at Versailles would not have made their tragic errors.[52]

No saint-like priest he, Friedrich Adler never quit making trouble. He infuriated the communists by lambasting the Stalinist purges, he protested against the Popular Front, he re-joined the Second International, then blasted them for refusing to oppose Hitler during the opening salvos of World War II. During one of Adler's speeches, Léon Blum leaned over to his neighbor and whispered, "he shoots better than he speaks."[53] After he fled from Europe to New York in 1940, the Nazi capitulation found him in apartment 14C on West 106th Street, just off Central Park West. "Dear Friend," he wrote to Einstein on Mercer Street in June 1945, "I was very happy to see you personally again." Not missing a beat, Adler plunged into a discussion of an American socialist and Max Born's latest on special relativity. The world changed utterly and not at all. Adler moved back to Zurich, where he worked on a biography of his father and lived until January 1960.[54]

THE ASSASSIN OF RELATIVITY

Friedrich Adler is a troubling figure, an ever-recurring tertium quid in historical set-pieces we are used to staging as two-part Manichean struggles. He appears, infuriatingly, to stand for the right to the left and, enragingly, for the left to the right. He stood with clear-sighted sanity and obsessive insanity; the best, most efficient party official and dutiful son, but also the schismatic, patricidal opponent of the ameliorist Austrian Socialist establishment that his father had devoted his life to building. Fritz was a hero to the revolutionaries Lenin and Trotsky, nominated to a prestigious place in the Third International, then rescuer to the Second International, then marching alone with the "Second and a Half" into the political night. From his cell came *Ernst Mach's Conquest of Mechanical Materialism*, which began with an homage to Mach, grappled with Mach's sworn enemy Max Planck, and then pulled back from both—neither anthropocentric-relativist nor trans-humanly absolutist. Finally, prison-author of a book litigating between his old friend Einstein's work and that of Lorentz and yet one that ends up throwing both accounts into the trash, Adler saw himself as the discoverer of the fatal flaw in the relativisms of both Einstein and Mach: Adler, the truest relativist with the anti-relativist, privileged frame K.

Here was the pacifist killer, the socialist individualist, the sideways gear in the motor of modernity. How to make sense of this wrongness in the proportion? Some have tried to find a psychodynamic explanation—a task in a certain sense launched by Friedrich Adler's psychiatrist-father long before Fritz gunned down Count Stürgkh. Melancholia, mania, fears of a nervous collapse—all these figure in the letters between son and father. One recent psychohistorical account makes Fritz Adler's gunshots into an attempt to impress his father. That seems, on the face of it, rather unlikely, given the lifetime Victor Adler had devoted to working within the Austro-Hungarian establishment. A step up might be to track the event as a symbolic (as well as real) killing. Even better: an account that took the fundamental cycle of ambivalence through his many relations of attachment and rebellion. Retainer and regicide: Fritz's deep friendship and fury with Einstein, with his filial devotion to and up-ending of Mach, and his ferocious devotion to the Party then his denunciation of it—all double actions of attraction and repulsion, need and hate.

But even corrected, refined, made more multivalent, the reductively psychological account won't do. It won't do for Adler and it won't do for Einstein—not because there is nothing to psychological explanations, but because here, in this case, they are too broad, too crude, too unmatched to the phenomena we want to understand. In Adler's case the post-hoc psychopathologizers are actually joining, after the fact, one side in the 1917 trial: that of the yellow press, struggling to depoliticize Adler's act. For both Einstein and Adler, reductive psychohistories too crudely make a bilayer system of the real (psychological) and the symbolic (physics, philosophy, politics).

Another possibility: We could take politics as our base—either the explicit politics that Friedrich Adler held or the politics of which he was a part but which were beyond his control. Then we could see his commitment to Machism as an expression of his "fundamental" Austro-Marxism, a stance he had taken (according to his biographers) since late childhood, growing up as he did in the heart of the movement. His killing of Stürgkh, his philosophy, his commitment to a relativity of physics—all could be seen as part of his "underlying" politics. Even his psychological state could be taken to emerge from a view that Adler claimed to have held since childhood, that justice meant freedom from raw domination—a view not that far from Einstein's.

Or flip the account a third way: perhaps Adler's philosophical beliefs lie deepest, to the necessary inclusion of sensory data at the beginning of our all-too-human knowledge and beliefs. Because he was trained by Mach, we could make Adler's politics and physics an epiphenomenal description of underlying views about how we secure knowledge.

None of these "true ground" accounts rings true to me. Much as we might like to "solve" Friedrich Adler by classing him in familiar dichotomous categories, maybe we should not. Maybe this strange, disquieting cipher helps us not so much by his typicality, but as a kind of anti–case study, a disturbing and forgotten (though once more famous) *Doppelgänger* to Einstein, an exemplar of nontypicality in a profoundly untypical moment. Maybe Adler could remind us of the limits of our categories, and in so doing help us understand a bit better who Einstein was by setting him in relief and showing us the explosive confluence of philosophy, politics, and physics from 1905 to 1919. After all, Einstein too drew heavily on Mach—*and* his nemeses Boltzmann and Planck. Einstein too could both sympathize with the assassin and take pleasure in the company of royalty and his cordial exchanges with Franklin Roosevelt. Einstein too wanted a physics of relative lengths of times but also of absolute speed of light and one set of physical laws—Machian relativism *and* Planckian absolutes. In these tensions, perhaps, lies a breakdown of comforting structures we too easily adopt from the early 20th century.

Of course, Einstein won his relativity battles and left to us a legacy unmatched in the history of science. True too, Friedrich Adler lost his wars—for a way to skew the dictatorships left and right in pre–World War II Austria; for an alternative to relativism and absolutism in philosophy; for a relativistic path to a privileged frame of reference. But both Adler and Einstein endlessly challenged simplistic splits. If the hybrid of philosophy, physics, and politics presses us to rethink the tumultuous years of the 20th century outside the realm of simple dichotomies, that might be a good last shot from the assassin of relativity.

CHAPTER THIRTEEN

14

SPACE, TIME, AND GEOMETRY: EINSTEIN AND LOGICAL EMPIRICISM

Michael L. Friedman

EINSTEIN'S GENERAL THEORY OF RELATIVITY revolutionized our conceptions of space, time, and motion. Einstein understood this theory, in particular, as casting an entirely new light on the problem of absolute versus relative motion, which had plagued the foundations of mechanics since the time of Newton. Even more dramatically, however, Einstein's new theory successfully applies a non-Euclidean geometry to represent physical space for the very first time. Specifically, it uses a non-Euclidean geometry of variable curvature (depending on the distribution of mass and energy) to represent the phenomenon of gravitation, where idealized bodies or "test particles" ("freely falling" under no other influence than gravity) are not affected by an externally impressed gravitational force, but rather follow geodesics or straightest possible paths in a variably curved, four-dimensional space-time manifold.

This four-dimensional space-time geometry has implications for three-dimensional, purely spatial geometry as well. One can visualize the purely spatial geometry of the solar system in two dimensions, for example, by imagining that the sun is a massive ball placed on a rubber sheet—the resulting "Schwarzschild geometry" is then circularly (spherically) symmetric with increasing negative curvature as one radially approaches the central sun. Gravity is not an external force acting on the planets (now viewed as comparatively much less massive test particles), but is rather a direct manifestation of the strongly non-Euclidean variable curvature increasing in the neighborhood of the sun: only at great distances from the sun does the geometry (the rubber sheet) approach a state of Euclidean flatness or zero curvature.

Einstein's revolutionary application of non-Euclidean geometry in physics had a decisive influence on 20th-century philosophy of geometry as well. In particular, logical empiricist philosophers like Reichenbach, Schlick, and Carnap appealed to Einstein's theory, in the context of earlier mathematical work by Riemann, Helmholtz, Poincaré, and Hilbert, in articulating a new view of the nature and character of geometrical knowledge. This view took its starting point from a rejection of Kant's conception of the necessary and synthetic a priori character of specifically Euclidean geometry—and, indeed, from a rejection of any role at all for spatial intuition within pure mathematics. We must sharply distinguish, on this view, between pure or mathematical geometry, which is an essentially uninterpreted axiomatic system making no reference whatsoever to spatial intuition or any other kind of extra-axiomatic or extra-formal content, and applied or physical geometry, which then attempts to coordinate such an uninterpreted formal system with some domain of physical facts given by experience. Since, however, there is always an optional element of decision in setting up such a coordination in the first place (we may coordinate the uninterpreted terms of purely formal geometry with light rays, stretched strings, or rigid bodies, for example), the question of the geometry of physical space is not a purely empirical one, but rather essentially involves an irreducibly conventional and in some sense arbitrary choice. In the end, there is therefore no fact of the matter whether physical space is Euclidean or non-Euclidean, except relative to one or another essentially arbitrary stipulation that coordinates the uninterpreted terms of pure mathematical geometry with some or another empirical physical phenomena.

Reichenbach's *Philosophie der Raum-Zeit-Lehre*, first published in 1928, is the most developed expression of this view, and it is particularly clear and explicit about the connection between Einstein's general theory of relativity and a conventionalist conception (inspired by Poincaré) of the question of whether space is Euclidean or non-Euclidean:

> Considered alone, the statement that a certain geometry holds for space is therefore without any meaning. It acquires a content only if we add which coordinative definition is to be used in the comparison of distant lengths.... [Such] statements only become characterizations of objective states of affairs if the coordinative definitions to be used are added, and [they] must be changed if other coordinative definitions are used. This [situation] signifies the relativity of geometry. The conception of the problem of geometry just sketched is essentially the result of the works of Riemann, Helmholtz, and Poincaré and has become known as conventionalism.... Einstein's achievement here consists only in having put this theoretical insight into practice in physics.[1]

In particular, this conception echoes Poincaré's famous thesis, in *Science and Hypothesis* (1902), that the question of whether space is really Euclidean or not is strictly meaningless—analogous to a purely conventional choice of coordinate system or units of measure.

Now Einstein himself made a celebrated contribution to 20th-century philosophy of geometry: his paper *Geometrie und Erfahrung* was first delivered in 1921. This paper begins with a very clear and sharp version of the distinction between "pure" and "applied"—mathematical and physical—geometry. Mathematical geometry, according to Einstein, derives its certainty and purity from its "formal-logical" character as a mere deductive system operating with "contentless conceptual schemata." The primitive terms of mathematical geometry—such as "point," "line," "congruence," and so on—do not refer to objects or concepts antecedently given (by some sort of direct intuition, for example), but rather have only that purely "formal-logical" meaning stipulated in the primitive axioms. These axioms serve, therefore, as "implicit definitions" of the primitive terms, and all the theorems of mathematical geometry then follow purely logically from the stipulated axioms. Applied or physical geometry, by contrast, arises when one gives some definite extra-axiomatic interpretation of the primitive terms via real objects of experience. But now the purity and certainty of mathematical geometry (which, in the end, rests simply on the purity and certainty of logic) is irrevocably lost, and we end up with one more empirical science among others:

> In so far as the propositions of mathematics refer to reality they are not certain; and in so far as they are certain they do not refer to reality. Full clarity about the situation appears to me be have been first obtained in general by that tendency in mathematics known under the name of "axiomatics." The advance achieved by axiomatics consists in having cleanly separated the formal-logical element from the material or intuitive content. According to axiomatics only the formal-logical element constitutes the object of mathematics, but not the intuitive or other content connected with the formal-logical element.[2]

Thus, these famous words from Einstein's paper, which were clearly intended and standardly taken as a refutation of the Kantian conception that mathematics is the paradigm of synthetic a priori truth, are a vivid expression of the modern axiomatic conception of geometry we now associate with the work of Hilbert. Indeed, Einstein himself takes the notion of "implicit definition," to which he appeals here, from Schlick's generalization of Hilbert's point of view to all of empirical science in his 1918 treatise *Allgemeine Erkenntnislehre*.[3]

So far, then, Einstein's own view of the matter is perfectly in harmony with the canonical 20th-century philosophy of geometry enunciated by logical empiricism. In articulating his view of specifically physical geometry, however, Einstein takes an entirely different tack. For he here views physical geometry, in particular, as a straightforward empirical theory of the actual physical behavior of "practically rigid bodies," and he claims, in a striking passage, that "without [this conception] I would have found it impossible to establish the [general] theory of relativity."[4] Immediately thereafter, in the same passage, Einstein considers Poincaré's geometrical conventionalism—apparently as the only real alternative to his own view—and suggests that "if one [following Poincaré] rejects the relation between the practically rigid body and geometry, one will in fact not easily free oneself from the convention according to which Euclidean geometry is to be held fast as the simplest."[5] Einstein concedes, in a continuation of this passage, that "Sub specie aeterni Poincaré in my opinion is correct," for "practically rigid bodies" are in fact unsuitable to play the role of "irreducible elements in the conceptual framework of physics"—nevertheless, they must provisionally "still be called upon as independent elements in the present stage of theoretical physics"—when, in particular, we are still very far from an adequate micro-theory of the structure of matter.[6] And where such rigid bodies must "still be called upon as independent elements," it is clear, is precisely in the foundations of the general theory of relativity.

In *Geometrie und Erfahrung*, therefore, Einstein decisively rejects Poincaré's conventionalist philosophy of geometry. He does so because, if one follows Poincaré, "one will ... not easily free oneself from the convention according to which Euclidean geometry is to be held fast as the simplest"—so that, in particular, Einstein had to adopt the competing, more straightforwardly empiricist conception of physical geometry as a theory of the factual behavior of rigid measuring rods in order successfully to formulate the general theory of relativity. But now we have a serious historical and philosophical puzzle: How could the logical empiricists have so badly misinterpreted the philosophical significance of Einstein's theory? And how could they have done this with a significant amount of encouragement from Einstein himself (as in Einstein's explicit endorsement of Schlick's work, for example)? How, in particular, could the logical empiricists (starting with Schlick) have read Einstein's *Geometrie und Erfahrung* and Poincaré's *Science and Hypothesis* as perfectly compatible, when Einstein himself explicitly says that he had to reject Poincaré in order to arrive at his new theory? What, more generally, is the relationship between Einstein's conception and Poincaré's? The key to understanding these questions, as we shall see, is an appreciation of the very subtle and unexpected way in which Einstein, in creating the general theory of relativity, connected the 19th-century discussion of the foundations of geometry by Helmholtz and Poincaré with his own revolutionary reinterpretation of the problem of motion.

I begin with Helmholtz's initial formulation of the problem of physical geometry circa 1860–70.[7] In particular, I begin with what we now call the Helmholtz-Lie theorem, which was first articulated by Helmholtz in connection with his psychophysiological researches into space perception. Helmholtz was inspired by Riemann's (1854) work on what the latter called "n-fold extended manifolds" to attempt to derive Riemann's fundamental assumption or "hypothesis"—that physical space is distinguished from all other such manifolds (such as color space or sound space, for example) by being not only three-dimensional, but also metrically infinitesimally Euclidean—from what Helmholtz took to be the fundamental "facts" generating our perceptual representation or intuition of space. Helmholtz's starting point was that this representation is in no way immediately given or "innate," but instead arises by a process of perceptual accommodation or learning based on our experience of bodily motion. Since our representation of space arises kinematically, as it were, from our experience of moving up to, away from, and around the objects that "occupy" space, the space thereby constructed must satisfy a condition of "free mobility" that permits arbitrary continuous motions of rigid bodies. In Euclidean geometry, for example, such "rigid motions" are expressed by a transformation group consisting of translations (along any straight line) and rotations (around any point). And, more generally, if we know that a transformation group of "rigid motions" exists on a space (or "manifold"), we can conclude that this space is at least infinitesimally Euclidean.

In fact, the Helmholtz-Lie theorem not only fixes the geometry of space as infinitesimally Euclidean, it also implies that space has constant (rather than variable) curvature as well; and it thereby fixes the geometry of space as one of the three classical geometries of constant curvature: Euclidean (zero curvature), hyperbolic (negative curvature), or elliptic (positive curvature). (The surface of a sphere, for example, which has constant positive curvature, has a group of rigid motions and satisfies the condition of free mobility; but an egg-shaped surface, which has variable positive curvature, does not: one cannot continuously move a triangle from the flatter end of the egg to the pointy end while keeping it rigid or congruent to itself.) But how do we know, on Helmholtz's view, which of the three classical geometries then holds? At this point, he says, we investigate the actual behavior of rigid bodies (rigid measuring rods, for example) as we move them around in accordance with the condition of free mobility. That physical space is Euclidean (which Helmholtz of course assumed) means that physical measurements carried out in this way are empirically found to satisfy the laws of this particular geometry to a very high degree of exactness (so the sum of the angles of a triangle is always equal to 180°, for example, and the ratio of the circumference to the diameter of a circle is always equal to π). Thus, Helmholtz's view is Kantian insofar as space has what Helmholtz calls a "necessary form" expressed in the condition of free mobility, but it is empiricist insofar as which of the three possible geometries of constant curvature actually holds is then determined by experience.

It was precisely this Helmholtzian view of physical geometry that set the stage, in turn, for the contrasting "conventionalist" conception articulated by Poincaré. Indeed, Poincaré developed his philosophical conception immediately against the background of the Helmholtz-Lie theorem, and in the context of his own mathematical work on group theory and models of hyperbolic geometry. Following Helmholtz and Lie, Poincaré viewed geometry as the abstract study of the group of motions associated with our initially crude experience of bodily "displacements." So we know, according to the Helmholtz-Lie theorem, that the space thereby constructed has one and only one of the three classical geometries of constant curvature. Poincaré disagreed with Helmholtz, however, that we can empirically determine the particular geometry of space simply by observing the behavior of rigid bodies. No real physical bodies exactly satisfy the condition of geometrical rigidity, and, what is more important, knowledge of physical rigidity presupposes knowledge of the forces acting on the material constitution of bodies. But how can one say anything about such forces without first having a geometry in place in which to describe them? We have no option, therefore, but to stipulate one of the three classical geometries of constant curvature, by convention, as a framework within which we can then do empirical physics. Moreover, since Euclidean geometry is mathematically the simplest, Poincaré had no doubt at all that this particular stipulation would always be preferred.

We know that Einstein was intensively reading Poincaré's *Science and Hypothesis* as he was creating the special theory of relativity in 1905, and it seems very plausible, accordingly, that Poincaré's conventionalism played a significant role in philosophically motivating this theory. More specifically, whereas Poincaré had argued, against both Kant and Helmholtz, that the particular geometry of space is not dictated by either reason or experience, but rather requires a fundamental decision or convention of our own, Einstein now argues, similarly, that simultaneity between distant events is not dictated by either reason or experience, but requires a new fundamental definition based on the behavior of light. (Indeed, Poincaré had himself argued that distant simultaneity requires a convention or definition—also involving the velocity of light—in his 1898 article, "The Measurement of Time.") Moreover, Einstein proceeds here, in perfect conformity with Poincaré's underlying philosophy in *Science and Hypothesis*, by "elevating" an already established empirical fact—the invariant character of the velocity of light in different inertial reference frames—into the radically new status of what Poincaré calls a convention or "definition in disguise" (here, a definition of simultaneity).

As we also know, however, Einstein tells us in *Geometrie und Erfahrung* that he needed to reject Poincaré's geometrical conventionalism in order to arrive at the general theory of relativity. In particular, Einstein here adopts a

Helmholtzian conception of (applied or physical) geometry as a straightforward empirical theory of the actual physical behavior of "practically rigid bodies," and he claims that "without [this conception] I would have found it impossible to establish the [general] theory of relativity." Here, as Einstein explains in the same passage, he has in mind the following line of thought:[8] According to the principle of equivalence (equating gravitational and inertial mass), gravitation and inertia are essentially the same phenomenon. So, in particular, we can model gravitational fields by "inertial fields" (involving centrifugal and Coriolis forces, for example) arising in non-inertial frames of reference (that is, accelerating or rotating frames of reference). If we now consider a uniformly rotating frame of reference in the context of special relativity, we then find that the Lorentz contraction differentially affects measuring rods laid off along concentric circles around the origin in the plane of rotation (because of the variation in tangential linear velocity at different distances along a radius), whereas no Lorentz contraction is experienced by rods laid off along a radius. Therefore, the geometry in such a rotating system will be found to be non-Euclidean (the ratio of the circumference to the diameter of concentric circles around the origin in the plane of rotation will differ from π and depend on the circular radius).

The importance of this line of thought for Einstein is evident in virtually all of his expositions of the general theory, where it is always used as the primary motivation for introducing non-Euclidean geometry into the theory of gravitation. Moreover, as John Stachel has shown, this particular thought experiment in fact constituted the crucial breakthrough to what we now know as the mathematical and conceptual framework of general relativity.[9] For, generalizing from this example, Einstein quickly saw that what he really needed for a relativistic theory of gravitation is a four-dimensional version of non-Euclidean geometry (comprising both space and time). He understood that a variably curved generalization of what we now call the flat Minkowski metric of special relativity should serve as the representative of the gravitational field, and turning to the mathematician Marcel Grossmann for help, he then discovered the Riemannian theory of manifolds. Einstein's repeated appeal to the example of the uniformly rotating frame of reference in his official expositions of the theory therefore appears to reflect the actual historical process of discovery very accurately, and to explain, in particular, how the idea of a variably curved, four-dimensional space-time geometry was actually discovered in the first place.

Einstein's introduction of non-Euclidean geometry into physics thus followed a remarkably circuitous route, and it involved, in particular, a process by which Einstein delicately positioned himself within the debate on the foundations of geometry between Helmholtz and Poincaré. In creating the special theory of relativity in 1905, Einstein took inspiration, as we have seen, from

Poincaré's conventionalist philosophy of science in *Science and Hypotheses*—not, however, as applied to spatial geometry, but rather to what we now call the (four-dimensional) geometry of space-time. Yet, when the special theory was in place, Einstein then faced a radically new situation in the theory of gravitation, for the Newtonian theory of universal gravitation—based, as it was, on an instantaneous action at a distance (and thus on absolute simultaneity)—was itself incompatible with the new conceptual structure of the special theory based on a relativized conception of simultaneity. Einstein was therefore faced with the problem of adjusting the theory of gravitation to this new relativistic conceptual structure, and he addressed this problem, in the first instance, by appealing to the already familiar fact that gravitational and inertial mass are equal. This fact led him to his principle of equivalence—the idea that gravitation and inertia are the very same physical phenomenon—which he then applied, as we have seen, to non-inertial frames of reference (accelerating and rotating frames) within the conceptual structure of the special theory (what we now call the structure of Minkowski space-time). This led Einstein, in turn, by the example of the uniformly rotating frame, to a non-Euclidean spatial geometry (now linked to the action of a gravitational field), which he was then able, finally, to generalize to the non-Euclidean space-time geometry of general relativity.

Moreover, the crucial thought experiment of the uniformly rotating frame of reference essentially involved, as Einstein tells us in *Geometrie und Erfahrung*, a naively Helmholtzian, rather than a sophisticated Poincaré-inspired, perceptive on the relationship between the behavior of rigid bodies and physical geometry. Indeed, it was necessary for Einstein to have already rejected the more sophisticated perspective on rigid bodies suggested by Poincaré in creating the special theory of relativity. For, as is well known, Poincaré was actually the first person to discover what we now know as the Lorentz group governing inertial reference frames in special relativity, and Poincaré had formulated, accordingly, a Lorentzian version of the mathematics of special relativity still in some sense committed to a classical æther. For Poincaré, we might say, the Lorentz group (and therefore the Lorentz contraction) thus operated at the level of electrodynamics—governing the microscopic electromagnetic forces ultimately responsible for physical rigidity—but not, as in Einstein, at the more fundamental kinematical level governing the basic concepts of space, time, and motion formulated prior to and independently of any particular dynamical theory. Just as, in the special theory, Einstein takes the Lorentz contraction as a direct indication of fundamental kinematical structure, independently of all dynamical questions about the microphysical forces actually responsible for physical rigidity, here, in the example of the uniformly rotating reference frame, Einstein similarly takes the Lorentz contraction as a direct indication of fundamental geometrical structure. And without this remarkably circuitous procedure of delicately situating himself, as it were, between Helmholtz and Poincaré, it is indeed hard to imagine how Einstein could have ever discovered the idea of a variably curved, four-dimensional space-time geometry in the first place.

There is an important sense, however, in which this same idea of a variably curved space-time geometry, once discovered, renders the preceding debate between Helmholtz and Poincaré quite irrelevant. For, as we have seen, this debate is itself framed by the Helmholtz-Lie theorem and is therefore limited to spaces of constant curvature. But the general theory of relativity, of course, employs a space-time of variable curvature—dependent on the distribution of matter and energy—and here Helmholtz's principle of free mobility therefore fails. In the purely spatial, two-dimensional rubber sheet model of the Schwarzschild geometry in the neighborhood of the sun described at the beginning, for example, a (breadthless) "measuring rod" can be rigidly transported on the circumference of a circle at a given radial distance from the sun (due to circular [spherical] symmetry), but it cannot be rigidly transported away from or off of the given circumference. As a result, the common conceptual framework for discussing the problem of physical geometry within which both Helmholtz and Poincaré developed their (conflicting) solutions is simply inapplicable to the general theory.

Now Einstein begins *Geometrie und Erfahrung*, as we have seen, by invoking Schlick's elaboration of the notion of "implicit definition" in his *Allgemeine Erkenntnislehre*, which is itself based, in this context, on Schlick's earlier work on the philosophical significance of the general theory of relativity, *Raum und Zeit in der gegenwärtigen Physik*, first published in 1917. And it is precisely here, in particular, that Schlick finds that Poincaré's conventionalist philosophy of geometry still holds in general relativity.[10] We still have a choice, that is, about whether to use a non-Euclidean or a Euclidean physical geometry, even within the context of Einstein's new theory; however, in the latter case, we would have to introduce further complications into the simple and "natural" physical coordination effected by Einstein's principle of equivalence (which directly coordinates freely falling "test particles" affected only by gravitation to the four-dimensional space-time geodesics of a variably curved, non-Euclidean, space-time geometry), and we would thereby introduce onerous complications into our total system of geometry plus physics. But Poincaré had maintained that only mathematical simplicity explains our preference for Euclidean (three-dimensional) spatial geometry, and all we are now doing, in the context of Einstein's new theory, is extending Poincaré's viewpoint to a non-Euclidean (four-dimensional) space-time geometry.

This attempt to link Poincaré's conventionalist philosophy of geometry and the general theory of relativity is certainly plausible, but it is subject to very deep difficulties from our present point of view. Most importantly, Poincaré's conception of space and geometry is entirely based, as we have seen, on the principle of free mobility first formulated by Helmholtz and then brought to precise mathematical fruition in the Helmholtz-Lie theorem. It was precisely this principle, moreover, that forms the indispensable link between pure or mathematical and applied or physical geometry for both Poincaré and Helmholtz. In particular, the principle of free mobility, in this context, serves as our crucial coordinating principle—our crucial link between pure and applied geometry—setting up a relationship of coordination or designation between purely geometrical notions, like that of geometrical equality or congruence, for example, and the idealized behavior of physical rigid bodies.

And it was precisely here, for Poincaré, that a remarkable conceptual situation then arises: it turns out that there are three and only three possible geometries compatible with the principle of free mobility—and thus compatible, as we have seen, with our fundamental coordinating principle linking pure and applied geometry. Our fundamental coordinating principle leaves the choice of Euclidean or non-Euclidean geometry entirely open, and it thus makes perfect sense, in this very special conceptual situation, for Poincaré to maintain that the choice of Euclidean geometry, in particular, is then determined by a convention or stipulation based on the greater mathematical simplicity of specifically Euclidean geometry. In the radically new conceptual situation created by the general theory of relativity, however, this particular view of the matter is no longer applicable. Not only is the space-time structure of general relativity incompatible with the fundamental presupposition of the Helmholtz-Lie theorem, the principle of free mobility, but, in the general theory, there is one and only one way to effect the required coordination between our purely mathematical formulation of the theory and concrete physical reality—namely, the principle of equivalence, which directly coordinates the purely mathematical notion of a (semi-)Riemannian space-time geodesic with the behavior of freely falling bodies ("test particles") affected only by gravitation. Here, unlike in the very special situation addressed by Poincaré, we are not faced with what we might call a common or generic coordinating principle, leaving the specific choice of geometrical structure for physical space still open, but rather with a singular or unique coordinating principle (the principle of equivalence), compatible with one and only one geometrical structure—the geometrical structure for physical space-time given by Einstein's formulation of the general theory. Here the idea of an arbitrary or conventional choice of physical geometry has itself lost all real meaning and application—both from a mathematical and an empirical point of view.

CHAPTER FOURTEEN

By contrast, Einstein's own engagement with the problematic of conventionalism, and, in particular, with the debate on the foundations of geometry between Helmholtz and Poincaré, was an especially timely and fruitful one. It allowed him, as we have seen, to take the critical step, via the principle of equivalence, from the interpretation of three-dimensional, non-Euclidean spatial geometry to that of four-dimensional, non-Euclidean space-time geometry. And it is in precisely this sense, we might say, that Einstein himself made the crucial transition from a 19th- to 20th-century philosophy of physical geometry. One unforeseen consequence, however, was that the fundamentally new perspective on the foundations of geometry actually created by Einstein in this way has proved much more difficult to grasp than it otherwise might. In particular, in the canonical 20th-century philosophy of geometry bequeathed to us by logical empiricism, we remained preoccupied with the problematic of conventionalism, and with a concern for the behavior of rigid bodies, long after these had lost all specific relevance to physical theory—where a new concern for space-time geometry, and with the problem of motion, can now be seen as the true successor to the late 19th-century tradition in the mathematical and philosophical foundations of geometry that was subject to a far-reaching and radical transformation in the work of Einstein.

What we are now in a position to see, finally, is that the crucial step in this transformation is strikingly and accurately depicted in the passage on the uniformly rotating frame of reference from *Geometrie und Erfahrung*.[11] For it is precisely here that Einstein joins two previously independent problems involving the physical interpretation of space, time, and geometry in a quite remarkable and unexpected fashion. Since Newton's original creation of a coherent mathematical-physical theory of space and time in the 17th century, the problem of the relativity of motion had persistently afflicted its conceptual foundations. The discovery of the concept of inertial reference frame in the late 19th century introduced considerable clarity into this situation, and Einstein had already put this concept to fruitful new use in his creation of the special theory of relativity. In view of the principle of equivalence, however, Einstein now sees that an extension of the special principle of relativity is called for and, in particular, that non-inertial frames of reference, in the context of the mathematical structure of special relativity (what we now call the structure of Minkowski space-time), can be used to model the gravitational field. Here Einstein encounters the decisive example of a uniformly rotating frame of reference, whose three-dimensional, purely spatial geometry turns out to be non-Euclidean. And Einstein is able to take this example as a direct physical indication of fundamental geometrical structure (and not as a mere field of force causing apparently rigid bodies to be distorted) by delicately positioning himself, as we have said, within the late 19th-century debate on the foundations of physical geometry between Helmholtz and Poincaré. But the upshot of Einstein's radical transformation (and eventual transcendence) of this debate is that the conceptual foundations of geometry can no longer be fruitfully pursued in independence of the problem of motion. The new problem of physical geometry is precisely the problem of space-time: space, time, motion, and geometry now constitute one indissoluble whole.

15

EINSTEIN AS A STUDENT

Dudley Herschbach

INTRODUCTION

Late in life, when reflecting on his uncanny papers of 1905, Einstein liked to say, "Nobody expected me to lay golden eggs." A century later, nobody can expect to comprehend fully how he did it. Yet much is known about Einstein as a student, enabling us to trace his maverick path as a fledgling scientist. It was a rough path; even with his talent and dedication, help from others was crucial. If Einstein were reincarnated as a graduate student today, it seems unlikely that he would complete a Ph.D. His saga offers perspectives that should embolden current students and prod faculty to reform doctoral programs.

Einstein revered the memory of his great predecessors in physics. In his home at Princeton, he had a portrait of Newton above his bed and portraits of Faraday and Maxwell in his study.[1] At commemorations, Einstein felt homage to scientists was best rendered by striving to understand what they "were aiming at, how they thought and wrestled with their problems." This accords with a kindred credo of Gerald Holton[2] that I have long advocated to students. It thus seemed apt, when invited to take part in this conference, to focus on Einstein as a student.

My paper is addressed to three presumed audiences. First, to many earnest students who find the transition from neophyte to Ph.D. scientist a daunting journey. May you be heartened to see that it was rough even for Einstein, despite his immense talent and passion for science. Second, to earnest faculty concerned that typical Ph.D. programs have evolved into an ill-defined and often demoralizing grind, poorly suited to fostering the creativity of students. May you find that aspects of the Einstein story prove useful in pressing for reforms. Third, to anyone curious about the gestation of ideas that enabled Einstein to bring forth in 1905 those epochal "golden eggs," along with his Ph.D. thesis. May you discern clues to approaches that can be emulated by lesser mortals.

Einstein's own commentaries and correspondence tell us much about his student years, and a host of outstanding historians of science and biographers have added many insightful books and essays. In this paper I merely sample morsels from this ample smorgasbord. Rather than construct a narrative essay, I invite readers to consult chronologies,[3] annotated in four stages: Einstein's schoolboy era, extending to age seventeen; his four years in a teacher

TABLE 15.1   Chronology, 1879–1896

Age	Year	
0	1879	March 14: AE born in Ulm, Germany. Ancestors for two centuries are Swabian Jews, but parents irreligious.
1	1880	Family moves to Munich. Father and uncle Jakob partners in manufacturing firm for plumbing and electrical apparatus.
5	1884	AE enchanted by a compass. Has private lessons at home (too young for admission to public primary school).
6	1885	AE starts violin lessons, continues to age 13. October: Enters Catholic primary school; also begins private Jewish religious instruction (required by law).
9	1888	October: AE enters first year of nine-year *Gymnasium* program.
10	1889	Uncle tells AE of Pythagorean Theorem; AE devises a proof. Max Talmud (later Talmey) begins regular visits, which continue for six years, and brings AE popular science books.
11	1890	AE's intense "religious paradise," lasting about a year. Reading science, including Darwin, disenchants AE with religion; he becomes a "fanatic freethinker."
12	1891	AE enthralled with "holy little geometry book" (from Talmud); finds Euclid's axiomatic-deductive method a trustworthy "road to paradise." Over next four years, learns analytic geometry and calculus outside school.
13	1892	AE disdains bar mitzvah. Captivated by Mozart sonatas ("love is a better teacher than duty"); reads Kant.
15	1894	Family moves to Milan. Intends AE to stay in Munich to complete *Gymnasium*, but in December, AE quits school, joins family in Italy.
16	1895	Summer: Sends uncle essay on the state of ether in magnetic fields. October: Fails entrance exam for ETH; enrolls in cantonal school in Aarau, boards with Winteler family. Ponders *Gedanken* ride on light wave.
17	1896	January: Renounces German citizenship. September: Passes exam for Aarau diploma. Essays on Goethe and on future plans.

training program at the Swiss Federal Polytechnic Institute in Zurich;[4] his first two years as a graduate student, desperate to find a steady job; and his first three years as a patent examiner in Bern, including his *annus mirabilis*. For each stage, I enliven the chronology with some observations and a few quotations, chiefly from Einstein or his contemporaries. After commenting on some aspects of his early papers, I offer my views about lessons that today's academic enterprise should take from the Einstein saga.

Schoolboy: Munich, Milan, and Aarau

A charming source for Einstein's childhood is an affectionate memoir by his younger sister Maja.[5] She describes worries of their parents that little Albert was retarded because he was unusually slow to talk. Einstein himself, decades later, said that one of his earliest memories was "the ambition to speak in whole sentences . . . so I would try each sentence out . . . saying it softly. Then, when it seemed alright . . . say it out loud." Up to at least age seven, the habit of softly and slowly rehearsing his words persisted. Maja reports that Albert avoided play with other children and had wild temper tantrums. At five, "he grabbed a chair and with it struck the woman tutor, who . . . ran away in fear and was never seen again." Another time, "he used a child's hoe to knock a hole in [Maja's] head." Fortunately, his "violent temper disappeared during his early school years." Today, a child with such unusual speech and antisocial behavior might have to contend with therapists and be put on drugs in order to attend preschool or kindergarten.

Maja describes also Albert's early traits of self-reliance, persistence and tenacity. He was fond of puzzles and "building many-storied houses of cards . . . as high as fourteen stories." In primary school, Albert was "self-assured and . . . confidently found the way to solve difficult word problems." He did well, both in primary and high school, but "the style of teaching [by rote learning] in most subjects was repugnant to him." Especially galling at the *Gymnasium* was the "military tone . . . the systematic training in the worship of authority."

Albert's intellectual growth was strongly fostered at home. His mother, a talented pianist, ensured the children's musical education. His father regularly read Schiller and Heine aloud to the family. Uncle Jakob challenged Albert with mathematical problems, which he solved with "a deep feeling of happiness." Most remarkable was Max Talmud, a poor Jewish medical student from Poland, "for whom the Jewish community had obtained free meals with the Einstein family." Talmud came on Thursday nights for about six years, and "invested his whole person in examining everything that engaged [Albert's] interest." Talmud had Albert read and discuss many books with him. These included a series of twenty popular science books that convinced Albert "a lot in the Bible stories could not be true," and a textbook of plane geometry that launched Albert on avid self-study of mathematics, years ahead of the school curriculum. Talmud even had Albert read Kant; as a result Einstein began preaching to his schoolmates about Kant, with "forcefulness" [F: 25].

By law, a male German citizen could emigrate only before the age of seventeen without having to return for military service. This was impetus for Albert's decision, made without consulting his parents, to conspire with a doctor for a medical release from the Munich Gymnasium, join his family in Italy, and renounce German citizenship. For nine months, he enjoyed freedom from school. He also wrote an essay on his "naive and imperfect" analysis of the state of the ether in a magnetic field, and studied diligently on his own to prepare for the ETH entrance examinations. He was permitted to take the exams for the engineering department, although he was much younger than the prescribed entrance age of eighteen. He failed because he did poorly in modern languages and descriptive sciences, but he did very well in mathematics and physics. That led the ETH Director to urge Albert to enroll in the cantonal school in Aarau, a small town near Zurich, whose graduates were directly admitted to the ETH.

At Aarau, Albert had a happy year, both in the school and lodging in the home of Jost Winteler, one of the teachers. Long after, in contrasting Aarau with Munich, Einstein wrote:

> By its liberal spirit and by the simple seriousness of its teachers . . . this school . . . made me realize how much superior an education towards free action and personal responsibility is to one that relies on outward authority and ambition. True democracy is no empty illusion. [F: 38]

The Wintelers welcomed Albert into their large family. His congenial ties to them proved durable: Maja married a Winteler son, and Michele Besso, one of Einstein's best friends, married a Winteler daughter. In an essay for his graduation exam, written in "execrable French," Albert described with blithe confidence his ambitions:

> A happy person is too content with the present to think much about the future. On the other hand, young people in particular are fond of making bold plans. Besides, it is natural for a serious young man to form as precise an idea of the goal of his strivings as possible.
>
> If I am lucky enough to pass my examinations, I will attend the Polytechnic in Zurich. I will stay there four years to study mathematics and physics. My idea is to become a teacher in these fields of natural science and I will choose the theoretical part of these sciences.

CHAPTER FIFTEEN

PLATE 1 Salvador Dalí (1904–1989), *The Persistence of Memory*, 1931; oil on canvas 9½×13". The Museum of Modern Art, New York. Digital image © The Museum of Modern Art. Licensed by SCALA/Art Resource, New York. © 2006 Salvador Dali, Gala-Salvador Dali Foundation/Artists Rights Society (ARS), New York.

PLATE 2 Andy Warhol, "Albert Einstein," from *Ten Portraits of Jews of the Twentieth Century*, 1980; screenprint. Licensed by Corbis.

PLATE 3   Tony Robbin, *Lobofour*, 1982; acrylic on canvas with painted wire rods. Private Collection. © Tony Robbin.

PLATE 4   Claude Monet (French,
1840–1926), *Rouen Cathedral Façade and
Tour d'Albane (Morning Effect)*, 1894; oil
on canvas, 41¾×29⅛ in.) Museum of Fine
Arts, Boston, Tompkins Collection—Arthur
Gordon Tompkins Fund. Photograph ©
Museum of Fine Arts, Boston.

PLATE 5   Claude Monet, *Rouen Cathedral,
West Façade, Sunlight*, 1894; oil on
canvas, (39½×26 in.) Chester Dale Collection,
image © 2007 Board of Trustees, National
Gallery of Art, Washington, D.C.

PLATE 6   Claude Monet (French, 1840–1926), *Rouen Cathedral, Façade, 1894;* oil on canvas (39⅝×26 in.). Museum of Fine Arts, Boston, Juliana Cheney Edwards Collection. Photograph © Museum of Fine Arts, Boston.

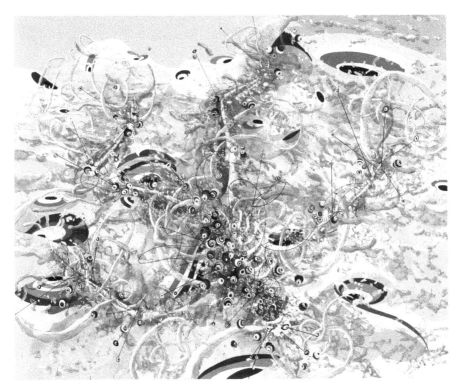

PLATE 7   Matthew Ritchie, *The Eighth Sea*, 2002; oil and marker on canvas, 99×121 in.
Private Collection. Courtesy of Andrea Rosen Gallery. Photograph by Oren Slor.

PLATE 8　Carsten Höller, *Neon Circle*, 2001. Courtesy of Carsten Höller and Casey Kaplan Gallery, New York. Collection of the Henry Art Museum, Seattle, WA.

PLATE 9　Doug Aitken, *Interiors.* Courtesy 303 Gallery, New York.

PLATE 10 *Proposition Player,* overview. Matthew Ritchie, Andrea Rosen Gallery.

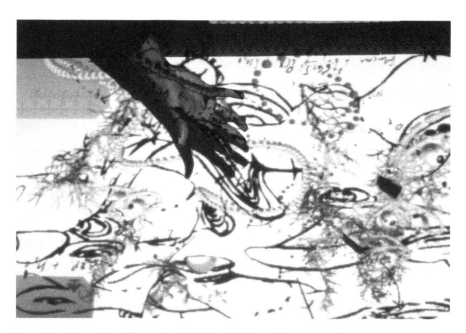

PLATE 11 *Proposition Player*, detail. Matthew Ritchie, Andrea Rosen Gallery.

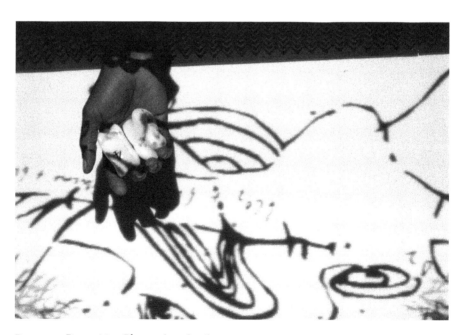

PLATE 12 *Proposition Player*, close detail. Matthew Ritchie, Andrea Rosen Gallery.

Plate 13 *The Fine Constant*, upper view. Matthew Ritchie, Andrea Rosen Gallery.

Plate 14 *The Fine Constant*, lower view. Matthew Ritchie, Andrea Rosen Gallery.

PLATE 15 *The Fine Constant*, side view. Matthew Ritchie, Andrea Rosen Gallery.

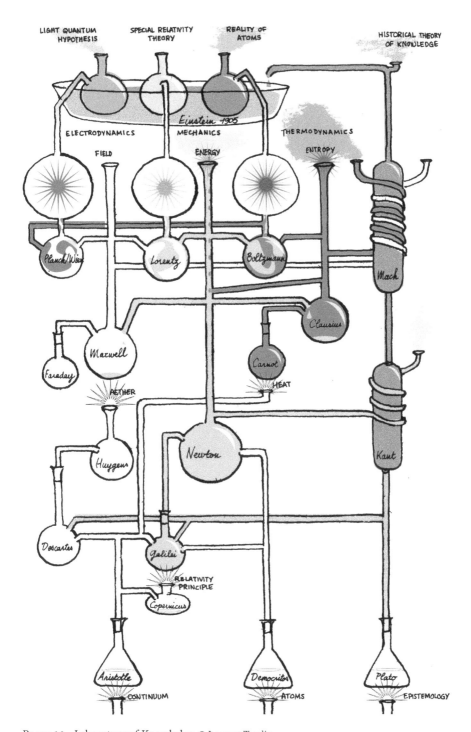

PLATE 16   Laboratory of Knowledge. © Laurent Taudin.

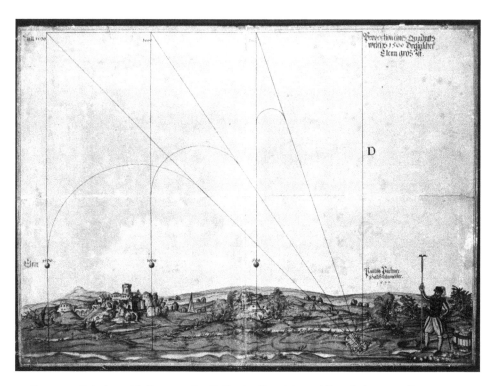

PLATE 17   Painting with Construction Reflecting Practitioners' Knowledge according to
Paulus Puchner. Mathematisch-Physikalischer Salon, Staatliche Kunstsammlungen, Dresden.

PLATE 18 Schematic of a particle trajectory in bound motion. The gradient of the action $S(\mathbf{q},\mathbf{q}_0,E)$ gives the momentum at point $\mathbf{q}$; however, this gradient is in general multivalued because bound motion loops back on itself an infinite number of times. Einstein showed that, for type A motion, this multivalued vector field in real space can be mapped onto a single-valued vector field on a torus.

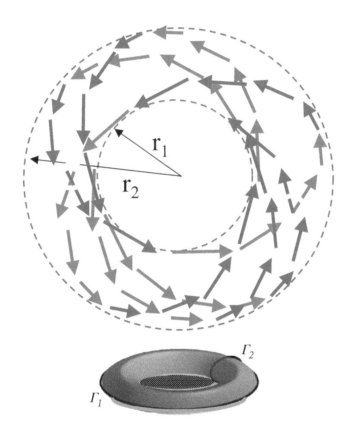

PLATE 19 Momentum vector field generated by a particle moving under a central force between inner and outer turning points r_1, r_2 (top). Einstein argued that such a vector field will be double-valued at all points in the allowed region of motion (see yellow circles) and can be mapped to an irrotational vector field on a torus (bottom). The red vectors describe the parts of the trajectory moving inward radially, and the blue those moving outward; these map to the top and bottom of the torus, as indicated. There exist two nontrivial loop integrals of such a field on a two-torus, illustrated by Γ_1, Γ_2 in the schematic. The Einstein quantization rule, Equation 18.3, requires that these two loop integrals yield an integer times Planck's constant h.

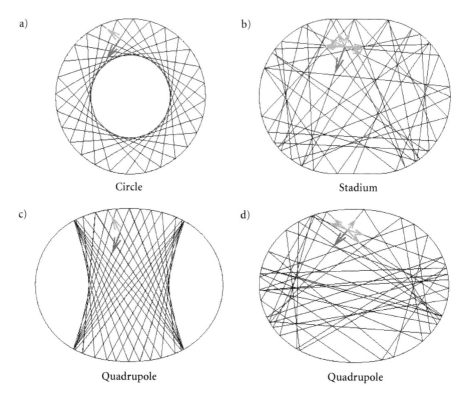

a) Circle

b) Stadium

c) Quadrupole

d) Quadrupole

PLATE 20 (a) A quasiperiodic trajectory in the circular billiard. Each trajectory visits any neighborhood (e.g., the yellow circle) of the allowed region of motion an infinite number of times, but in a finite number of momentum directions. This system is integrable (angular momentum and energy are conserved) and exhibits only Einstein's type A motion. (b) A generic trajectory in the stadium billiard. Each trajectory returns to any neighborhood of the allowed region in an infinite number of momentum directions. The system is non-integrable and highly chaotic; it exhibits only Einstein's type B motion. (c, d) Two trajectories starting in the same neighborhood of quadrupole billiard with different initial momenta. The first leads to regular (type A) motion; the second to chaotic (type B) motion. This is the generic behavior of Hamiltonian dynamical systems, a mixture of quasiperiodic and chaotic dynamics.

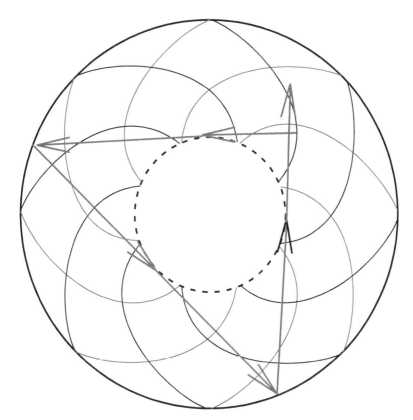

PLATE 21 Red and blue curves denote the wavefronts (curves of constant action S_n) for the EBK solution to the circular billiard. The rays are the normals to the wavefronts and travel in straight lines until they reflect from the boundary, switching from one set of wavefronts to the other. Note that the wavefronts are such that the rays specularly reflect at the boundary; therefore, an initial ray starting in the yellow neighborhood, normal to the red wavefronts, will follow an actual ray trajectory in the billiard, switching between red and blue wavefronts as shown. When it returns to the yellow neighborhood, it must be on one of the allowed wavefronts. This will be the case for the circular billiard (Einstein type A motion); it will not be the case for the stadium billiard because of the generic chaos in its ray dynamics (Einstein type B motion). Therefore, no finite set of wavefronts exists for the stadium billiard, and one cannot construct EBK solutions.

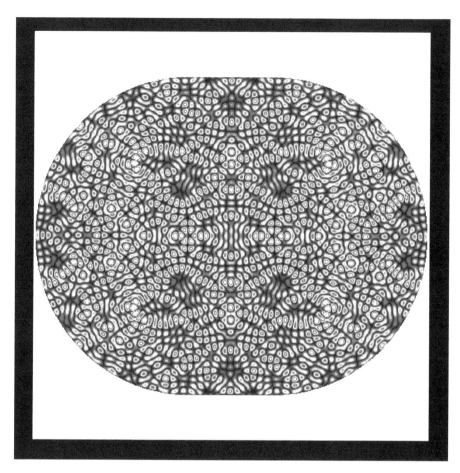

PLATE 22   False color plot of the probability density for a "chaotic" wavefunction of the stadium billiard. Red areas are regions of high density; blue are regions of low density. Note that there are few regions of parallel nodal lines, indicating the lack of smooth wavefronts as are required for the EBK-WKB method to work.

These are the reasons that have led me to this plan. It is primarily a personal gift for abstract and mathematical thought and a lack of fantasy and practical talent. Moreover, my hopes lead me to the same resolution. This is quite natural; one always wishes to do the things one has the most talent for. Moreover, there is a certain independence in the profession of science that greatly appeals to me. [P: 40; C: D22]

College Years: 1896–1900 in Zurich

Perhaps with Jost Winteler as a model, Einstein entered ETH as a candidate for a specialist teacher diploma rather than as an engineering student. He enrolled in Department VI A, which dealt with mathematics, physics, and astronomy. It had only 23 students, 11 of them freshmen—a small fraction of the ETH student body of over 800 students, most of them in engineering fields. Each of the VI A students had an individual study plan, decided at the start of each semester in consultation with the department head. The program comprised a few core courses, which were graded, plus ungraded elective courses. At least one elective each semester had to be taken outside the student's department.

Einstein's transcript shows he followed, at least nominally, a standard program within VI A. About half those courses were mathematics, extending

TABLE 15.2 At Swiss Polytechnic Institute, 1896–1900

Age	Year	
17	1896	October: AE to Zurich; enrolls in ETH program for diploma to teach high school math and physics. Among ten other students in that program are Marcel Grossmann (math) and Mileva Marić (physics).
18	1897	Through music, meets Michele Besso (engineer), who urges AE to study Mach.
19	1898	January: AE dismayed by father's bankruptcy. October: AE passes midway oral exams, using Grossmann notes to cram. Gets highest score among math and physics candidates (5.7 out of 6).
20	1899	August: In a letter to Mileva (MM), AE first proposes to discard ether. October: AE applies for Swiss citizenship.
21	1900	Spring: AE and MM write diploma essays, both on heat conduction, for Prof. Weber (scores 4.5/6 and 4/6, respectively). July: AE graduates from ETH. On final diploma exams, of five candidates in math and physics, AE scores fourth (4.9/6); MM scores lowest (4/6) and fails. August: AE is the only one of the four passing diploma candidates not appointed as an assistant at ETH.

beyond theory of functions, projective and differential geometry, and partial differential equations, to number theory and elliptic functions. About a quarter of the VI A courses were laboratory work. Einstein took more than the required minimum of electives outside VI A, among them courses on Goethe's works and worldview, Kant's philosophy, prehistory of man, geology of mountains, politics and cultural history of Switzerland, banking and stock exchange, social consequences of free competition, statistics and personal insurance, and foundations of national economy. To graduate, VI A students had to pass two sets of oral exams on the core math and physics curriculum. One set was usually taken after the first two years, the other at the end of the fourth year. The final required also a written diploma thesis.

Einstein's view of his ETH years was bittersweet. Over fifty years later, he stressed that he lacked qualities expected of a "good" student, easy comprehension and docile focus on what was offered in lectures, but benefited from the freedom allowed by the Swiss system. Among his typical comments are these:

> [Gradually I learned] to arrange my studies to suit my intellectual stomach and my interests. Some lectures I would follow with intense interest. Otherwise I "played hooky" a lot and studied the masters of theoretical physics with a holy zeal at home. [F: 50]
>
> I really could have gotten a sound mathematical education. However, I worked most of the time in the physical laboratory, fascinated by the direct contact with observation. [E: 15]
>
> In all there were only two examinations; for the rest one could do what one wanted . . . a freedom, which I thoroughly enjoyed . . . up to a few months before the examinations. [E: 17]

In his self-study of theory, Einstein embraced works almost entirely missing from the VI A courses. He read books by Kirchhoff, Hertz, Helmholtz, Mach, Boltzmann, and Drude, learned Maxwell's electromagnetism from a recent text, and studied papers by Lorentz. In the electrotechnical lab, Einstein's performance matched his zeal; he then "still expected to approach the major questions of physics by observation and experiment" [P: 132]. However, he was not allowed to construct an apparatus that he designed to measure the earth's movement against the ether; "the skepticism of his teachers was too great." In coping with the exams, Einstein gratefully received crucial help from his classmate Marcel Grossmann, who had prepared superb notes on the core courses. Einstein deeply resented the exams:

> It is, in fact, nothing short of a miracle that the modern methods of instruction have not yet entirely strangled the holy curiosity of inquiry; for this delicate

little plant, aside from stimulation, stands mainly in need of freedom. . . . It is a very grave mistake to think that the enjoyment of seeing and searching can be promoted by means of coercion and a sense of duty. . . . I believe that it would be possible to rob even a healthy beast of prey of its voraciousness, with the aid of a whip, to force the beast to devour continuously, even when not hungry. [E: 17]

Despite his devotion to study, Einstein enjoyed some special social outlets in Zurich. He was regularly invited for weekly lunch or dinner by families (as Max Talmud had been by his family); in a letter of thanks, he wrote that "I often came to you in a dejected or bitter mood and there invariably found joy and an inner equilibrium" [F: 52]. On Saturday nights and holidays he played his violin with chamber music groups.

Most special was his romance with Mileva Marić, his future wife. They studied together, and when apart exchanged many letters. In those from Einstein, usually both playful and ardent, he often reports what he is reading or research ideas he is hatching. His liaison with Mileva was opposed by his parents and puzzled friends, but that did not faze Einstein. Even when Mileva, having failed the diploma exam, was required to repeat it next year, Einstein presumed that she would also go on to get a doctorate. In letters to her during summer vacation in 1900, he wrote:

> Even my work seems to me pointless and unnecessary if I am not telling myself that you are happy with what I am and what I do. . . . I am also looking forward very much to our new studies. You must now continue with your investigation—how proud I will be when maybe I'll have a little Ph.D. for a sweetheart while I am myself still a totally ordinary man. [C: D75]

> So courage, little witch! I can hardly wait to be able to hug you and squeeze you and to live with you again. We'll happily get down to work right away, and money will be as plentiful as manure. [C: D71; F: 71]

Graduate Student: 1900–1902, "The Gypsy Years"

Einstein's antic hope for plentiful money was soon frustrated. During his four years at ETH, although his parents could not give him regular support, he had a monthly allowance from affluent relatives in Genoa, but that ended when he graduated. He expected to be appointed at ETH as a teaching assistant. Because Department VI A provided service courses for the large flux of engineering students, several assistants were needed. The few graduates in math or physics who wanted to be assistants usually were promptly appointed. In 1900, Einstein was the only exception. He felt insulted when Professor Heinrich

TABLE 15.3   Graduate Student, 1900–1905

Age	Year	
21	1900	October: AE starts on thermoelectricity as thesis project, in Weber's lab.
		December: AE submits first paper (on capillarity) to *Annalen der Physik*.
22	1901	February: AE becomes Swiss citizen, completing process begun October 1899.
		March: AE exempted from military service: "varicose veins, flat and sweaty feet."
		March–April: AE seeks assistant jobs in several countries, in vain.
		April: AE criticizes Planck's radiation theory in letter to MM.
		April 13: AE gets letter from Grossmann about possible job at patent office.
		May: AE and MM enjoy Lake Como; learn MM is pregnant.
		May 15–July 15: AE is a substitute teacher of math at Winterthur. AE excited about Lenard's observation of photoelectric effect.
		July: MM again fails diploma exam.
		September: AE becomes private tutor at Schaffhausen.
		November 23: AE submits proposed thesis on gas kinetics theory to Prof. Kleiner.
		December 11: Patent office job advertised; AE applies on December 18.
23	1902	January: Daughter Lieserl born; in response, AE wishes to give birth himself.
		February 1: AE withdraws thesis, moves to Bern, advertises private lessons.
		April: AE forms "Olympia Academy" with Maurice Solovine and Conrad Habicht.
		April 30: AE submits paper 2 (on electrochemistry) to *Annalen*.
		May: AE interviewed for patent office job.
		June 23: AE starts provisional job at patent office as "Expert III Class."
		June 26: AE submits paper 3 (on statistical thermodynamics).
		October: AE's father dies in Milan; he is shocked and desolate.
24	1903	January: AE and MM married, with Habicht and Solovine only witnesses.
		AE writes Besso that he'll not pursue Ph.D.: "comedy."
		AE submits paper 4 (on foundations of thermodynamics).
		September: Lieserl survives scarlet fever; likely given up for adoption.
		December: AE talks on electromagnetic waves at *Naturforschende Gesellschaft*.
25	1904	March 29: AE submits paper 5 (on fluctuations).
		May: son Hans Albert born.
		Summer: Besso takes job at patent office as "Expert II Class."
		September: AE made "definitive" in patent office; still "Expert III Class."
		Late October: Habicht leaves Bern for teaching post in Schaffhausen.
26	1905	March–November: AE publishes 21 reviews in *Beiblätter zu den Annalen*.
		March–June: AE completes papers 6, 7, 8, and 9.
		July 20: AE submits paper 7 as his Ph.D. thesis; accepted July 27.
		August–December: AE submits slightly revised paper 7, and papers 10 and 11.
		November: Solovine leaves Bern for University of Lyon.
	1906	January 15: All formalities completed, AE becomes a Ph.D.

Weber, who had supervised Einstein's *Diplomarbeit*, hired two mechanical engineers as assistants. He was also rebuffed by other ETH faculty.

Despite these disappointments, Einstein returned to the ETH for the fall term to pursue an experimental doctoral thesis under Weber. Shortly before, Einstein added this postscript in a letter to Mileva:

> For the investigation of the thermoelectric Thomson effect I have again resorted to another method, which has some similarities to yours for the determination of the dependence of heat conduction on temperature and which indeed presupposes such an investigation. If only we could already start tomorrow! With Weber we must try to get on good terms at all costs, because his laboratory is the best and the best equipped. [C: D74]

While preparing for her repeat try at the diploma exam, Mileva also intended to carry on experiments toward a doctoral thesis. Ever optimistic, Einstein expected her experiments would aid his project and he would complete his thesis by Easter of 1901 [C: D85].

The ETH did not grant doctoral degrees (until 1911), but ETH graduates could obtain a doctorate from the University of Zurich without further ado by merely submitting a dissertation. This policy naturally encouraged graduate students to work on projects proposed or endorsed by their faculty advisor, often related to their *Diplomarbeit*. The plans of Einstein and Mileva conformed to this pattern. By spring, however, both had abandoned Weber's lab, and Einstein was convinced that a "poor reference" from Weber would foreclose prospects elsewhere [C: D94].

Meanwhile, Einstein had submitted in December his first scientific paper, a theoretical analysis of capillarity as a means of characterizing attractive intermolecular forces in liquids. This most likely was stimulated by a lecture by Hermann Minkowski that Einstein had attended at ETH in the spring before his graduation. Minkowski had then just published an encyclopedia article on the subject and gave out reprints to his audience. Ruefully, Einstein remarked to another student, "This is the first lecture on mathematical physics we have heard at the Poly" [F: 58]. He apparently undertook this work entirely on his own initiative. He was proud and excited at contributing results that he took to be "entirely new despite their simplicity and might yield a law of nature" [C: D79]. He had high hopes that this first paper, published in *Annalen der Physik* in early March of 1901, would help him get a job. Within a few weeks, he had also decided to make intermolecular forces the subject of his doctoral thesis [C: D100].

Einstein sent out during March and April more than a dozen letters or postcards (with postpaid return) pursuing job possibilities, all in vain. This went to Wilhelm Ostwald:

Esteemed Herr Professor! Because your book on general chemistry inspired me to write the enclosed article, I am taking the liberty of sending you a copy of it. On this occasion permit me also to inquire whether you might have use for a mathematical physicist familiar with absolute measurements . . . because I am without means, and only a position of this kind would offer me the possibility of additional education. Respectfully yours, Albert Einstein [gives Milan address of parents]. [C: D92]

Three weeks later, Einstein followed up with:

Esteemed Herr Professor! A few weeks ago I took the liberty of sending you from Zurich a short paper which I published in Wiedemann's *Annalen*. Because your judgment of it matters very much to me, and I am not sure whether I included my address in the letter, I am taking the liberty of sending you my address hereby. Respectfully . . . [C: D95]

Unknown to Einstein, on April 13, 1901, his father Hermann also wrote Ostwald:

Esteemed Herr Professor! Please forgive a father who is so bold as to turn to you in the interest of his son. I shall start by telling you that my son Albert is 22 years old, that he studied at the Zurich Polytechnikum for 4 years, and that he passed his diploma examinations in math and phys with flying colors last summer. Since then, he has been trying unsuccessfully to obtain a position as an Assistant, which would enable him to continue his education in theoretical & experimental physics. All those in position to give a judgment in the matter, praise his talents; in any case, I can assure you that he is extraordinarily studious and diligent and clings with great love to his science.

My son therefore feels profoundly unhappy with his present lack of position, and his idea that he has gone off the tracks with his career & is now out of touch gets more and more entrenched each day. In addition, he is oppressed by the thought that he is a burden on us, people of modest means.

Since it is you . . . whom my son seems to admire and esteem more than any other scholar currently active in physics . . . I make the humble request to read his paper . . . and to write him, if possible, a few words of encouragement, so that he might recover his joy in living and working. [C: 99]

Ostwald did not respond. Only nine years later, Ostwald made amends: right after he received the Nobel Prize, he became the first to nominate Einstein for it [P: 45].

On the same day as his father's letter, Einstein received a hopeful message from Marcel Grossmann: his father, a friend of the director of the Swiss patent office in Bern, had recommended that Einstein be considered for the next vacancy. Einstein responded gratefully:

> Dear Marcel! When I found your letter yesterday, I was deeply moved by your devotion and compassion which did not let you forget your old luckless friend. . . . I would be delighted to get such a nice sphere of activity and I would spare no effort to live up to your recommendation. I came here to my parents three weeks ago in order to search from here for an assistant's position at a university. I could have found one long ago had Weber had not played a dishonest game with me. All the same, I leave no stone unturned and do not give up my sense of humor. . . . God created the donkey and gave him a thick hide. . . .
>
> As for science, I have conceived a few marvelous ideas, which only have to be properly hatched. I now firmly believe that my theory of attractive forces between atoms can be extended also to gases . . . and that the characteristic constants for nearly all elements will be determined without major difficulties. Then the question of the inner kinship of molecular forces and Newtonian forces will move a big step closer its solution. It is possible that experiments already done by others for other purposes will suffice for testing the theory. In that case I shall utilize everything achieved so far about molecular attraction in my doctoral thesis. It is a glorious feeling to recognize the unity of a complex of phenomena, which appear to direct sense perception as quite distinct things. [C: D100]

As things turned out, it was another fourteen months before Einstein was actually hired at the patent office. During that stretch, he had only six months of salaried income from temporary jobs as a substitute teacher and as a private tutor. In May, he learned he had gotten Mileva pregnant and, in late July, that she had again failed the diploma exam. A few weeks before the exam, Einstein had sent her an earnest pledge:

> Rejoice now in the irrevocable decision I have made! About our future I have decided the following: I'll look for a position *immediately*, no matter how modest. My scientific goals and personal vanity will not prevent me from accepting the most subordinate role. As soon as I have such a position I will marry you and take you to live with me. . . . Then no one can cast a stone upon your dear head. . . . your parents and mine [will] just have to reconcile themselves to it as best they can. [HP: 184]

In December, this pledge as yet unmet, Einstein made another earnest proposal that he would not manage to fulfill. Mileva was at home with her parents, waiting to give birth; she anticipated the baby would be a girl and had even named her Lieserl. Einstein, having just heard that the patent office job was about to be advertised, wrote:

> I'm even happier for you than for myself. . . . We'll be students as long as we live and won't give a damn about the world. . . . The only problem that still needs to be solved is the question of how we can take our Lieserl to us; I do not want us to have to give her up. [C: D127]

Einstein did carry out his plan to prepare a thesis dealing with molecular forces in gases. In November, he submitted this to Alfred Kleiner, the professor of physics at the University of Zurich, and wrote Mileva that "he won't dare reject my dissertation" [C: D126]. It was January before Kleiner read the thesis, and he did reject it, supposedly because Einstein had sharply criticized Boltzmann, but likely also because theoretical results offered in the thesis lacked experimental confirmation [C: 175]. (No copy of the thesis is extant.) Soon after withdrawing his thesis (to avoid forfeiting half the submission fee), Einstein heard from her father that Mileva had given birth to a daughter, Lieserl, after an exhausting labor. His letter in response includes curious comments:

> you must suffer enormously if you cannot even write me yourself. . . . our dear Lieserl too must get to know the world from this aspect right from the beginning! . . . I would like once to produce a Lieserl myself, it must be so interesting! [C: D134]

Graduate Student and Patent Examiner: 1902–1905 in Bern

Einstein more than once gave up on the prospective patent office job [C: D126]. When it was finally advertised, he applied immediately, soon resigned incautiously from a salaried tutorial post, and in early February of 1902 moved to Bern. He advertised "private lessons . . . given most thoroughly . . . trial lessons free." There were few takers, but one led to a lifelong friendship. Maurice Solovine, a Romanian student, found Einstein was eager to discuss philosophy and literature as well as physics. Soon Conrad Habicht, a mathematics student Einstein had known in Zurich, joined their discussions. The trio decided to meet regularly as a book and debate club, which they dubbed the "Akademie Olympia" as a spoof of pompous societies. For about two and a half years, they met regularly, often several evenings a week, had a modest dinner ("sausage, cheese, fruit and tea"), then indulged in intense, typically boisterous debate that sometimes "went on far into the night, to the annoyance of the neighbors" [HD: 38]. The reading was systematic and eclectic; in a memoir Solovine lists Karl Pearson's *Grammar of Science*, Ernst Mach's

Mechanics, John Stuart Mill's *Logic*, David Hume's *A Treatise on Human Nature*, Baruch Spinoza's *Ethics*, as well as Sophocles' *Antigone*, lectures by Hermann von Helmholtz and André-Marie Ampere on physics, and discussions by Bernhard Riemann of the foundations of geometry and by Richard Dedekind of the concept of number. Special attention was devoted to Henri Poincaré's *Science & Hypothesis*, which "held us spellbound for weeks."[6]

While waiting five more months for the patent office job, Einstein completed two more papers, again without a mentor. In an extension of his first paper, which he had anticipated a year earlier [C: D101], he applied his theory of intermolecular forces to systems comprising metal electrodes immersed in dilute salt solutions. However, he concluded with an apology for "only setting out a meager plan for a demanding investigation" that required experimental solution and the hope his paper would "induce some researcher to attack the problem" [F: 100]. In the other paper, possibly related to his rejected thesis, and likewise alluded to in a letter the previous year [C: D122], his aim was to fill what he saw as a gap in Boltzmann's kinetic theory by providing a sounder derivation of the laws of thermal equilibrium and the second law of thermodynamics from statistical mechanics.[7]

In late June, Einstein finally began work at the patent office. He was to continue there for more than seven years, eight hours a day, six days a week. Einstein enjoyed deciphering drawings and elaborate descriptions to decide whether the invention would work and was actually new. He found congenial the instructions issued to his dozen or so "patent slaves" by the director, Friedrich Haller: "When you pick up an application, think that anything the inventor says is wrong. . . . You have to remain critically vigilant" [F: 104]. Indeed, from his boyhood on, Einstein was much interested in the design of machines and experiments, and he patented several devices of his own invention.[8] He became so adept at processing patents that he had time in the office to do some of his own calculations and writing, "guiltily hid in a drawer when footsteps approached" [HD: 39]. Ten years after his Bern era, Einstein recalled fondly "that temporal monastery, where I hatched my most beautiful ideas" [F: 102].

Soon after he entered his temporal monastery, Einstein was stunned when at age fifty-five his father suffered a fatal heart attack. A few months later, Albert and Mileva were married in Bern, with no family members present. People in Bern did not know about Lieserl, and she was not brought there, perhaps for fear of offending propriety because Einstein's appointment was still provisional [F: 114]. The Olympia Academy continued to meet, now usually in the Einstein's apartment. Solovine noted that Mileva, "intelligent and reserved, listened to us attentively without ever intervening in our discussions."[9]

Einstein also now had congenial scientific interactions with colleagues at the University of Bern and the Natural Science Society, as well as the patent office. Only three weeks after his wedding, Einstein wrote to Besso about completing his fourth paper, which carried further his treatment of the foundations of thermodynamics:

> On Monday I finally sent off my work, after many changes and corrections. Now it is perfectly clear and simple, so that I am quite satisfied with it. . . .
>
> I have now decided to become a Privatdozent, provided of course I can get away with it. On the other hand, I won't become a Ph.D., as this doesn't help me much and the whole comedy has become a bore to me. [F: 112]

He had learned that the University of Bern had an unusual policy, allowing a shortcut to the status of *Privatdozent* (unsalaried, but with the privilege to lecture at the university and collect fees from subscribing students). Scholars with "other outstanding achievements" could skip both a doctoral and habilitation thesis, submitting instead other published work. Einstein applied, offering his third and fourth papers. This ploy failed and soon he wrote Besso again: "The university here is a pigsty. I won't lecture there, it would be a waste of time" [F: 112].

More than a year later, Einstein submitted a fifth paper, completing a trilogy on statistical thermodynamics. He evaluated fluctuations of the internal energy about its average value. This, he emphasized, brought out the significance of Boltzmann's constant, which "determines the thermal stability of the system" because it sets the scale of the fluctuations.[10]

Significant fluctuations soon occurred both at home and work. In May, 1904, Mileva gave birth to a son, Hans Albert. That summer, Einstein was joined at the patent office by his friend Michele Besso, whom he had encouraged to apply. In the fall, Einstein's provisional appointment, after twenty-seven months, was made permanent. However, only after another twenty months was he promoted to Class II, in contrast to Besso, an engineer, who started at that rank. In late October, Habicht departed Bern and the Olympia Academy ended.

In another few months, Einstein erupted in his *annus mirabilis*. During 1905, he submitted six papers. The three most celebrated papers were submitted within a span of fifteen weeks. Within that span he also completed a fourth paper, which became his Ph.D. thesis. Moreover, during the year Einstein contributed twenty-one reports to a review journal, *Beiblätter zu den Annalen der Physik*. His reviews, in the category "theory of heat" [*Wärmelehre*], summarized and commented on papers published in German, French, Italian, and English journals. Evidently he interleaved writing his own papers with preparing these reviews, as eight of his reviews appeared in March, six in June, three in September, and four in November. How much work Einstein had to do at the patent office is not known. In accord with policy, eighteen years later his patent assessments were destroyed. However, it seems likely he would have had more

TABLE 15.4 First Five Papers, 1900–1904

Title	Received	Published	No. Pages
1. Conclusions Drawn from the Phenomena of Capillarity	Dec 16, 1900	Mar 1, 1901	11
2. On the Thermodynamic Theory of the Difference in Potentials between Metals and Fully Dissociated Solutions of Their Salts and on an Electrical Method for Investigating Molecular Forces	Apr 30, 1902	Jul 10, 1902	17
3. Kinetic Theory of Thermal Equilibrium and of the Second Law of Thermodynamics	Jun 26, 1902	Sep 18, 1902	17
4. A Theory of the Foundations of Thermodynamics	Jan 26, 1903	Apr 16, 1903	18
5. On the General Molecular Theory of Heat	Mar 29, 1904	Jun 2, 1904	9

These papers were all published in *Annalen der Physik*. Paper 1 was submitted from Zurich; papers 2–4 from Bern. Paper 1 mentions the work of R. Schiff and compiles data from *Allgemeine Chemie* of W. Ostwald (1891) and the Landolt-Börnstein tables (1894). Papers 1, 2, and 4 have no footnotes or explicit literature references. Paper 3 mentions Maxwell and Boltzmann in the text and has two footnotes to sections of Boltzmann's *Gastheorie*; Paper 5 mentions Boltzmann and Planck in the text and has two footnotes, one to AE's paper 4, the other to Boltzmann's *Gastheorie*.

to do in 1905 than earlier. As nearly all the other examiners were mechanical engineers, Einstein probably had to contend with a flood of electrical engineering patents, generated by the rapid development of electrical industry [G: 248].

Einstein's creative outburst in 1905 is often said to be comparable only to that achieved by Newton in 1666. Both soared in their mid-twenties, but otherwise, what a contrast! With Cambridge University closed by the plague, Newton had retired to his mother's estate and, as a bachelor, was free to concentrate totally on science and mathematics. He is thought to have conceived several of his great ideas during the plague recess, but Newton published little until many years later. Einstein, with much else to do, must have labored mightily to bring forth so quickly his golden eggs. He never identified accelerating factors, but two speculative possibilities seem to me plausible. After the arrival of Hans Albert and the departure of Habicht, Einstein might have begun devoting more of his evenings to putting his ideas in writing. In lieu of the exuberant verbal sparring he'd enjoyed with his Olympia Academy, he had calmer stimulation in talking with Besso at the patent office and on the way home. Also, like the birthing urge induced by Lieserl, the presence of his infant son perhaps spurred Einstein to deliver his intellectual progeny.

Comments on Einstein's Early Papers

The content, antecedents, and response to Einstein's early papers have been amply discussed in editorial commentaries included in the Collected Papers and in several other excellent sources.[11] Here I note only a few aspects related to his transition from student to scientist.

His first two papers, and presumably also his rejected (and vanished) thesis of late 1901, employed a dubious conjecture about intermolecular forces.[12] In analogy with gravitation, Einstein assumed the potential energy for interaction of a pair of molecules involved a universal function of the distance between them. This was not compatible with the then well-known theory of fluids developed by van der Waals, nor with experimental data indicating that distance dependence does vary for different pairs of molecules. Remarks in Einstein's letters show that, three months before submitting his first paper (treating capillarity), he had finished reading Boltzmann's *Gastheorie*, which presents the theory of van der Waals [C: D75]. A year later, shortly after submitting his ill-fated thesis, Einstein had derived from his own theory a consequence that contradicted the results of van der Waals. Blissfully confident, Einstein then thought his result, if confirmed by experiment (as he seemed to expect), "would be the end of the molecular-kinetic theory of liquids" [C: D127].

He did not realize until several years later that his conjecture about distance dependence was wrong, and then declared his first two papers "worthless beginners' works" [P: 57]. This must have been a disillusioning experience. He had been proud of his first paper; he had sent a copy to Boltzmann as well as to Ostwald [C: D85]. Misled by seeming success in fitting considerable data (but of limited sensitivity) with adjustable parameters, he had hoped to find a "general law." Never again did he succumb to such an approach. Decades after, without mentioning his "beginners' works," he wrote:

> I despaired of the possibility of discovering the true laws by means of constructive efforts based on known facts. The longer and the more despairingly I tried, the more I came to the conviction that only the discovery of a universal formal principle could lead us to assured results. The example I saw before me was thermodynamics. [E: 53]

Indeed, as well as making much direct use of thermodynamics, Einstein soon came to emulate its character.[13] Rather than pursuing a "constructive theory" that attempts "to build a picture of complex phenomena out of some relatively simple propositions," he strived for "a theory of principle" that starts "from empirically observed general properties of phenomena" and infers from them results "of such a kind that they apply to every case which presents itself, without making assumptions about hypothetical constituents"[E:53]. This became his distinctive approach, especially prominent in his 1905 papers.

Einstein finally wrapped up the Ph.D. comedy in July of 1905. He completed the paper that he submitted for his doctoral thesis (#7) at the end of

TABLE 15.5   Papers of the Annus Mirabilis, 1905

Title	Received	Published	No. Pages
6. On a Heuristic Point of View Concerning the Production and Transformation of Light	Mar 18, 1905	Jun 9, 1905	17
7. A New Determination of Molecular Dimensions	Aug 19, 1905	Feb 8, 1906	17+1
8. On the Movement of Small Particles Suspended in Stationary Liquids Required by the Molecular-Kinetic Theory of Heat	May 11, 1905	Jul 18, 1905	12
9. On the Electrodynamics of Moving Bodies	Jun 30, 1905	Sep 26, 1905	31
10. Does the Inertia of a Body Depend on Its Energy Content?	Sep 27, 1905	Nov 21, 1905	3
11. On the Theory of Brownian Motion	Dec 19, 1905	Feb 8, 1906	11

These papers were all submitted from Bern and published in *Annalen der Physik*. Paper 6 has footnotes citing work of Drude, Planck, Lenard, and Stark, plus some enlarging on details. Paper 7 is AE's doctoral thesis, dedicated to Marcel Grossmann and published separately as a booklet (Buchdruckerei K.J. Wyss, Bern, Jan 1906); both the *Ann. Phys.* version and the booklet are dated April 30, 1905, although the thesis was not submitted until July 20, 1905, and the paper not until mid-August 1905. Both versions have one footnote to Kirchhoff and one to AE's paper 8. The *Ann. Phys.* paper has some minor changes and a short supplement dated Jan 1906 (requested by Drude, the editor) that uses updated data from the latest edition of Landolt-Börnstein (1905). Paper 8 cites AE's paper 3, paper 4, and Kirchhoff. Paper 9 has no literature citations, but has three explanatory footnotes and thanks his "friend and colleague Michele Besso for loyal support and valuable stimulation." Paper 10 cites only paper 9. Paper 11 cites paper 8, Gouy (1888), Planck, and Kirchhoff.

April, but set it aside in favor of turning out an offshoot of it, his Brownian motion paper (#8) in early May, then his relativity paper (#9) during June. Maja reported that Einstein had first submitted the relativity paper as his thesis, only to have it turned down because it "seemed a little uncanny to the decision-making professors" [F: 123]. He then submitted his April paper, which provided a means to determine both molecular size and Avogardro's number from experimental diffusion rate and viscosity data for sugar solutions. It was quickly endorsed by Kleiner and approved by the Zurich faculty. According to a story Einstein liked to tell, after an objection that his thesis was too short, he added a sentence and it was then accepted [P: 88]. The Brownian motion paper was published just before Einstein submitted the thesis, and a footnote citing it was added ("fuller explanation can be found . . ."). That footnote may be the now legendary but unidentified final sentence. When printed as customary in a booklet, the text portion of the thesis was only 16 pages.

Einstein did not have to contend with either mentors or reviewers in producing his papers. Today it seems astonishing that an unknown graduate student could readily publish in a leading physics journal, and that the papers usually appeared in print only two to three months after submission. In the *Annalen* volumes for 1902–1905, each monthly issue typically had fifteen papers, nearly all by single authors. About 20 percent were purely theoretical papers. Those were handled either by the editor, Paul Drude, himself a theorist, or by the only theorist among the five-member advisory board, Max Planck.

Publication appeared to be equally prompt whether or not the author was distinguished, or had an academic affiliation, or whether the paper reported experiments or theory.[14] Nowadays, if a new Einstein were to appear as a graduate student and try to publish in a major journal anything as audacious as most of his early papers, the student would surely be squelched by a withering gauntlet of reviewers. Fortunately, new options are now becoming available in cyberspace.

A curious aspect of Einstein's papers is his meager citation of relevant work by others (itemized in notes to Tables 15.4 and 15.5). This was quite atypical in the *Annalen* even a century ago. Pais suggests he "simply did not much care," and quotes a striking statement made by Einstein in a 1907 paper:

> What is to follow might already been partially clarified by other authors. However, in view of the fact that the questions under consideration are treated here from a new point of view, I believed I could dispense with a literature search which would be very troublesome for me, especially since it is to be hoped that other authors will fill this gap.

Pais comments: "This statement is not arrogant if, and only if, arrogance is a mark of insecurity. To me these lines express ebullience, total self-assurance, and a notable lack of taste" [P: 165]. Most odd is Einstein's failure to acknowledge, in 1905 or later, seminal essays by Henri Poincaré. His *Science and Hypothesis* appeared in German translation in 1904 and kept the Olympia Academy "spellbound for weeks." That edition included an excerpt from his *Measure of Time*, wherein Poincaré explicitly treats issues directly relevant to the nascent theory of special relativity [G: 238]. Moreover, Poincaré also points out, as important unsolved problems, Brownian motion and the photoelectric effect, which soon became Einstein's other two golden eggs [R: 10]. In words that must have gripped the Olympia Academy, Poincaré wrote:

> Let us notice, however, the original ideas of M. Gouy [in 1888] on the Brownian movement. According to this scientist, this singular movement does not obey Carnot's principle [the second law of thermodynamics]. The particles which it sets moving would be smaller than the meshes of that tightly drawn net; they would thus be ready to separate them, and thereby to set back the course of the universe. One can almost see Maxwell's demon at work.[15]

In the introduction to his paper (#8), Einstein likewise featured the challenge to the second law:

> If the movement discussed here can actually be observed . . . then classical thermodynamics can no longer be looked upon as applicable with precision to bodies even of dimensions distinguishable in a microscope: an exact determination of

actual atomic dimensions is then possible. On the other hand, had the prediction of this movement proved to be incorrect, a weighty argument would be provided against the molecular-kinetic conception of heat. [C2: D16]

Einstein does not mention Gouy until a later paper (#11). Actually, thermal fluctuations such as those that produce Brownian motion do not violate the second law; that was shown in 1922 by Leo Szilard in his Ph.D. thesis, cursed by Maxwell's demon and at first doubted, then blessed by Einstein.[16]

In contrast to his original journal articles, elsewhere Einstein was unstinting in his appreciation:

> A hundred times a day I remind myself that my inner and outer lives are based on the labors of other people, living and dead, and that I must exert myself in order to give in the same measure as I have received and am still receiving.[17]

Lessons for a New Century

Efforts to improve science education and literacy face a root problem: science and mathematics are regarded not as part of general culture, but rather as the province of priest-like experts. Einstein is seen as a towering icon, the exemplar *par excellence* of lonely genius. That fosters an utterly distorted view of science, both in students considering a career and in the wider public.

Among many others, I've tried to combat that distortion by emphasizing two key aspects.[18] First, science enjoys a tremendous advantage over other human enterprises: The goal—understanding nature—*waits patiently to be discovered*. That is why ordinary human talent, given sustained effort and freedom in the pursuit, can achieve marvelous advances. It is also why it is vital to have some maverick scientists willing to explore unorthodox paths, as unanticipated roadblocks often obstruct routes favored by consensus. Second, science is intrinsically a *cooperative, democratic social enterprise*. In contrast to hierarchical organizations, governed by a chain of officers, science relies on independent units free to pursue their own interests. Despite seemingly chaotic freedom, the enterprise is efficiently "coordinated by an invisible hand" because each unit can observe and apply results found by others.[19]

These special aspects of science, and the great variety of its subdomains, make it congenial for people with a huge range of talents and temperaments. That point was nicely made by Enrico Fermi. He was asked if he could think of something his fellow Nobel laureates in physics had in common. After a pause, he replied: "No, I can't think of anything they have in common. Not even intelligence."[20] Thus we certainly cannot expect to divine from the Einstein saga what is necessary or sufficient to produce boldly creative scientists. But his maverick journey to his doctorate highlights aspects that should compel attention in current discussions of science education, from grammar to graduate school. In summary:

1. *Early enchantment.* At the age of ten to twelve, Einstein's passion for science was aroused. Thanks chiefly to Max Talmud, Einstein read many popular science books and became enamored with geometry and "the purity of pure thinking." He went on, outside of school, to learn calculus on his own. That and his violin taught him the pleasures of taking ownership of subjects by self-study ("love is a better teacher than duty").

2. *Confident sense of mission.* By seventeen (in his Aarau essay), Einstein felt he had a "personal gift" for theoretical science, which offered him a "goal for striving" and "certain independence." These traits were reinforced by disdain of authority, exercised throughout his academic experience, and by his critical, questioning attitude, honed in the patent office.

3. *Freedom for solo initiatives.* The four-year course at ETH required only two major exams. Einstein attended only lectures that interested him and spent much time performing experiments and in intense self-study of theory. After graduation, papers he wrote and submitted on his own were promptly published. The Ph.D. required no further courses, only a single paper of modest length, on a topic he chose without any prior approval.

4. *Supportive friends.* Einstein enjoyed conversation and correspondence, social and scientific, with many friends. At ETH, Einstein often studied with Mileva or Marcel Grossmann. At Bern, the Olympia Academy was a major indulgence. Conversations with Michele Besso helped Einstein recognize the crucial issue in special relativity. He credited his musical friends, Besso among them, for preventing him from "getting sour."

5. *Cultural resonance.* Holton has explored "how the cultural milieu in which Einstein found himself resonated with and conditioned his science."[21] In gymnasium and university, Einstein received a broad humanistic education. In this, Holton finds the cultural roots of Einstein's urge to generalize and unify, as well as the tension seen in his rebellion against and reverence for traditional ideas. Goethe appeared an especially strong influence—on Boltzmann and Planck, as well as on Einstein—to pursue "a Faustian drive toward a *Weltbild* encompassing all phenomena."

The exhilaration of taking ownership at a young age, outside of school, is testified to in memoirs of many scientists. Edward Wilson's love affair with ants[22] and Oliver Sacks' with chemistry[23] are notable recent examples. Arnold Sommerfeld remarked: "Mathematics is like childhood diseases. The younger you get it, the better."[24] That likely applies to much else.

If young Einstein were reincarnated today, he might not have a Talmud as mentor, but he would find abundant opportunities to explore science. I'll be parochial and assume he appears in Princeton, just so I can point to activities of Science Service, a small nonprofit outfit in Washington, DC, that I know about firsthand.[25] It publishes a weekly, *Science News*, written for laypeople. As it covers all fields, it would serve young Einstein like the popular science books he read a century ago and later praised for giving him a broad, qualitative perspective. Science Service also conducts premier science fairs at the middle school and high school levels. Best known is the Science Talent Search (long sponsored by Westinghouse and now by Intel). Largest is the International Science and Engineering Fair (ISEF, now also sponsored by Intel). This brings together more than 1,200 students from fifty countries (as yet over 90 percent from the United States), winners of hundreds of local, state, and regional fairs in which about a million students take part. *Science News*, as well as the projects displayed at the STS and ISEF fairs, are now on the Web; recently a delightful edition has been added, designed for pre–high school kids.

In high school and college today, Einstein would certainly be unhappy about myriad tests and the loss of freedom to cut classes. In graduate school, he might resent having to take some courses rather than concentrate solely on research. On the other hand, unlike his situation in Zurich, he would now find on the faculty many theoretical physicists offering advanced courses. (Perhaps too many!)

Most galling to him as a graduate student, I expect, would be the almost certain loss of opportunity to publish solo papers, and to get a doctorate with a single brief paper, all the while gladly working at a "temporal monastery."

Today, the time to complete a Ph.D. in American universities has expanded to a norm of six or seven years, even in the sciences. For those who seek faculty positions, a postdoctoral stint of two or three years is expected. In my view, how academic science has come to be funded is a major factor: funding is for particular projects, not for people. Most graduate students and postdoctoral fellows, therefore, are supported chiefly by serving as hired hands on a project defined by a research grant. Veteran students are most useful in obtaining results to justify a grant renewal. That vital need works to extend the apprenticeship. Nowadays other options are rare. Mathematics is the only field I know of in which a single paper can earn a Ph.D. (at least at Harvard and some sister universities). There are very few fellowships for independent study. Today's Einstein, unless he could quench his strong yearning for independence, would rebel against this feudal system. I think he would not try for a Ph.D., unless perhaps in mathematics.

Shortening the time to the doctorate is extremely important, in my view. For a student entering graduate school with decent preparation, four years should be enough. Uncoupling support of graduate students from project grants would help them break loose. If support of students (and preferably also postdoctoral fellows) on grants to individual professors were abolished, the same money could be put into expanding greatly the number of fellowships that students can win for themselves, as well as into block training grants to university science departments. Winning a fellowship profoundly influences a student's outlook and approach to research; they are certified as national resources rather than as hired hands. Funding agencies should adopt a policy that a fellowship holder completing the Ph.D. in four years would be rewarded by receiving a postdoctoral stipend for a year to work at a laboratory of his or her choice. This and other reforms I've suggested[26] are surely quite reasonable. More simple is the proposal by Freeman Dyson to award students a Ph.D. "on the day they enter graduate school."[27]

Einstein would likely endorse two pieces of advice recently offered to students by Steven Weinberg: "Learn something about history of science. . . . [You] can get great satisfaction by recognizing that your work is part of history." Also, "Aim for rough water . . . that's where the action is."[28] Einstein advised a former student, "One must develop an instinct for what one can just barely achieve through one's greatest effort" [C8: D87]. Of Shakespeare it has been said that writing his plays "must have been easy or it would have been impossible." For Einstein, producing his golden eggs would have been impossible if it had not been difficult.

ACKNOWLEDGMENTS

I am grateful for the opportunity to learn much more about Einstein, at the conference and from the orgy of reading it induced before and after. I thank Gerald Holton and Peter Galison for helpful discussions and Bretislav Friedrich for compiling information about the *Annalen der Physik* a century ago.

LEARNING FROM EINSTEIN: INNOVATION IN SCIENCE

Jürgen Renn

Einstein's Revolution—A Fountain of Youth?[1]

Whenever we talk about innovation in connection with Einstein's contributions to science, it almost sounds like a magic spell invoking a "Fountain of Youth" from which even today's science and technology could be rejuvenated. It therefore comes as no surprise that all kinds of snake oil are on the market referring to Einstein and promising just such a rejuvenating effect. But most of the attempts to find current relevance in Einstein's revolution make superficial assumptions about its foundations, searching for roots in the individual circumstances of his creativity—in his brain, in his rejection of authority, in his supposedly childlike innocence, or in an attitude reflected by assertions such as "imagination is more important than knowledge."

From the point of view of an historical epistemology, such attempts make little sense. Reaching conclusions about the conditions of Einstein's revolution becomes possible only after one takes into account the entire intellectual laboratory from which that revolution emerged. Only thus can we be sure that Einstein's innovations in science were neither the product of charlatanism nor of alchemy. They actually turn out to be the result of a transformation of a system of knowledge in which the preservation of what has been handed down since the beginning of science is as important as the dramatic changes associated with Einstein's work.

The following begins by briefly presenting the main outlines of Einstein's revolution, focusing on the transformation of the concepts of space and time by the special theory of relativity from 1905 and the general theory of relativity from 1915. I will concentrate on the second phase of this revolution, which has been the main subject of my own research in the past years.[2] I shall then try to place this revolution within a panorama of the history of knowledge since the emergence of science.[3] By way of conclusion, I will draw some consequences of these historical reflections for the challenges represented by the transformative processes in today's science.[4]

The Relativity Revolution as Transformation of a System of Knowledge

Recent historical research has not only made profound changes in our image of Einstein, it has also recast our understanding of the structure of scientific revolutions. Einstein no longer appears as the isolated pioneer of 20th-century physics, but rather as the one who completed classical physics in a way that uprooted its foundations. A scientific revolution now resembles more a slow geological process than a dramatic surge of novelty, as what appears suddenly to break new ground can now be seen as actually having matured over time. With only slight exaggeration, one could say that Einstein just happened to scale the volcano when it finally erupted.

The relativity revolution was the final result of a long period of preparation; it also was in no way completed in 1905 when Einstein published his groundbreaking paper on the electrodynamics of bodies in motion. It started with his reinterpretation of Hendrik Antoon Lorentz's theory of electromagnetism in what one may call a "Copernicus process." In such a Copernicus process, technical aspects of a traditional system of knowledge are largely preserved while their physical interpretation changes. Indeed, in such a conceptual revolution, the architecture of knowledge is turned upside down in much the same way that Copernicus placed a previously peripheral element, the sun, at the center of his new system. Thus, when Einstein formulated his special theory of relativity, he kept most of the formal apparatus established by Lorentz, while reinterpreting a peripheral aspect of Lorentz's theory, an auxiliary variable for time, as a key element of his new system, namely as the time actually measured by clocks in a moving reference system. In short, the development of special relativity was clearly part of a larger development crowning and transforming the long tradition of classical electrodynamics. The best evidence for this is that Einstein in 1905 was not alone in being able to draw dramatic consequences from the development of classical physics up to this point. Thus, the French scientist and philosopher, Henri Poincaré, had had similar thoughts about a new interpretation of Lorentz's theory at around the same time as Einstein. That observation holds for the other revolutionary accomplishments of Einstein's *annus mirabilis* as well: They all resulted from reinterpretations of classical accomplishments, and they all found parallels in the work of other contemporary physicists. The light quantum hypothesis resulted from

FIGURE 16.1   Einstein with Lorentz in Leiden. Museum
Boerhaave Leiden.

a reinterpretation of Max Planck's radiation formula, and Einstein's explanation of Brownian motion was closely related to a new interpretation of Ludwig
Boltzmann's atomistic theory in the sense of a statistical mechanics.

There seems to be, however, a striking contrast between the first and the
second phase of the relativity revolution, which is represented by the formulation of general relativity in 1915. Whereas the emergence of the special theory of relativity was evidently the result of an activation of hidden potentials
of highly specialized 19th-century physics, potentials that could have been
used also by others, general relativity seems to be an example of science without context, created against the trends of the contemporary scientific establishment and with the initial support of only a few outsiders.

The basic premise of general relativity lay in the challenge to create a field
theory of gravitation compatible with the insights of special relativity. Although
Newton's classical action-at-a-distance theory of gravitation explained all
known phenomena up to that point, it was reasonable to look for a modification

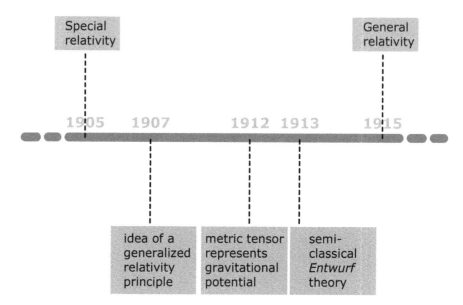

FIGURE 16.2   Timeline: 1907 Idea of a Generalized Relativity Principle.

because physical interactions according to special relativity cannot take place at speeds greater than the speed of light. But neither the path Einstein's research would take nor the theory that brought a solution to this problem in 1915 was in accordance with the expectations of his contemporaries (Fig. 16.2).

In 1907 Einstein questioned the barely established space-time framework of special relativity. In 1912, to the amazement of his colleagues, he abandoned the simple gravitational potential of Newtonian physics in favor of a sixteen-component object with whose mathematics he was barely acquainted. Despite this, he was able to formulate clear-cut heuristic criteria that would have to be satisfied by a gravitational field equation for this monster. However, in the winter of 1912/1913, he discarded what to modern eyes would have been the correct solution in favor of the semiclassical *Entwurf* theory. He even believed to have a proof that what later turned out to be the correct solution had to be ruled out. But in spite of his colleagues' growing skepticism, Einstein maintained his original heuristic agenda and returned in late 1915 to the correct solution he had discarded earlier, and finally formulated the general theory of relativity with its nonclassical consequences, a theory that essentially has withstood all later developments in physics and astronomy.

In recent years, all traditional arguments explaining this dramatic course of events in the framework of biographically oriented history of science have crumbled under closer scrutiny of the genesis of general relativity. Was Einstein's digression from the correct solution caused by his mathematical

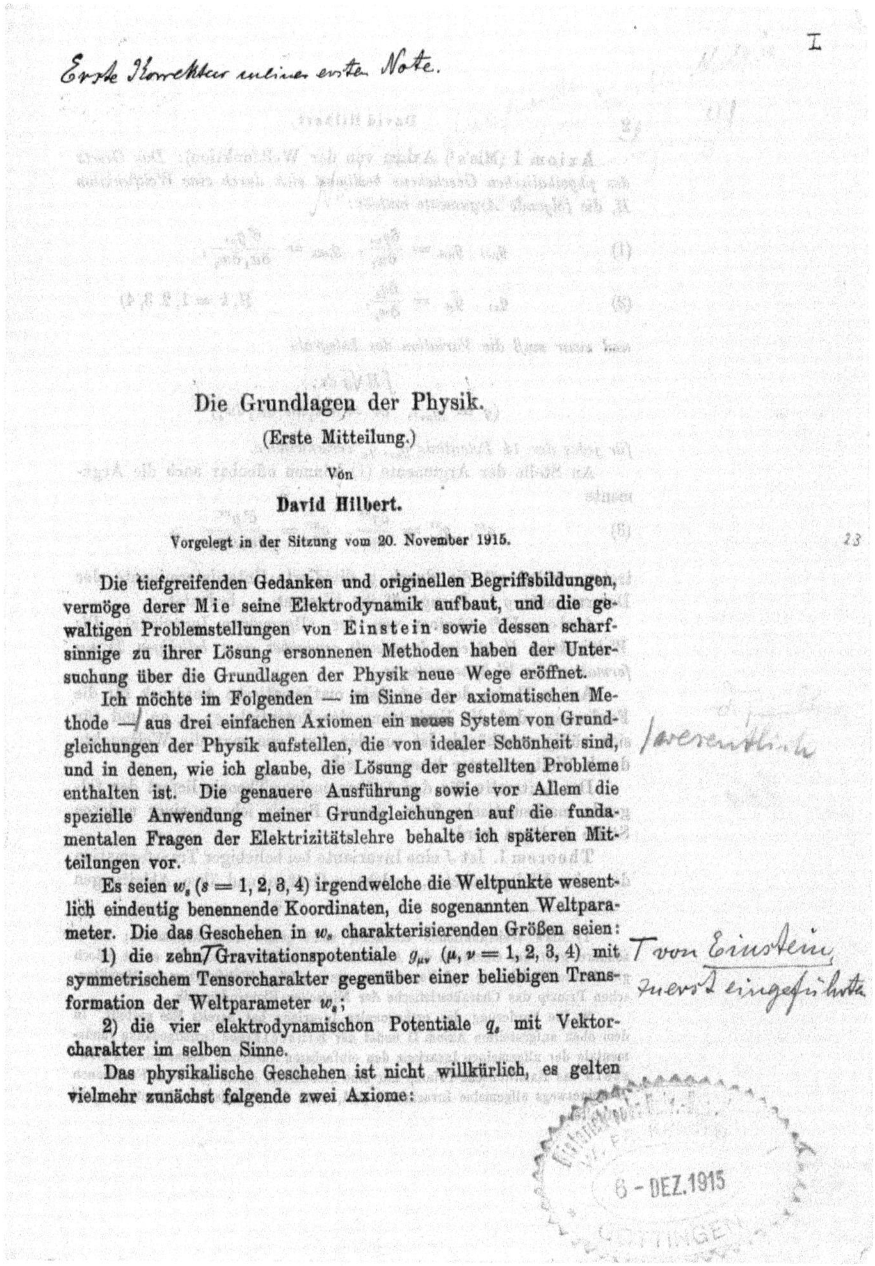

I

23

Die Grundlagen der Physik.

(Erste Mitteilung.)

Von

David Hilbert.

Vorgelegt in der Sitzung vom 20. November 1915.

Die tiefgreifenden Gedanken und originellen Begriffsbildungen, vermöge derer Mie seine Elektrodynamik aufbaut, und die gewaltigen Problemstellungen von Einstein sowie dessen scharfsinnige zu ihrer Lösung ersonnenen Methoden haben der Untersuchung über die Grundlagen der Physik neue Wege eröffnet.

Ich möchte im Folgenden — im Sinne der axiomatischen Methode — aus drei einfachen Axiomen ein neues System von Grundgleichungen der Physik aufstellen, die von idealer Schönheit sind, und in denen, wie ich glaube, die Lösung der gestellten Probleme enthalten ist. Die genauere Ausführung sowie vor Allem die spezielle Anwendung meiner Grundgleichungen auf die fundamentalen Fragen der Elektrizitätslehre behalte ich späteren Mitteilungen vor.

Es seien w_s $(s = 1, 2, 3, 4)$ irgendwelche die Weltpunkte wesentlich eindeutig benennende Koordinaten, die sogenannten Weltparameter. Die das Geschehen in w_s charakterisierenden Größen seien:

1) die zehn Gravitationspotentiale $g_{\mu\nu}$ $(\mu, \nu = 1, 2, 3, 4)$ mit symmetrischem Tensorcharakter gegenüber einer beliebigen Transformation der Weltparameter w_s;

2) die vier elektrodynamischen Potentiale q_s mit Vektorcharakter im selben Sinne.

Das physikalische Geschehen ist nicht willkürlich, es gelten vielmehr zunächst folgende zwei Axiome:

/ wesentlich

T von Einstein zuerst eingeführte

FIGURE 16.3 First page of David Hilbert's 1915 proofs. Niedersächsische Staats- und Universitätsbibliothek Göttingen.

incompetence? The alleged priority of the mathematician David Hilbert in having formulated the gravitational field equation before Einstein—albeit without giving it a detailed physical interpretation—seemed a strong case in point. However, this supposed triumph of mathematics over physics moldered under the evidence of the proofs of Hilbert's first paper (Fig. 16.3), which

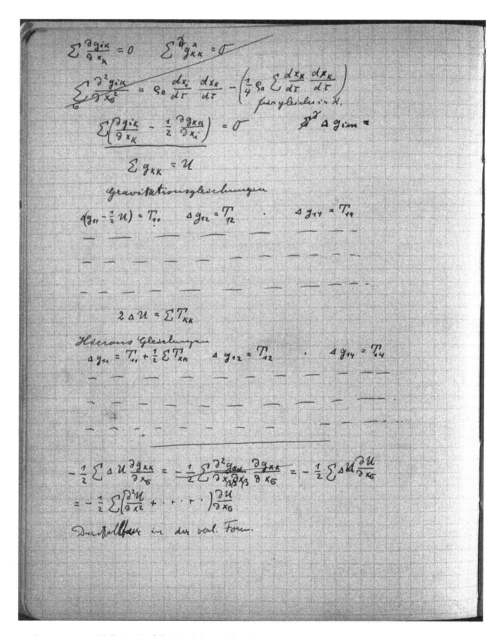

FIGURE 16.4   Folio 20L of the Zurich Notebook. © The Hebrew University of Jerusalem.

showed that his key insight into the possibility of a general covariant gravitational field equation came only after having seen Einstein's publication.

The deathblow for the argument that Einstein's zigzag path may have been a result of mathematical incompetence comes, however, from the fact mentioned earlier that he was already able to formulate the essentially correct field equation in the winter of 1912–13, as is evidenced by his Zurich notebook (Fig. 16.4).

Was Einstein's rambling therefore a consequence of physical prejudices preventing him from recognizing the correct solution even when it was under his nose? The natural candidates for such prejudices have equally failed, on closer inspection, to play quite the evil role ascribed to them. Refuting such legends has allowed us to concentrate on the real challenges of an historical understanding of the relativity revolution and the scientific innovations it has generated. These challenges can be formulated as three paradoxes.

1. *The paradox of missing knowledge*: How was it possible to create a theory such as general relativity that was capable of accounting for a wide range of phenomena—ranging from an expanding universe, via gravitational waves, to black holes—that were only later discovered in the context of several revolutions of observational astronomy or are at its edge even today? Which knowledge granted such stability to a theory that initially did not seem superior to its competitors, since no phenomena were known at the time that could not be explained with traditional physics as well?

2. *The paradox of deceitful heuristics*: How could Einstein have formulated the criteria for a gravitational field equation years before he established the solution? How could he establish a heuristic framework that would quickly lead him to a correct mathematical expression, and then to the conclusion that it was actually unacceptable, only to bring him back to essentially the same expression three years later?

3. *The paradox of discontinuous progress*: How could general relativity with its nonclassical consequences—such as the dependence of space and time on physical interactions—be the outcome of classical and special relativistic physics although such features are incompatible with their conceptual frameworks?

These paradoxes can hardly be resolved within a traditional history of ideas. Addressing the challenges they present requires taking into account dimensions that are usually neglected but are crucial to an historical epistemology of scientific knowledge; the long-term character of knowledge development, the complex architecture of knowledge, and the hidden mechanisms of the developmental dynamics of knowledge.

In fact, an adequate response to the missing-knowledge paradox can only be found if the long-term development of scientific knowledge is taken into account, and if Einstein's revolution is understood as the result of a successful integration of shared knowledge resources. The laws of planetary motion, for instance, including the curious perihelion advance of Mercury, which provided the first astronomical touchstone for the new theory, had long been established by extensive observations. The same is true for non-Euclidean geometry and absolute differential calculus—the mathematical language of general relativity. A crucial guiding principle of Einstein's investigative path—that all bodies, regardless of their nature, fall with the same acceleration—had been known to classical mechanics since Galileo's time. The knowledge on which Einstein based his famous thought experiments was also taken from classical physics, as well as from its critical revision by Ernst Mach. All such knowledge resources may be considered as part of the shared knowledge on the problem of gravitation that was available to scientists in the early 20th century. Their attempts to tackle this problem can be distinguished mainly by the perspective from which these knowledge resources were considered or disregarded. But even the spectrum of individual perspectives may be understood as an aspect of the knowledge system of classical science.

The answer to the second paradox, of how Einstein could have formulated the heuristic criteria for a gravitational field equation years before finding the solution, comes from considering the architecture of the shared knowledge resources available to him. These resources were part of a knowledge system with active components capable of providing heuristic guidance to his research. The characteristics of Einstein's search become indeed comprehensible once one realizes that this search was guided by the numerous knowledge representation structures inherited from classical physics. These structures figured, for instance, in his famous thought experiments, dealing with rulers, clocks, buckets, and elevators. A further characteristic of his search was his capability to outline a theory that had yet to be formulated. This capability was also rooted in the heritage of classical physics. The outline for a field theory of gravitation follows indeed from the mental model provided by the classical tradition of electromagnetic field theory, which was embodied in an exemplary way by Lorentz's electromagnetic field theory. This "Lorentz Model" provided Einstein not only with a useful overall structure but also with numerous default settings, which resulted from past experience and which allowed him to draw conclusions even in the absence of a fully developed theory. By their very nature, default settings are flexible and can be changed whenever new information becomes available. Conclusions can then be corrected without having to give up the entire interpretation of a situation in terms of a mental model.

The third paradox, of discontinuous progress, can only be resolved if one takes into account that the development of knowledge does not consist only of enriching a given architecture but also comprises processes of reflection by which this architecture can be transformed. In the case of the search for the gravitational field equation, the enrichment of the model provided by Lorentz's theory was guided by the relatively stable higher-order structures at the core of Einstein's heuristic principles, for instance, by the principle that the new theory must comprise the knowledge about gravitation represented by Newton's law. But Einstein's learning process was not determined solely by these higher-order structures, that is, by a top-down process of assimilation (i.e., the assimilation of new knowledge to the architecture characterized by these higher-order structures). It was also characterized by a reflection process modifying these higher-order structures as a consequence of the experiences made with their implementation, that is, by a bottom-up process of accommodation. The interaction between assimilation and accommodation—mediated by the elaboration of an appropriate mathematical formalism—has turned out to be the crucial process of knowledge dynamics that led to the emergence of general relativity as a nonclassical theory of physics. It is remarkable that this process, extending over eight years from 1907 to 1915, involved, just as the emergence of special relativity, a Copernicus process in which a preliminary theory providing most of the relevant technical apparatus was eventually subjected to a reinterpretation that gave rise to the crucial nonclassical concepts. Indeed, just as the emergence of the new space-time framework of special relativity had resulted from Einstein's elaboration and reinterpretation of Lorentz's theory, also the final version of the general theory of relativity, formulated in November 1915, emerged from Einstein's reinterpretation of a predecessor theory, the *Entwurf* theory, which in this case he himself had developed in 1913 with his mathematician friend Marcel Grossmann.

A Brief History of Knowledge

This brief review of the relativity revolution, which is based on results of our research project centered at the Max Planck Institute for the History of Science, may have offered an impression of what a theoretically guided history of science can reveal to us about the inner workings of a scientific revolution. But, to draw consequences that help us to address today's challenges, we need an even larger picture. It may thus be helpful for assessing what can be learned from Einstein to first determine our position within the long-term development of knowledge in which his revolution marks an important milestone that aids our orientation.

When one attempts to understand the long-term development of knowledge (Plate 16), it makes sense to look at the coevolution of material culture, of social and cognitive structures, in which individual and collective processes of

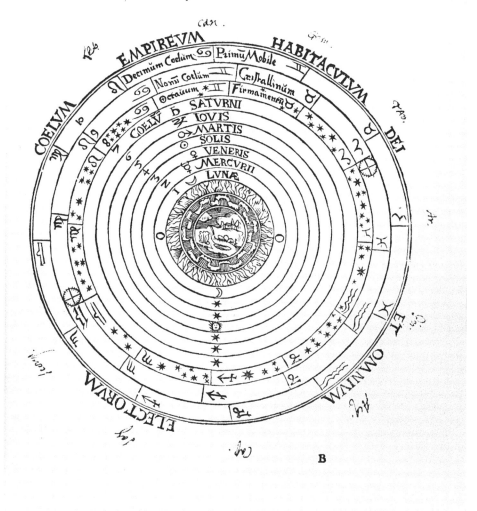

FIGURE 16.5 *Aristotle's Cosmologia*, from "Cosmographica" Petrus Apianus, 1540.
Kunstbibliothek, Staatliche Museum Berlin.

reflection mark the major turning points. A glance at the history of knowledge conceived in this way makes it evident that the themes of Einstein's revolution of physics—matter, space, motion, and radiation—are rooted in a development of knowledge systems going back to antiquity. Such systems typically comprise a rich architecture of knowledge, including intuitive, practical, and theoretical strands of knowledge. A full understanding of the historical place of Einstein's revolution requires taking into account this architecture and its development in the *longue durée* as indicated earlier in our discussion of the genesis of general relativity.

In antiquity, science emerged as a spin-off of the division of intellectual and physical labor. Its concepts were rooted in the intuitive and practical knowledge with which people mastered their environment. The ancient systems of natural philosophy distinguished themselves from mythological explanations by universalizing simple mental models of natural processes, as may be illustrated by Thales' reference to the fundamental role of water as a model for understanding nature. Also the great theories of antiquity continued to depend upon the reflection of elementary intuitive and practical knowledge that constituted their empirical bases. The concept of space in Aristotelian natural philosophy (Fig. 16.5), for instance, is connected with reflections about the different kinds of motions in our natural environment, ranging from the spontaneous "natural" fall of bodies downward via the "violent" motions caused by our intervention to the eternal motions of the celestial spheres. In contrast, the homogeneous and isotropic space of Euclidean geometry resulted from reflection on the practical, instrumental knowledge achieved by using compass and ruler.

At the dawn of modernity, Copernicus reinterpreted the knowledge accumulated by a long tradition of observation and calculation in astronomy, bringing an entirely new order to that body of knowledge. But Copernicus' heliocentric system not only questioned the Aristotelian image of the cosmos as onion-like space, it also constituted a challenge to the intuitive knowledge about motion incorporated into Aristotelian natural philosophy. The difference between real and apparent movement suddenly became a critical issue: its study would provide the basis for introducing the notion of relative motion, a crucial asset of Einstein's relativity revolution. It was, however, not only the new view of astronomy that sparked the scientific revolution of early modern times. There was also a general explosion of knowledge, much of which was created by the new class of engineer-scientists engaged in the workshops of the great cities, and which was helped along by new media such as printed books that served the integration of diverse strands of knowledge, theoretical as well as practical. This development also gave knowledge a new role as a guide for living and awakened hopes that science could improve life for everyone, rather than just remaining an intellectual consolation for a few experts, a transcendent aspect of science of which Einstein was still aware. It was virtually inevitable that such an expansion of knowledge would sooner or later conflict with the Aristotelian worldview of the Catholic Church.

In any case, the new challenging objects with which the great engineering endeavors of early modern times were confronted, from artillery via machine constructions to shipbuilding, triggered the transformation of the traditional Aristotelian system of knowledge into that of classical mechanics. But in spite of the dramatically increased significance of practical knowledge, the intuitive knowledge that was the source of the ancient theories of nature continued to shape some of the basic mental models of classical science right up to the time of Einstein. Galileo and Newton's atomism, as well as Descartes' or Huygens' concepts of a continuum, may in this sense be considered a direct continuation of ancient traditions. In the 19th century, these traditions culminated in the further elaboration of atomism and in attempts to found optics and electrodynamics on the idea of an all-pervading mechanical ether, essential to the vision of an all-encompassing mechanical worldview (Plate 17).

In spite of the successes of the mechanical foundation of classical science in the 18th and 19th centuries, the seemingly universal validity of the mechanical worldview eventually came under attack. As knowledge kept accumulating, it became more and more difficult to formulate mechanistic mental models that were free of contradictions when used in an ever-increasing variety of applications. Ether and atoms grew to suffer from a "Figaro" problem—they had to serve a multitude of functions in areas ranging from optics to thermodynamics, functions that were becoming ever more irreconcilable within the framework of classical physics. At the same time, electrodynamics and thermodynamics acquired—in the mid-19th century—an increasingly independent status as conceptually autonomous subdisciplines of physics with central concepts such as "field" and "entropy" no longer permitting a purely mechanical explanation. Philosophers and historians of science like Mach, who were able to look beyond the trenches of disciplinary specialization, took this situation as an occasion to examine critically the foundations of mechanics, as well. In summary, at the end of the 19th century, what in hindsight was called classical physics actually comprised three subsystems of knowledge, mechanics, electromagnetism, and thermodynamics. These subsystems of classical physics may be compared to the shifting continental plates in geology, in the sense that borderline areas of tension were developing between them. In the case of systems of knowledge, these tensions resulted from problems of compatibility among partly independent conceptual domains brought into contact by borderline problems situated on the fault lines between them. Just as in the geological case, such fault lines represent potential sites for tremors, which may either hardly be noticeable or else may lead to major structural rearrangements.

CHAPTER SIXTEEN

Masters of classical physics such as Planck, Lorentz, and Boltzmann were at the end of the 19th century exploring such borderline problems whose potential impact on the transformation of classical science though was at first hardly perceptible to them. Planck dedicated himself to the problem of the thermal equilibrium of radiation, which was situated between the domains of thermodynamics and electromagnetism; Lorentz worked on the problem of the electrodynamics of moving bodies, which was situated at the borderline between mechanics and electromagnetism; and Boltzmann dealt with the kinetic theory of gases, which bridges mechanics and thermodynamics. These masters of classical physics tended to look at these borderline problems primarily from the point of view of their own knowledge continents associated with the domain-specific organization of classical science. The truly explosive potential of these problems became visible, however, only from a more encompassing perspective which would include a critical reflection on the basics of classical physics like the one found in the works of Mach.

During the industrial age, which at the same time was the age of disciplinary specialization of science, the triumph of science had turned into an ever more powerful secularization process, in which not only its original promise as a life guide, but more generally its reflective dimension had faded into the background. The unity of science became less and less a question of its content than of its institutionalized methods. Apart from the works of philosophically oriented scientists like Mach or Poincaré, the old ideal of a unity of content in science survived mainly in the popular science literature of the 19th century. Despite its tendency toward unreflective mythmaking, popular science was also able to encourage

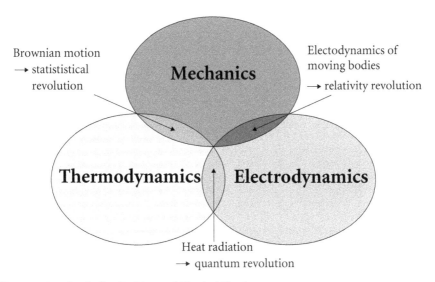

FIGURE 16.6 Borderline Problems of Classical Physics.

critical reflection and attempts to gain an overview of knowledge, as proven in the case of the young Einstein, a passionate reader of such literature. It was from the perspective he thus acquired that the borderline problems of classical physics (Fig. 16.6) and their solutions by the masters of classical physics could become the starting points for the scientific revolution of 1905.

Einstein's autodidactic studies as well as the informal discussions with his Bohemian *Denkkollektiv*, in particular within the "Akademie Olympia (Fig. 16.7)," represent the intellectual laboratory in which these solutions would be transformed into the basic principles of a new physics.

In short, the substance of Einstein's revolution was not new, but rather the result of an accumulation and integration of knowledge extending over ages. What was new was the conceptual architecture into which this knowledge was transformed in a kind of Copernican revolution. The place of models of intuitive knowledge that were fundamental to the mechanistic worldview, such as the ether, was now taken by new, nonclassical models of space and time. The classical understanding of light based on the mental model of a wave carried by a medium was replaced by a dualism of wave and particle aspects no longer expressible in intuitive mechanistic terms. And in the place of the classical atoms and their movements came new particles—the existence of which was no longer hypothetical, but which could only be described with the help of statistical laws. The conditions for Einstein's revolutionary renewal of physics

FIGURE 16.7 "Olympia Academy": Conrad Habicht, Maurice Solovine, and Albert Einstein.
Bibliothek der Eidgenössischen Technischen Hochschule, Zurich.

lay as much in the way he succeeded in preserving the received knowledge as in the way he was able to perceive its limits. One of the important aspects of his reflective approach to science was his awareness of the long-term development of knowledge, evidence of which can also be found in his own philosophical and popular writings, for instance in his book with Leopold Infeld on the evolution of physics. This encompassing perspective was clearly also responsible for the fact that Einstein did not stop critically examining conceptual assets of classical physics after his breakthrough of 1905. He saw at the same time that the revolution was not over while it was still worth preserving insights of classical physics. There is no better proof of this than what was outlined in the beginning, the genesis of the general theory of relativity.

Innovation and Guidance for Living

Now that we have looked at the structure of Einstein's relativity revolution as well as at its context in the long-term development of knowledge, what can we say about the conditions for scientific innovation to occur? The history of science in the 20th century should warn us in any case against a simplistic, technocratic answer to this question that abstracts from the cultural, social, and ethical contexts of science. Whoever speaks about innovation without thinking about its nature as a reflective dimension of our existence in society renounces the freedom to decide consciously upon the priorities and direction of innovation. The ever-graver consequences of scientific progress for our self-definition as a culture, as a society, and as a biological species—in short, for the definition of our lives—makes it imperative that such considerations enter our discussions of innovation.

The insights we have gained from our review of the history of Einstein's revolution and its roots allow two conclusions that may be relevant to the present subject: First, that science is only the tip of the iceberg, the substance of which is constituted by a world of knowledge subject to continuous transformations. The knowledge of the sciences and the humanities, practical and technical knowledge, as well as the intuitive knowledge of our everyday life are all part of interconnected systems of knowledge. Second, reflection—that is, thinking about thinking—is the crucial mechanism enabling structural changes in such systems of knowledge and is therefore also the only way to guide those transformations, to make them as we would like them, insofar as such guidance is possible at all.

From the perspective of an historical theory of knowledge, therefore, it is worth asking the question whether the leaden fog of unavoidable necessity that seems occasionally to hang over science—and which has given science its negative reputation as an instrument of blind rationality—is not perhaps the result of unexploited freedom; for instance, the result of a division of labor that has become self-perpetuating, or subordinate to other unreflected constraints. Either possibility would be evidence of deficits in the conditions under which science develops, the conditions under which it may give rise to innovative perspectives, or the conditions under which science might serve as a guidance for living.

Today's grand challenges are represented by problems encountered in the aftermath of the great civilizing ventures that were initiated in the early modern period. They no longer concern only the local fate of city-states or nations but unavoidably our civilization as a whole, such as the problems of global climate, nutrition, or disease. Against the background of these present challenges, some of the key notions of an historical epistemology, as outlined here with reference to Einstein's revolution, take on new meaning, such as the notions of borderline problems, of the integration of knowledge, of challenging objects, of shared knowledge versus individual perspectives, and of the reflective reorganization of knowledge. In a rapidly expanding world of science that is still characterized by growing specialization, the notion of borderline problems demanding the integration of differently structured areas of knowledge has, if anything, gained in relevance. But in contrast to the perspective of classical science, the perspective appropriate to today's challenges is no longer one that can take potentially infinite horizons and a fundamentally stable cosmos for granted, but one that must deal with the limits of evolving systems of either an ecological, cognitive, or cosmological nature. In this sense it may be characteristic for the scientific challenges of our time that a borderline problem such as quantum gravity situated between quantum theory and general relativity is intimately linked to issues concerning the limits and boundary conditions of the largest evolving system, the universe.

Whatever the perspectives capable of transforming today's challenges into tomorrow's scientific innovations may be, their effectiveness will depend on overcoming the division of labor between primary and reflective levels of knowledge that characterizes the still prevailing classical image of science. Only when this boundary is overcome can reflective thinking about science affect the structure of its contents. Einstein's implementation within his theory of gravitation of the historically based philosophical criticism of science by Mach provides a striking example of how reflective knowledge can actually transform science. Even at the time, both Mach and Einstein belonged to the small minority of those who took such reflections at all seriously. I would like to make the surprising claim that in view of the new information technologies our situation may have become more hopeful because they have not only accelerated the information explosion, but have also offered us new means for coping with it.

This brings me back to another key concept of our analysis of Einstein's revolution from the perspective of historical epistemology, the integration of

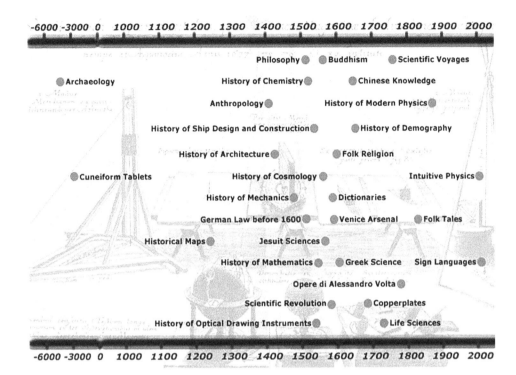

-6000 -3000 0 1000 1100 1200 1300 1400 1500 1600 1700 1800 1900 2000

Philosophy ● Buddhism ● Scientific Voyages
● Archaeology History of Chemistry ● ● Chinese Knowledge
Anthropology ● History of Modern Physics ●
History of Ship Design and Construction ● ● History of Demography
History of Architecture ● ● Folk Religion
● Cuneiform Tablets History of Cosmology ● Intuitive Physics ●
History of Mechanics ● ● Dictionaries
German Law before 1600 ● ● Venice Arsenal ● Folk Tales
● Historical Maps Jesuit Sciences ●
History of Mathematics ● ● Greek Science Sign Languages ●
Opere di Alessandro Volta ●
Scientific Revolution ● ● Copperplates
History of Optical Drawing Instruments ● ● Life Sciences

-6000 -3000 0 1000 1100 1200 1300 1400 1500 1600 1700 1800 1900 2000

FIGURE 16.8   ECHO: European Cultural Heritage Online; http://echo.mpiwg-berlin.mpg.de.

knowledge as a presupposition for its reflective reorganization. The Internet has indeed profoundly changed the conditions for the integration of knowledge, both in a material and in an intellectual sense. It has actually become conceivable to develop the representation of information on the Web, which at the moment resembles an enormous jigsaw puzzle, into a global, interactive, and transparent representation of human knowledge that is structured by the meaning of its content (Fig. 16.8). Interactive networking could become part of a conscious structuring of the knowledge represented on such an "epistemic Web." A thoroughgoing understanding of the architecture of knowledge, as offered by an historical theory of knowledge, could contribute much to improve the conditions for realizing this important developmental step.

Regarding this future development merely as a problem of greater efficiency amounts to an absurd underestimation of the immense potential for innovation in a nontechnocratic sense that is harbored by the new information technologies. As we may also learn from Einstein's revolution, such a transformation, which amounts to the restructuring of a global system of knowledge perhaps definitively transcending traditional disciplinary specialization, will not simply happen by itself, no matter how urgently the dynamics of knowledge development pushes for it. The potential inherent in new forms of knowledge integration will only be realized if we succeed in overcoming the resistance inherent in the established compartmentalization of scientific information flow, for instance, against open access to scientific knowledge and to cultural heritage—a crucial precondition for taking the step toward an epistemic Web.

The conceptual novelties that could emerge from a global, interactive, and transparent representation of human knowledge, with its enormously enhanced potential for self-organization, are as difficult to predict as it would have been to foresee Einstein's revolution of the concepts of space, time, matter, and radiation from the perspective of classical physics. What is clear even now is that future transformations of science will depend ever more on a culture of sharing knowledge, to which belong its accessibility to all just as much as its openness for change in a collective process of reflection (Fig. 16.9). Considering how much a single individual like Einstein could accomplish even under rather poor conditions of sharing knowledge, we can read the history of his revolution also as an encouragement to contribute to a reflective science that is adequate to today's global challenges and may serve as an orientation for our lives.

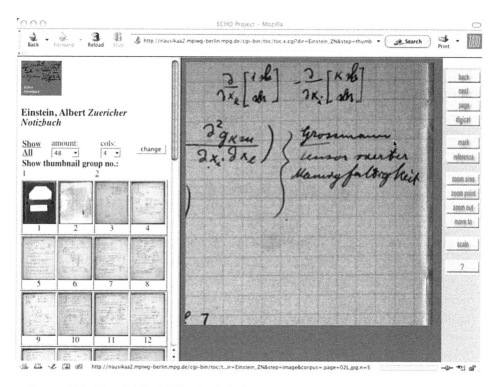

FIGURE 16.9   Einstein's Zurich Notebook Online.

EINSTEIN AND ℏ: ADVANCES IN QUANTUM MECHANICS

Jürg Fröhlich

Anyone who is not shocked by quantum mechanics has not understood it.
　　　　　　—*Niels Bohr*

I have thought a hundred times as much about the quantum problems as I have about the general theory of relativity.

　　　　　　—*Albert Einstein*

I recall having committed myself to you to give a report on quantum theory at this conference. After much reflection back and forth, I have come to the conviction that I am not competent to give such a report in a way that really corresponds to the state of things. The reason is that I have not been able to participate as intensively in the modern development of the quantum theory as would be necessary for this purpose. I beg you not to be angry with me! Perhaps, Herr Gross from Santa Barbara or Zeilinger will do a good job.

Ladies and gentlemen, what I have just read to you is a slightly modified version of the text of a letter Einstein wrote to Lorentz in connection with the 1927 Solvay Conference. Given the composition of the audience to which my lecture is addressed and my lack of scholarly erudition, the paraphrasing of Einstein's letter[1] accurately reflects my own anxieties.

Let me begin this lecture with yet one more Einstein quotation, one that is particularly pertinent to the discussion that follows: "Every physicist thinks that he knows what a photon is. I have spent my whole life trying to find out what a photon is and I still don't know it."

I am convinced that, after fifty more years of research, we are in a much more comfortable position than Einstein, at least with regard to photons—but to indicate why this is so within the compass of such a lecture is nigh impossible. My presentation will necessarily be sketchy.

A great deal has been written about Einstein, and about Einstein's contribution to quantum mechanics.[2] I refer the reader to Abraham Pais' fine intellectual biography[3] of Einstein and to Albrecht Fölsing's recent more general biography[4] for a more detailed presentation of Einstein's contributions to the quantum theory.

My aim in this lecture is to trace some of the developments in quantum theory initiated by Einstein's original contributions. It is not enough, I believe, to extol Einstein's remarkable insights. Much has happened since then, and I hope that historians of physics will take up the challenge of describing what we understand, at present, and how we came to where we are, with some of the important technical details. On one hand, this would be the story of how the reductionists among physicists have been able to give a "foundational"—if not fundamental—description of the elementary entities out of which matter, as we know it, can be thought to be constituted. This is the story of how quarks, gluons, leptons, W- and Z-bosons were put together in the form of the "standard model" of particle physics, and of various attempts to encompass gravity within a quantum-mechanical description. On the other hand, it would be the story of numerous impressive developments, brought forward by emergentists, in, for example, condensed matter physics: how we have come to understand, in a mathematically rigorous fashion, the stability of nonrelativistic matter as well as some of the elastic, electric, magnetic, and thermal properties of different forms of condensed matter in terms of the interactions among the microscopic constituents, electrons and nuclei. It would also involve an account of the statistical mechanical foundations of thermodynamics and of transport theory, of how, for example, the Navier-Stokes equations that describe liquid flow can be derived from microscopic dynamical laws. An important part of the narrative—in both the reductionist and the emergentist story—would be to describe the diverse and fructiferous interactions between theoretical and mathematical physics and pure mathematics.

Of course, I will not even begin to implement this project in this lecture!

PLANCK, EINSTEIN, AND THE AGENDA OF PHYSICS IN THE 20TH AND 21ST CENTURIES

I would like to put Einstein's work in a general context by recalling how Max Planck set the agenda of physics in the 20th century (and, perhaps, of much of physics in the 21st century) by noticing that, dimensionally, all physical quantities can be expressed in terms of four fundamental constants of nature:

1. Boltzmann's constant: $k_B \cong 1.38 \times 10^{-16}$ erg/K
2. Planck's constant: $h \cong 6.63 \times 10^{-27}$ erg sec

3. The speed of light: $c \cong 3 \times 10^{10}$ cm/sec

4. The Planck length: $L_p = \sqrt{\hbar G_N / c^3} \cong 1.6 \times 10^{-33}$ cm, which is related to Newton's gravitational constant, G_N

These constants appear in two fundamental laws of nature (see Fig. 17.1), Newton's law of universal gravitation and Planck's law for the spectral energy density of black body radiation.[5] Planck held the view that these were universal laws, equally valid and applicable on earth as on distant stars. This is why he used the term *"absolute constants"* to describe the three constants that we now call

$$\text{Planck mass: } M_p = \sqrt{\hbar c / G_N} \cong 2.2 \times 10^{-5} \text{ gr}$$

$$\text{Planck length: } L_p = \sqrt{\hbar G_N / c^3} \cong 1.6 \times 10^{-33} \text{ cm}$$

$$\text{Planck time: } T_p = L_p / c \cong 5.4 \times 10^{-44} \text{ sec.}$$

Newton's Law of Universal Gravitation

$$F(r) = G_N \frac{mM}{r^2}$$

Planck's Law of Black Body Radiation

$$\rho(T, \nu) = 2 \frac{4\pi\nu^2}{c^3} \frac{h\nu}{e^{h\nu/kT} - 1}$$

FIGURE 17.1   Two Fundamental Laws.

Further constants of nature, such as the mass of the proton or the electric charge of an electron, can be expressed in terms of the Planck mass, the Planck length, and the Planck time as pure dimensionless constants. Einstein held the opinion that a fundamental theory of nature ought to predict such dimensionless constants, but we might disagree with him on this particular issue. Some of these constants might, as Dirac first suggested, be dynamical and evolve in time or, at least, in energy scale.

Each of the constants k_B, h, c, and L_p stands for a *revolution* in the physics of the 20th century and for revolutionary developments in pure mathematics triggered by problems of physics. (For mathematicians, the constants k_B, h, c^{-1}, and L_p are the parameters of four different *deformations*.) Taken together, these constants stand for a revolution yet to take place. Let me make some brief comments and indicate what these four revolutions have been about.

1. Boltzmann's constant and the *atomism* it is connected with, through the equations

$$\frac{R}{k_B} = N \text{ and } \frac{F}{N} = e$$

where R is the universal gas constant,[6] 8.3×10^7 erg/K; N is Avogadro's number, 6×10^{23} (the number of atoms or molecules in a mole of a pure substance); F is Faraday's constant (the amount of charge necessary to deposit one mole of a monovalent substance in electrolysis); and $-e$ is the charge of an electron; stand for the transition from continuum mechanics to the Hamiltonian mechanics of point particles,[7] and from thermodynamics to the statistical mechanics of atoms and molecules. The foundational hypothesis of equilibrium statistical mechanics is Boltzmann's postulate that the entropy s per particle of a system of N particles in thermal equilibrium, with a density $\rho = 1/v$ and an average energy u per particle, is given by

$$s(u,v) = \lim_{N \to \infty} \frac{k_B}{N} (\ln \Omega \, (Nu))$$

where $\Omega(E)$ is the number of microstates accessible to the system at energy E.[8] (Note the appearance of Boltzmann's constant k_B.)

2. The representation of the atomism of electrically charged particles interacting with electromagnetic radiation is inconsistent without *quantum theory*—the most revolutionary change in the *Weltbild* of physics in the 20th century. Planck's constant h stands for the transition from Hamiltonian mechanics to matrix mechanics, and from the symplectic geometry of classical phase spaces to the noncommutative geometry of quantum theory.

3. *Special relativity*—in which the speed of light c plays a special role as a limiting velocity of signal propagation—brings about the transition from Newton's absolute space and universal time to the four-dimensional space-time continuum of Einstein, Poincaré, and Minkowski. The inverse of the speed of light is the deformation parameter in the deformation of the Galilei group to the Poincaré group.

4. *General relativity*, with the Planck length L_p, ties the geometry of space-time to the distribution of matter and energy and provides us with a purely *geometrical* description of gravitational interactions.

The special and general theories of relativity profoundly changed our conception of space, time, and symmetry. We are awaiting an even more profound change originating from a unification of quantum theory with the theory of gravitation. One might expect that this change will make space-time appear as an emergent structure. And the relation between Einstein's field equations and a more fundamental theory of space-time-matter might be expected to be vaguely reminiscent of the relation between the Vlasov equation and the Hamiltonian equations of motion for point particles: Einstein's field equations might correspond to some kind of *mean-field* (large "N") *limit* of a more fundamental law of nature, just like the Vlasov equation corresponds to the mean-field limit of an N-particle Hamiltonian system, as $N\rightarrow\infty$.

None of the four revolutions in our understanding of nature was immediately and universally approved and accepted by the scientific community—least so, quantum theory.

As late as 1913, Ernst Mach commented on two of them as follows: "I must . . . assuredly disclaim to be a forerunner of the relativists as I withhold from the *atomistic belief* of the present day" (emphasis added). It is astonishing that as perspicacious and insightful a physicist as Mach would doubt the existence of atoms as late as 1913.

The quantum theory of light faced the stiffest opposition. Einstein's discovery of photons was based on the observation that, at high frequencies, where Wien's law for black body radiation is valid, the spectral entropy of thermal radiation obeys the same law as the entropy of an ideal gas of N particles, if one identifies N with the quotient $\dfrac{U(v)}{hv}$, where $U(v)$ is the internal energy of thermal radiation of frequency v. Einstein thus concluded: "Monochromatic radiation of low density behaves, thermodynamically, as if it consisted of mutually independent quanta of energy of magnitude hv."[9] However, Einstein's quanta, i.e., the photons, were not taken seriously. Neither experimentalists, such as Robert Millikan, who succeeded in experimentally verifying Einstein's law for the photoelectric effect,[10] nor leading theorists around Bohr accepted Einsteins's photon hypothesis before Debye and Compton did their famous experiments in 1922.

Einstein took part in all four revolutions described above as a leading or the key contributor. For Einstein, h and k_B were always fellow travellers. I would have liked to include in this presentation a discussion of the following exciting and very topical problems of statistical physics that have evolved from his work:

- Derivation of the fundamental laws of thermodynamics from the (quantum) statistical mechanical description of open systems
- Quantum Brownian motion
- Transport theory in open (quantum) systems
- Dissipation-free, "quantized" transport

However, the technical complexity of these topics makes it impossible to give a brief, yet understandable account.

I conclude this introduction with a representation of the Planck-Einstein hypercube (Table 17.1), which summarizes succinctly the revolutionary

TABLE 17.1   Planck–Einstein Hypercube

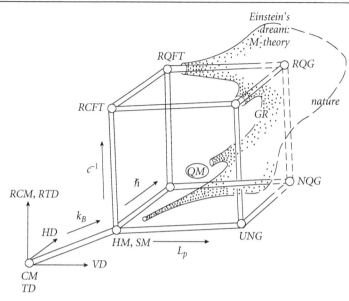

$k_B, \hbar, c^{-1}, L_p$:   Symbols of revolutions in physics

Parametrize deformations of mathematical structure

CM, TD: Continuum mechanics, thermodynamics

RCM, RTD: Relativistic CM, relativistic TD

HD: Hartree dynamics

VD: Vlasov dynamics

HM, SM: Hamiltonian mechanics, statistical mechanics

UNG: Universal Newtonian gravity

QM: Quantum mechanics

RCFT: Relativistic classical field theory

RQFT: Relativistic quantum field theory

GR: General relativity

NQG: Newtonian quantum gravity

RQG: Relativistic quantum gravity, M-theory

The Tetrahedron Favored by the Younger Generation

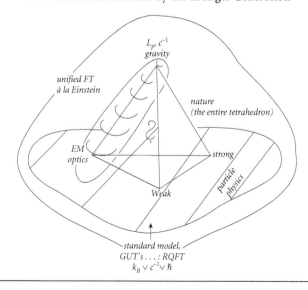

developments in the physics of the 20th century and draws attention to some of the fundamental tasks of physics in the 21st century. I expect that these tasks will not be accomplished unless patience, intellectual courage, and generous funding of fundamental science will prevail! I should add that I am deeply worried that this may not be the case.

Einstein's Contributions to Quantum Theory

Let me continue with a lightning sketch of the history of some of the discoveries in the quantum theory of nonrelativistic matter, i.e., of atoms and molecules, and of electromagnetic radiation. This is the field in quantum theory to which Einstein made his most fundamental contributions. It is the field underlying optical spectroscopy, quantum optics, laser physics, photovoltaics, and many other areas of physics that keep us busy and fascinated nowadays.

Whereas Planck's law of black body radiation was given a firm theoretical foundation within seventeen years of its discovery in 1900, it took the better part of eighty years until the theoretical explanation of the data of atomic and molecular spectroscopy—processes like spontaneous emission of light from an atom, Compton scattering of photons and electrons, laser physics, the theory of resonances, the stability of matter, among others—was put on a mathematically firm basis, going beyond approximate treatments with uncontrolled error terms. When talking about the quantum theory of radiation and matter, it is appropriate to remind the audience that we are also celebrating eighty years of matrix mechanics, as originally discovered by Heisenberg. Heisenberg is a more controversial figure than Einstein; but he is undoubtedly one of the great geniuses of physics without whom quantum mechanics might not have been formulated in the way we presently use it.

Quantum theory is the only field of physics of which Einstein said that his contributions were "revolutionary." Table 17.2 gives a brief chronology of developments in quantum theory until the early 1930s.

As already mentioned, Einstein's contributions to quantum theory start in 1905 with the discovery of the photon, that is, of the idea that monochromatic electromagnetic radiation of frequency ν is made up of quanta of energy of magnitude $h\nu$. It took a few years until these so-called photons, the field quanta of electromagnetic radiation, were also understood to carry momentum. Curiously, the name of Johannes Stark comes up in this connection. Stark was, I believe, an impressive physicist but, as you may know, he became a nasty anti-Semite and a Nazi. He did make, however, an important contribution to understanding the momentum of photons.[11]

It should be stressed that all of Einstein's contributions to quantum theory were actually contributions to *quantum statistical mechanics*, more particu-

TABLE 17.2   Chronology of Discoveries in Two Strands

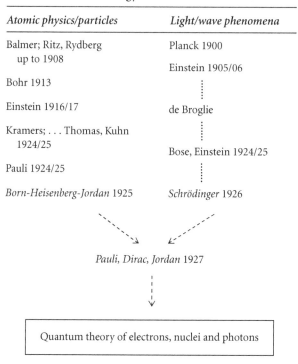

Atomic physics/particles	Light/wave phenomena
Balmer; Ritz, Rydberg up to 1908	Planck 1900
	Einstein 1905/06
Bohr 1913	⋮
Einstein 1916/17	de Broglie
Kramers; . . . Thomas, Kuhn 1924/25	⋮
	Bose, Einstein 1924/25
Pauli 1924/25	⋮
Born-Heisenberg-Jordan 1925	*Schrödinger* 1926

Pauli, Dirac, Jordan 1927

Quantum theory of electrons, nuclei and photons

larly to the quantum theory of thermal equilibrium (Table 17.3). Many of his contributions grew out of meditations on Planck's formula. What Einstein studied most intensely was black body radiation and the interactions between radiation and matter in a regime, where the motion of material particles can be treated nonrelativistically. In 1907, he formulated a quantum description of the harmonic vibrations of atoms arranged in a crystal that enabled him to explain the behavior of the specific heat of solids at very low temperatures.[12] (His theory was important for Nernst's discovery of the third Law of Thermodynamics.) In 1915, in collaboration with Wander J. de Haas, Einstein proposed to test Lorentz' conjecture that electrons were the charged particles in Ampère's hypothesis that ferromagnetism is caused by the microscopic circular motion of electric charges.[13] Although this hypothesis is wrong, the effects studied by Einstein and de Haas and by Barnett (the Barnett-Einstein-de Haas effect) are *real*. In 1917, he presented a new derivation of Planck's formula, using ideas that led to the foundation of spectroscopy, and introduced the concept of stimulated emission of light from atoms[14]; in 1924, having understood the counting method that Bose had used to derive Planck's law, Einstein applied this same method to derive the properties of quantum (Bose) gases in thermal equilibrium at very low temperature.[15] Along the way, he formulated the first semirigorous microscopic theory of a phase transition. All these researches involved h and k_B.

The analysis of thermal equilibrium in quantum systems and quantum field systems, that is, in systems with an infinite number of degrees of freedom, later led to extraordinarily deep insights in quantum physics and in pure mathematics. The modern developments started with the work of Kubo, and of Martin and Schwinger on quantum statistical mechanics. These advances were complemented by the (only *seemingly* unrelated) 1957 work of Jost in axiomatic quantum field theory that concerned the celebrated CPT theorem, and then by the 1967 work of Haag, Hugenholtz, and Winnink concerning the general analysis of equilibrium states of quantum systems. Why quantum statistical mechanics and the CPT theorem have a common basis is understood only by "insiders": It is a reflection of the common mathematical structure underlying these results. The work of Jost and of Haag et al. gave rise to major developments in pure mathematics, namely in the theory of von Neumann algebras. I am referring to the work of Tomita and Takesaki, and of Connes and others. It also led to a better understanding of general space-time symmetries of quantum systems (including a certain anti-unitary involution, CPT, which I will consider shortly). Perhaps these discoveries offer a glimpse of how geometry and quantum theory may ultimately be united. The unification of geometry and quantum theory is an area where, so far, success has been rather limited.

When studying relativistic quantum systems with an infinite number of degrees of freedom, one is dealing with quantum field theory. Local relativistic quantum field theory (RQFT) emerges when one incorporates into quantum theory the relativistic notion of causality: namely, that if one performs experiments in causally disconnected regions of space-time, regions that cannot be connected by light beams, these experiments cannot influence one another. When the insights gained in quantum statistical mechanics are combined with the basic principles of RQFT, in particular with locality and Einstein causality, one concludes that, besides Lorentz invariance, there always exists a general anti-unitary symmetry

$$J = CPT$$

which is the composition of the operations C (charge conjugation), P (space reflection), and T (time reversal). This symmetry maps matter to antimatter, (e.g., electrons into positrons).

That there might be a general symmetry between matter and antimatter is an idea that first came up in Wolfgang Pauli's 1919 article on general relativity, and it was explicitly mentioned in a paper by Einstein in 1925. But Pauli and Einstein studied classical theories, and they perceived this symmetry as

TABLE 17.3   Statistical Mechanics and RQFT

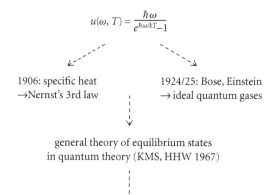

1905 ("sehr revolutionär"): light composed of
photons = "particles" with

$$E = \hbar\omega, \ \vec{P} = \hbar\vec{k} \ (1909)$$

$$\hbar \vee k_B$$

atomism, quantum theory, thermal equilibrium

$$u(\omega, T) = \frac{\hbar\omega}{e^{\hbar\omega/kT} - 1}$$

1906: specific heat   1924/25: Bose, Einstein
→Nernst's 3rd law   → ideal quantum gases

general theory of equilibrium states
in quantum theory (KMS, HHW 1967)

operator algebras, Tomita–Takesaki theory → anti-unitary
symmetries, noncommutative geometry (Connes)

$$\hbar \vee k_B \vee c^{-1}$$

RQFT, QED

Heisenberg, Pauli, . . . , Feynman, Schwinger, Tomonaga, Dyson (1930–1950 & ever since)
Locality & "Einstein causality" → vacuum is equilibrium state for Lorentz boosts →
anti-unitary symmetry J = CPT: matter → anti-matter; spin ↔ statistics
Generalizations to curved space-times: black holes (Hawking radiation), dS_4, AdS_5, . . .

$$\hbar \vee k_B \vee c^{-1} \vee L_p$$

gravitational instability against black-hole formation;
cosmological constant problem; space-time uncertainty relations:

$$\Delta t_{event} \times diam_{event} \gtrsim L_p^2/c$$

→ RQFT's not fundamental but effective
low-energy quantum theories ($L_p = 0$)

a problem they wanted to get rid of. The modern ideas about antimatter start with Dirac's work on his relativistic electron equation, and his subsequent postulate that the vacuum is the state where all the negative-energy one-particle states that this equation exhibits are occupied. The final form in which this CPT symmetry is currently understood came in the work of Res Jost. The mathematical framework underlying Jost's work also leads to an understanding of how the spin of particles—the intrinsic angular momentum of particles—is related to their quantum statistics: half-integral spin ↔ Fermi-Dirac statistics, integral spin ↔ Bose-Einstein statistics—with interesting complications in two spatial dimensions!

I mention this because all these results are connected, mathematically, to the general quantum theory of thermal equilibrium so dear to Einstein.

By the way, the ideas just alluded to generalize to other space-times (i.e., other than the Minkowski space-time, the usual space-time of RQFT), such as the space-times surrounding black holes, where they lead to the prediction of Hawking radiation, or to some more esoteric space-times considered by string theorists and cosmologists.

Unfortunately, RQFTs describe too many degrees of freedom at very high energy scales or, put differently, in arbitrarily tiny space-time cells. This feature leads to the well-known ultraviolet divergences of RQFTs. The analysis of these divergences has given rise to the development of renormalization theory, of renormalization group methods, and of conformal field theory, which have profoundly changed the way we analyze many problems in physics and mathematics. The Nobel prizes of Gerard 't Hooft and Martinus Veltmann, and of David Gross, H. David Politzer, and Frank Wilczek have been awarded in recognition of fundamental work in this area.

Another consequence of the high-energy properties of RQFTs are various *no-interaction theorems*, which I mention here because they are ultimately based on an analysis of gases of interacting Brownian paths, and this feature connects the analysis to Einstein's work on Brownian motion of 1905.

If one attempts to cure the ultraviolet divergences of relativistic quantum field theories by coupling RQFT to gravity, one discovers that these theories have severe gravitational instabilities. These problems will presumably be cured, ultimately, only after another revolution in our conception of space-time and matter will have occurred.

But let me return to my discussion of Einstein's contributions to quantum theory. In 1916–17, Einstein returned to the study of interactions between atoms and electromagnetic radiation and, in a sense, laid the foundation for the modern theory of atomic spectroscopy. He rederived Planck's law by introducing his famous A and B coefficients, important to determine the transition rates for spontaneous and stimulated emission of radiation by an atom. This work laid the foundation for the general statistical rules regarding radiative transitions between the stationary states of an atom. It marks the introduction of probabilistic concepts into quantum theory. The phenomenon of stimulated emission of photons, when complemented by a quantum mechanical treatment of atoms, made possible, much later, the design of masers and lasers. Einstein also made a contribution to the Bohr-Sommerfeld-Epstein quantization rules: he put them in the natural, general context of integrable Hamiltonian systems.

Why did Einstein succeed in discovering important laws about the interactions between matter and radiation on the basis of relatively crude theoretical arguments? One reason is that the fine structure constant is so small. The laws discovered by Einstein are neither fundamental nor exact, but hold with great precision, just because the fine structure constant $\cong {}^1/_{137}$ is very small.

To conclude, I would like to point out that most major revolutions in 20th century physics have grown out of *very mathematically oriented analysis of physical problems*. Einstein had a powerful mathematical intuition, and Erwin Schrödinger was a very talented mathematician and was in touch with mathematicians such as Herman Weyl, because he understood the role of mathematics in modern theoretical physics—to mention but two major mathematical physicists.

For a survey of Einstein's work on quantum theory, see Table 17.4.

Beyond Physics

In his younger years, Einstein succeeded very well in constructing a productive coexistence between his efforts in fundamental and in applied research. We have much to learn from him in this respect!

Einstein had many dreams outside of physics—of nuclear disarmament, world government, peace, human rights, and unbureaucratic higher education.[16]

I hope it will be part of Einstein's legacy for the 21st century that fundamental science will not be undermined but will be generously supported, and that the future will see the establishment of a productive balance and a kind of unity between fundamental and applied research. At many universities throughout the world, it appears that the situation in teaching fundamental science and in carrying out fundamental research in physics and mathematics is quite precarious. Bottom-line payoffs set the agenda in many of the sciences, and speculations of what might be useful and applicable merge with speculations about what might boost the price of shares on the stock market.

Several colleagues working at universities in Berlin are particularly concerned about the situation of fundamental science at their institutions. One would hope that those who bear responsibility for the universities in Berlin will keep excellent traditions alive!

I conclude with the wish that we all learn something useful from Einstein's opinions and ideas about human rights, disarmament, and peace, and that we *do* what we can to promote them![17]

TABLE 17.4   Einstein's Work on Quantum Theory

	Matter & Radiation
1905	photons, photoelectric effect

$$\frac{m_{e^-}}{2}v^2 = \hbar\omega - W$$

1916/17	atomic spectroscopy & Planck's law

$$\hbar\omega_{nm} = E_n - E_m \; \text{(Bohr 1913)}$$

A & B coefficients; absorption, spontaneous & induced
emission of γ's $\rightarrow$ lasers, photovoltaics, sensors, . . . 2005!

Condensed matter, cold atoms

1906/07	specific heat ($\rightarrow$ Nernst's 3rd law)
1915	Einstein–de Haas
	(Maxwell, Hertz, Richardson, Barnett) magnetism, g-factor of e^-
	Coriolis force $\sim$ Lorentz force $\rightarrow$ London eq., vortex quantization, superfluidity, superconductivity
1924/25	Theory of ideal quantum (Bose) gases, BEC: First understanding of a phase transition $\rightarrow$ continuous symmetry breaking

superconductivity   superfluidity

phase transitions
& critical phenomena

1932/34	Semi-vectors (2-component spinors) complex Lorentz group (with W. Mayer)

Twistors

1935	Einstein–Podolsky–Rosen entanglement, Bell's inequalities
	(Does Ψ describe reality or only our best guess of how reality will reveal itself in a measurement?)

Acknowledgment

I would like to thank Silvan Samuel Schweber for his notes on my paper.

EINSTEIN'S UNKNOWN CONTRIBUTION TO QUANTUM THEORY

A. Douglas Stone

ALBERT EINSTEIN MADE MANY SEMINAL contributions to quantum theory, starting with his famous 1905 paper on light quanta, which he himself termed "revolutionary," and continuing through his 1925 work on the quantum ideal gas in which he predicted the phenomenon now called Bose-Einstein condensation. In 1909 he identified clearly the necessity of a "fusion" of the wave and particle theories of radiation, and in 1916 he laid the foundation of the quantum theory of radiation and introduced the A and B coefficients that determine equilibrium between quantized atomic levels and thermal radiation. In the 1916 paper he first recognized and expressed the intrinsic randomness of quantum processes, which he later decided was unacceptable in a complete theory of mechanics: "The weakness of the theory lies . . . in the fact that it does not bring any nearer the connection with the wave theory and . . . it leaves the moment and direction of the elementary processes to 'chance.'" It is interesting that, at the time he wrote those words, he did not regard this weakness as a fatal flaw, adding "all the same, I have complete confidence in the reliability of the method used here."[1]

The extent of Einstein's contributions, both conceptual (the photon, wave-particle duality of light, intrinsic randomness) and mathematical (the law of the photoelectric effect, the A and B coefficients, the statistical mechanics of the ideal Bose gas and of lattice vibrations) made him arguably the leading contributor to the old quantum theory, surpassing even Niels Bohr and Max Planck. Indeed T. S. Kuhn has argued that Planck never really interpreted the "quantum of action" (as he called it) to imply the quantization of energy in individual mechanical oscillators until after Einstein's 1907 paper on the specific heat of solids,[2] in which Einstein makes this statement.

For although one has thought before that the motion of molecules obeys the same laws that hold for the motion of bodies in our world of sense perception . . . we must now assume . . . that the diversity of states that they can assume is less than for bodies within our experience. For we make the additional assumption that . . . the energy of elementary structures can only assume the values 0, hν, 2hν, etc.[3]

Moreover, it is well known that Planck (and others) did not accept the photon concept until a decade or more after its proposal by Einstein in 1905. Finally, despite Einstein's much stronger public identification with relativity theory, he himself frequently emphasized his lifelong interest and focus on quantum theory, commenting at one point to Otto Stern: "I have thought a hundred times as much about the quantum problems as I have about General Relativity theory."[4] In this chapter I will discuss a brilliant but relatively little-known 1917 paper in which Einstein both attempted to put the old quantum theory on a firmer mathematical footing and concurrently identified a basic conceptual shortcoming of this theory.[5]

Most of Einstein's known work on quantum theory relates to the quantum theory of radiation, and all of it before 1917 is based on the Planck relation

(18.1) $$\varepsilon = nh\nu$$

where ε is the energy of some physical degree of freedom, n is an integer, h is Planck's constant, and ν is the frequency associated with periodic motion of that degree of freedom. In the case of electromagnetic radiation, ν is the frequency of the radiation, and each light quantum (photon) would carry energy hν. In the case of a mechanical oscillator such as a vibrating molecule, ν is the frequency of the vibration, and the energy of oscillation can be any integer n times hν. However, unlike an ideal Newtonian oscillator, the energy of a quantum oscillator can *only* take these quantized values. This was the first key new idea of the quantum theory and one that, as noted above, Einstein was the first to state clearly (although it was implicit in Planck's work).

However, the Planck relation only holds for simple harmonic oscillations or for phenomena, like light, that satisfy a linear wave equation with simple harmonic solutions (i.e., oscillations of any amplitude have the same frequency). The allowed quantum energy levels for *arbitrary* mechanical systems are not given by the Planck relation. For example, the allowed energy levels for motion of an electron around a hydrogen atom are not given by multiplying the frequency of an electron orbit by Planck's constant. That is one reason why the spectrum of hydrogen was not immediately understood following the acceptance of the Planck relation. For the thirteen years after the introduction of Planck's constant, the correct quantization rule for mechanical systems other than harmonic oscillators was not known, until the work of Bohr showed that the correct rule for the atomic levels of hydrogen was derived from the principle of quantization of angular momentum, L, in units of Planck's constant

(18.2)
$$2\pi L = nh$$

The angular momentum is a special case of a general quantity of importance in classical mechanics, called the action integral. The basic Lagrangian equations of motion, which are a general statement of Newton's Laws, are derived from William Hamilton's principle of minimizing the action integral along a trajectory. Following the work of Bohr, Arnold Sommerfeld realized that Bohr's condition (2) could be generalized by requiring that, for *any* mechanical system undergoing bound periodic motion, the action integrals associated with that motion must be set equal to an integer times $h/2\pi$. (Bound periodic motion refers to motion that is restricted to some finite region of space and repeats itself after some period of time, T; only bound motion leads to quantized energy levels in quantum theory, although as we shall see, the bound motion need not be periodic to get quantized levels). The Sommerfeld condition is always applicable for bound motion in one dimension, which must be periodic; however, Einstein recognized that its generalization to the more realistic situation of motion in all three dimensions of space was problematic. Einstein, who had not contributed previously to the problem of the correct quantization rule for mechanical systems, addressed this issue for the first and only time in his career at the May 11, 1917, meeting of the German Physical Society.

The lecture and subsequent paper were entitled "On the Quantization Condition of Sommerfeld and Epstein."[6] It contained an elegant reformulation of the Bohr-Sommerfeld quantization rules of the old quantum theory that extended and clarified their meaning; the technique that Einstein invented, now known as the Einstein-Brillouin-Keller (EBK) method, is still used in contemporary research.[7] Even more impressive, however, the paper contained a brilliant insight into the limitations of the old quantum theory when the mechanical system in question is non-integrable (or in modern terminology, chaotic). The paper was cited by Louis de Broglie in his historic thesis on the wave properties of matter and by Erwin Schrödinger in the second of his seminal papers on the wave equation for quantum mechanics,[8] but was then ignored for twenty-five years until the Einstein quantization scheme was independently discovered by Joseph Keller in the 1950s.[9] However, its significance was not fully appreciated until the early 1970s when theorists, led by Martin Gutzwiller, finally began to address the fundamental difficulty that Einstein had identified in 1917, founding a subfield of research now known as

Quantum Chaos. Even today, Einstein's brilliant insight into the failure of the Bohr-Sommerfeld approach is unknown to the large majority of researchers working in quantum physics. It seems appropriate in this centennial of Einstein's miracle year of 1905 to frame Einstein's achievement in this paper in a modern context and explain how he identified a simple criterion for determining if a dynamical system can be quantized by the methods of the old quantum theory.

Einstein's Motivation

Einstein's paper was written shortly after he completed his groundbreaking work on the quantum theory of radiation in which he introduced the famous A and B coefficients that determine the equilibrium between matter and radiation. Einstein had come back to the quantum theory after focusing intensely on the theory of General Relativity from 1911 until early in 1916. His deep involvement with general coordinate transformations and the geometry of curved space likely motivated him to search for a coordinate-independent formulation of the rules of the old quantum theory, i.e., a formulation that would give the same energy levels independent of the particular choice of coordinates to label points in space.

Einstein begins his paper by saying that, for a particle moving under an arbitrary force in one dimension, it is clear that the quantization condition is

$$(18.3) \qquad \int p\,dq = \int_0^T p\,\frac{dq}{dt}\,dt = nh$$

where the integral is over one period of the one-dimensional (bound) motion. In higher dimensions, bound motion is not in general periodic, but Sommerfeld had proposed the straightforward generalization for separable motion,

$$(18.4) \qquad \int_0^{T_i} p_i\,\frac{dq_i}{dt}\,dt = n_i h, \quad i = 1 \ldots d$$

where d is the number of degrees of freedom of the motion, and T_i is the period of the motion in each of the d separable coordinates. (A separable motion is one in which the motion described by each coordinate-momentum pair is independent; separable bound motion must be periodic in each such coordinate). Einstein comments:

> Notwithstanding the great successes that have been achieved by the Sommerfeld-Epstein extension of the quantum theorem for systems of several degrees of freedom, it still remains unsatisfying that one has to depend on the

separation of variables . . . because it probably has nothing to do with the quantum problem per se . . . the individual products $dp_i dq_i$ in a system of several degrees of freedom . . . are not invariants, therefore the quantization condition [Equation 18.4 herein] has no invariant meaning.

He then describes an alternative approach, which he modestly refers to as a "small modification," that allows a coordinate-independent formulation of the quantization rules.

EINSTEIN QUANTIZATION

As noted, the old quantum theory was a hybrid theory in which electron motion in the electric field of the nucleus was described by classical mechanics, but only certain discrete values of the classical variables were possible, for reasons that could not be understood within classical mechanics. The discrete values allowed were supposed to be determined by Equation 18.4, which, however, could give different values in different coordinate systems. Hence, Einstein set out to find classical-mechanical quantities that had coordinate-invariant values, which then could be quantized in an invariant manner. Einstein begins with the point, emphasized by Henri Poincaré at the 1911 Solvay Congress, that for a dynamical system $\Sigma p_i dq_i$ is such an invariant under coordinate transformations, and that hence the line integral $\int \Sigma p_i dq_i = \int p \cdot dq$ is also invariant. This quantity is familiar in classical mechanics as the phase space action

(18.5)
$$S(E, \mathbf{r}, \mathbf{r}') = \int_0^t L dt' - Et$$

where L is the usual Lagrangian function of a mechanical system, r', r are the beginning and end points of the trajectory, which requires time t to traverse.

Einstein is seeking to exploit the invariance property of certain line integrals that is well known, for example, in electrostatics. In electrostatics the electrical potential difference (voltage) between two points in space is represented as a line integral of the electric field along a path connecting the two points, and because the electric field is the gradient of this scalar potential, the value of this integral is independent of the path taken between the two points. It follows that the line integral of such a gradient field around any closed loop is zero, because one could just as well choose the path of zero length. Einstein develops an analogy between this situation and the action integral $\int p \cdot dq$ in mechanics, because the momentum in this formula can be derived from a scalar function, the action S of Equation 18.5; in particular $\partial S / \partial \mathbf{r} = \mathbf{p}(\mathbf{r})$, where $\mathbf{p}(\mathbf{r})$ is the momentum at the point $\mathbf{r}$. We can thus regard $\mathbf{p}(\mathbf{r})$ as the momentum "field" of the trajectory, generated by taking the gradient of the scalar function, $S(E, r, r')$. From Bohr and Sommerfeld, we know that for periodic trajectories this line integral must be quantized in units of $h/2\pi$. Since periodic trajectories would give a closed path, if the situation were identical to electrostatics, we would run up against the contradiction that such integrals must be zero. However, the situation in mechanics is subtly different. Although $p(r)$ is indeed the gradient of the scalar "function" $S(r)$, this function is not single-valued and neither is its gradient. This was clear to Einstein because typical trajectories for bound motion loop back on themselves and return to the neighborhood of the point r an infinite number of times (see Plate 18). Therefore the momentum at a point r can have more than one value (and in fact always does), and the action function S generated from integrating a trajectory is a multivalued function of its space coordinates. Far from being a drawback, Einstein recognized that this property avoided the paradox of zero action for closed paths and led to a natural invariant generalization of the Bohr-Sommerfeld quantization condition.

Einstein points out that, when the motion has certain properties (to be specified in the next section), the action S and its associated momentum field can be expressed by regarding $S(r)$ as defined on a doubly connected space (a torus or doughnut). Then he makes the crucial observation that a vector field derived from a scalar field on a torus does *not* have the property that all closed-loop integrals of the field are zero; in fact, for such a field on a torus in d dimensions, there exist exactly d independent irreducible loops that give a specific non-zero value to each line integral around the loop $\int p \cdot dq$ (see Plate 19). These values of course depend on the energy of the dynamical system, so by demanding that each of these invariant integrals are quantized, one arrives at an implicit formula for the quantized energy levels of the system without any reference to specific coordinate systems or to separability of the motion. Thus he proposes the generalized quantization rule:

(18.6) $$\oint_{C_i} p \cdot dq = n_i h, \quad i = 1 \ldots d$$

where C_i are the d irreducible closed loops for such a system with the topology of a torus in d space dimensions, and n_i are the quantum numbers associated with each energy level of the system.

EINSTEIN'S EXAMPLE

Einstein illustrates the construction he proposes with the example of a point mass moving in the plane perpendicular to its conserved angular momentum under the influence of a general central force (i.e., with a potential energy that depends only on the radial distance from the center). The electron moving around a hydrogen nucleus (proton) is an example of such a system; in this case, the force is simply e^2/r^2. Such problems are of course separable in polar coordinates and could be solved by using the generalized Sommerfeld condition, Equation 18.4; however, Einstein found it a convenient example to illustrate the approach. For any such trajectory with non-zero angular momentum, the "particle" will revolve around the center while oscillating in radial distance between its inner turning point r_1 and it outer turning point r_2. As long as the force law is not exactly proportional to $1/r^2$ or r, such orbits typically will not close on themselves. (In fact, they will "precess," in astronomical terminology, a fact of that we can be sure Einstein was aware from his work on the relativistic theory of the precession of the orbit of Mercury!) Therefore, one such orbit will define, after many revolutions, a momentum vector field everywhere in the annular region between the inner and outer turning points (see Plate 19). Inspection of such an orbit reveals that the vector field has exactly two possible values at each point in the allowed region, corresponding to the opposite values of the radial momentum that occur when the orbit passes through the point in question heading outward or inward (the two vector fields are colored red and blue in Plate 19). It is easy to convince oneself that another initial condition for the trajectory *with the same initial angular momentum and energy* would define the same vector field, so this double-valued vector field is unique. Now, Einstein says, one can just as well regard the red vectors as living on the "top" surface of a doughnut and the blue vectors on the "bottom" surface; i.e., one can think of this momentum vector field as being *single-valued* on a torus. One can calculate the momentum field everywhere on the torus if the force law and angular momentum are given; hence, one can calculate any loop integral of the type needed for Equation 18.6. Note here a crucial point: unlike the Sommerfeld rule, for which the integration follows an actual particle trajectory, here the loop integrals of Equation 18.6 are along *any* convenient loop on the torus, which is topologically non-trivial (i.e., it can't be shrunk to a point).

For the central force case that Einstein used as an example, one of these integrals is particularly simple. Consider a loop that travels around the center in a circle, just outside the inner turning radius (the circle of closest approach to the center). On this loop, the momentum p must be purely azimuthal by definition of the turning radius (the radial momentum is zero); dq is by choice azimuthal, so the loop integral is just $|\boldsymbol{p}|2\pi\, r_1 = 2\pi L$, where L is the angular momentum of the motion. From Equation 18.6, in order to quantize the motion, we must set $2\pi L = nh$, i.e., $L = n\,(h/2\pi)$, and we recover the familiar Bohr rule for the quantization of angular momentum for an arbitrary central force law. The integral over the second non-trivial loop (see Plate 19) will give a second condition that, combined with the first one, will quantize the allowed energies of the motion; its precise form will depend on the particular force law.

The Remarkable Insight

Up to this point Einstein's paper can be regarded as an elegant improvement on the Bohr-Sommerfeld quantization rules. However, in the next section Einstein introduces a deep physical insight through which he elucidates the nature and limitations of the old quantum theory in a manner that was only appreciated by physicists fifty years later: "We come now to a very essential point which I carefully avoided mentioning during the preliminary sketch of the basic idea [of Equation 18.6]." He reconsiders the multivalued vector field generated by a trajectory, as just discussed, but no longer assuming a central force law. He excludes the case of periodic closed trajectories as nongeneric, thus he focuses on a generic bound trajectory that passes through each small neighborhood dr of the classically allowed region in coordinate space an infinite number of times: "*A priori*, two types of orbits are possible, obviously of fundamentally different characteristics." Paraphrasing Einstein: An orbit of type A passes through dr an infinite number of times with only a finite number of different momentum directions; an orbit of type B passes through dr an infinite number of times with an infinite number of different momentum directions. In the latter case, the momentum p cannot be represented as a finite multivalued function of r, as it was in the example just discussed. He then remarks: "One notices immediately that type B [motion] excludes the quantum condition we have formulated. . . . On the other hand, classical statistical mechanics deals essentially *only* with type B; because only in this case is the microcanonical ensemble of one system equivalent to the time ensemble."

Here Einstein is referring to a basic principle of statistical physics: for complex and imperfectly known systems with many degrees of freedom, one may calculate their statistical properties, not by following an individual system over time, but by averaging over all similar systems with the same energy (the so-called microcanonical ensemble). This equivalence is referred to as the ergodic hypothesis. Einstein is completely correct that his type A motion does not satisfy the ergodic hypothesis, which is always assumed valid in statistical mechanics. However, by associating type B motion with statistical systems, Einstein might have led a reader to believe that type B motion only arises for many interacting particles and complex dynamical systems. He completes the main body of the article with no further general statements to clarify this point.

Fortunately Einstein appended to the article a note, added in proofs, in which he clearly recognizes that the difficulty with quantizing type B motion does *not* arise just for complex many-particle systems, but actually is associated with what we would now call "chaotic" motion, which occurs in quite simple systems as well. "If there exist fewer than *d* [constants of motion], as is the case, for example, according to Poincaré in the three-body problem, then the p_i are not expressible by the q_i, and the quantum condition of Sommerfeld-Epstein fails also in the slightly generalized form that has been given here." Einstein here has put his finger on the key idea: it is not the complexity or the number of degrees of freedom of the system that matter. The dynamics of systems with fewer constants of motion than degrees of freedom is fundamentally different from that of systems with one constant of motion for each degree of freedom. In modern terminology, the former type of system is non-integrable and the latter, integrable. Integrable systems are now known to be highly exceptional, and the generic case of a non-integrable system always has some regions of phase space where the dynamics are chaotic, i.e., where trajectories are exponentially sensitive to initial conditions. When trajectories are chaotic, they will cover the constant energy surface in phase space uniformly, and a phase space average of some quantity related to the trajectory will equal the time average of the quantity along the trajectory. There has been a tremendous amount of modern research into non-integrable dynamical systems and the implications of non-integrability for producing chaotic motion in the technical sense. Much of this research was aided by computer simulation; it is extraordinary that Einstein had a correct intuitive understanding of the crucial relationship between non-integrability and the ergodicity assumed in statistical mechanics. His specific appeal to the three-body problem showed that he had understood very well Poincaré's proof of non-integrability for this "simple" dynamical system.

But even more impressive, he had leaped ahead of his contemporaries and begun to struggle with the implications of non-integrable motion for quantum theory, coming to the correct conclusion that ergodic motion could not be quantized by any simple generalization of the Bohr-Sommerfeld quantization rules.[10] This was the brilliant insight of his 1917 paper that was completely lost to science for fifty years.

Einstein's Insight from a Modern Perspective

To clarify the relationship between Einstein's type A and type B motion, what is now called regular and chaotic motion, it is helpful to look at the example of dynamical billiards, a class of simple systems that have been extensively studied in this context. A billiard is a two-dimensional region of space in which a point mass can move freely between perfectly reflecting walls, the dynamics being completely determined by the shape of the boundary and the law of specular reflection (the rule that the angle of reflection from the boundary is equal to the angle of incidence). To make a connection with Einstein's example of central force motion, let us first consider a circular billiard of radius R [Plate 20(a)], for which the angular momentum is conserved and given by $L = |\boldsymbol{p}|R\sin\chi$, where χ is the angle of incidence for each reflection from the boundary. Such a billiard can have periodic orbits with the geometry of an inscribed polgyon (e.g., a square, equilateral triangle, etc.), but actually such orbits are so rare (measure zero in the technical sense) that they may be ignored in our general analysis. Generic orbits fill an annular region between the boundary and the inner caustic (the orbit shown in Plate 20(a) would completely fill the ring-shaped region if it were continued for a longer time). Any such *quasiperiodic* orbit will pass through any neighborhood of the allowed region of motion in exactly two directions, and thus the allowed energy levels of a particle confined to a circular billiard can be found (rather accurately) by the method developed by Einstein (described above) and refined by Keller (see discussion below).

In contrast, consider trajectories in the stadium billiard shown in Plate 20(b); all initial conditions (except a set of measure zero) for the stadium have been proven to lead to chaotic motion. From the example shown in Plate 20(b), one sees that a generic neighborhood is traversed in an infinite number of directions, displaying Einstein's pure type B motion. For the stadium billiard, this would be true for any initial position and direction of the point particle. However, based on the theory of the Hamiltonian transition to chaos (developed in the 1940s, 1950s, and 1960s by Andrei Kolmogorov, Vladimir Arnold, Jürgen Moser, and others), a generic billiard shape actually exhibits a mixture of the two types of motion exemplified by the circular billiard [Plate 20(a)] and the stadium billiard [Plate 20(b)]. A typical shape of this type and the resulting dynamics are shown in Plate 20(c,d). For this shape of billiard, known as the quadrupole billiard, *some* initial conditions lead to type A motion and others to type B motion; this situation is referred to as mixed dynamics. For the case of mixed dynamics, a subset of the quantum levels, those corresponding to regular motion, can be quantized by the Einstein method (now known as the Einstein-Brillouin-Keller or EBK method); whereas the subset corresponding to the chaotic region of phase space cannot be quantized by any similar analytic method. There is still (in 2005) *no* method for finding these individual "chaotic" states analytically, although there are well-developed methods for finding the contribution of the "chaotic" states to various quantities that characterize the energy level spectrum statistically (e.g., the density of states).

The question then arises, why is it impossible to use the Einstein method to find chaotic energy levels in closed mathematical form? We know that they exist and can be determined as a set of numbers calculated by a computer solution of Schrödinger's equation (e.g., for the stadium billiard). In fact, the average number of such levels in a given interval of energy is the same for a stadium billiard as for a circular (non-chaotic) billiard with the same area (this is known from a technique in quantum theory called the Weyl expansion). But for reasons to be discussed below, these levels and their associated quantum wavefunctions are too elusive to be determined by analytic methods. In fact, Einstein's insight about the failure of his method (and the Bohr-Sommerfeld method) for type B motion will reappear as the explanation for the elusiveness of chaotic states.

Keller's Improvement of the Einstein Method

Einstein's work, of course, preceded the Schrödinger equation by eight years, and he had no way of knowing that he was in essence trying to find the eigenvalues of a wave equation by means of his quantization condition. I have found no record that he reconsidered his approach after the wave equation was discovered. It is important to understand however that Bohr-Sommerfeld quantization and the more general EBK method constitute important analytic methods for solving Schrödinger's equation to a very accurate approximation. Hence, such approaches, known generically as "semi-classical methods" survive and are used in modern research in quantum physics.

For motion in one dimension (where none of the complexity that Einstein addressed arises), the relationship between the Schrödinger equation and the Bohr-Sommerfeld method of the old quantum theory can be simply demonstrated mathematically. The Bohr-Sommerfeld energy levels are obtained by solving the Schrödinger equation approximately using a technique known as the WKB (Wenzel-Kramers-Brillouin) method. This approximation, familiar in undergraduate quantum physics courses, is based on proposing a trial solution to the Schrödinger equation of the form

$$(18.7) \qquad \psi(x) = A(x)e^{\frac{2\pi i}{h}\int_x^{x_1} p(x')dx'} + B(x)e^{-\frac{2\pi i}{h}\int_x^{x_1} p(x')dx'}$$

where $p(x) = [2m(E - V(x))]^{1/2}$ is the local wavevector in the potential $V(x)$, which is assumed to be slowly varying, and x_1 is either one of the classical turning points in the potential. By insisting that this wavefunction is single-valued, one arrives at exactly the one-dimensional Bohr-Sommerfeld quantization condition, Equation 18.3 (above). Use of the so-called connection formula corrects this condition to include the zero point energy.

The EBK quantization formula, Equation 18.6, can be viewed as arising from the higher-dimensional version of WKB. Specifically, one introduces the same kind of trial wave function

(18.8)
$$\psi(x,y) = \sum_{n=1}^{N} A_n(x,y)e^{ikS_n(x,y)}$$

where I have specialized to the case of two-dimensional billiards discussed above and have defined $k = (2\pi|\mathbf{p}|/h) = (2\pi/h)[2mE]^{1/2}$. The functions $S_n(x,y)$ turn out to be the same as the action $S(\mathbf{r})$ defined in Equation 18.5, except that they have been divided by $|\mathbf{p}|$ to bring out the factor k in front of the exponent. The semiclassical limit is defined by $\lambda \to 0$ or $k = 2\pi/\lambda \to \infty$. Assuming that limit, we insert this trial function into the Schrödinger equation for the billiard and seek solutions to the leading order in λ. This yields two equations: the Hamilton-Jacobi equation for $S_n(\mathbf{r})$ (appropriately scaled) and the continuity equation $\nabla \cdot \mathbf{J} = 0$, where $\mathbf{J}$ is the usual particle current obtained from the wavefunction. The solutions of the two equations can be represented as a set of lines of constant $S_n(x,y)$, which we will call wavefronts, and the associated gradient fields, $\nabla S_n(x,y)$, which are the rays associated with these wavefronts. These rays correspond to the direction in which the particle is moving at that point on the wavefront. We will focus on the simple case of the circular billiard, which can be solved by the EBK method; in this case, one only needs two terms in Equation 18.6. It turns out that the amplitudes are equal, $A_1 = A_2 = A(x,y)$, so one needs to determine the three functions S_1, S_2, and A. The condition $\psi = 0$ on the boundary implies that, on the boundary, $kS_1 = kS_2 - \pi$ (the waves destructively interfere on the boundary) *and* that the particle/rays, ∇S_1, specularly reflect into the particle-rays ∇S_2 everywhere on the boundary. Functions S_1 and S_2, consistent with these requirements, can be found and are shown in Plate 21.

The final requirement is that the wavefunction (Equation 18.8) determined by these wavefronts must be single-valued. This will not hold true for an arbitrary value of the wavevector k; in fact, this requirement is only satisfied if

(18.9)
$$\oint_{C_i} \nabla S \cdot \mathbf{dq} = 2\pi n_i h + \frac{\pi}{2}m_i, \qquad i = 1 \ldots d$$

which is the Einstein quantization condition with a small modification, the presence of an additional "phase shift" on the right-hand side, which is a (fixed) integer times $\pi/2$. To see the connection to Equation 18.6, note that by the de Broglie relation, $p = h/\lambda$, and by the definition $k = 2\pi/\lambda$, we have $p = hk/2\pi$, and Equation 18.9 is just Equation 18.6 divided by Planck's constant h and multiplied by 2π, with the addition of an extra term to be discussed.

Equation 18.9 was derived from the EBK ansatz, Equation 18.8, by Keller in 1958,[11] who had no knowledge of Einstein's 1917 paper. The additional term in the equation arises from leakage of the wavefunction into classically forbidden regions beyond the turning points, the points beyond which conservation of energy forbids a particle to penetrate. The quantum tunneling effect allows some small probability of finding particles beyond these turning points in space; this effect leads to a small shift in their allowed energies from those predicted by Einstein's equation (18.6). Einstein could not have anticipated this effect because it arises from the failure of the old quantum theory near a turning point and involves the tunneling effect that had yet to be understood. Nonetheless we see that Einstein's topological quantization condition (with Keller's improvement) is a completely correct approximate solution of the relevant Schrödinger equation in the limit of short wavelengths, *if solutions of the WKB-EBK type can be found.*

THE PROBLEM OF TYPE B MOTION RETURNS

Where is Einstein's insight lurking to invalidate the WKB-EBK approach when the classical motion is chaotic? It is hidden in a very subtle place, in the sum over N terms in Equation 18.8: one needs to write an EBK trial wavefunction with a finite number N of actions $S_n(r)$ and their associated wavefronts. But imagine starting at an arbitrary point **r** on some wavefront and propagating the perpendicular ray at **r** along it until the ray hits the boundary. The ray will then switch from that set of wavefronts upon specular reflection to another set of wavefronts, and this process will repeat at each subsequent reflection (see Plate 21). The ray under this construction is at the same time tracing out a specific classical trajectory of a particle for the given billiard; hence by the recurrence property mentioned above, the particle/ray will return to the neighborhood of the initial point **r** an infinite number of times as the trajectory is followed for an arbitrarily long time. But by our previous argument, each time the ray passes through the neighborhood of **r**, it must be normal to one of the sets of wavefronts $S_n(r)$ of the EBK solution. If the billiard is one that generates *only* Einstein's Type B motion, then new wavefronts would be generated upon each return, and the required number of terms in the EBK solution would necessarily tend to infinity, implying that no solution of this form can be constructed. If the billiard generates a mixture of Type A and Type B motion [as for the quadrupole shown in Plate 20(c,d)] then some EBK solutions can be found, but other perfectly good solutions of the Schrödinger equation for that billiard are not of the EBK form.

Why does this innocent-seeming EBK expansion sometimes fail and other times succeed? The reason is that the expansion assumes that the wavefronts move in straight lines for distances much larger than the wavelength, i.e., that the radius of curvature of the wavefronts is much larger than the wavelength. This assumption is used explicitly in going from the Schrödinger equation and Equation 18.8 to the equations that determine $S_n(\mathbf{r})$.

Intuitively one can say that the EBK solutions consist of the sum of a fixed number of waves that superpose to form a standing wave. The number of waves in the solution is independent of the wavevector $k = 2\pi/\lambda$, which determines the energy of the solution, and each wave in the superposition looks locally like a plane wave, i.e., the wavefronts appear straight on the scale of the wavelength itself. But there is no reason to expect that all of the actual solutions of the Schrödinger equation *can* be constructed to satisfy this smoothness condition; it is only for certain very special billiard shapes that the smoothness condition can be satisfied. Those special shapes are precisely those in which only Einstein's Type A motion occurs. For any two billiards with the same area, there exists the same *number* of standing wave patterns in each interval of wavelength, but typically many of these patterns are not smooth in the EBK sense. Instead their "wavefronts" (no longer well-defined) bend and twist on the scale of the wavelength. These are the "chaotic" solutions, which must be found numerically. An example of such a solution is shown in Plate 22. In fact the numerical solution essentially consists of superposing a number of "smooth" EBK waves that grows inversely with the wavelength to form the "turbulent" standing wave patterns of the type shown. The fraction of solutions that are "chaotic" in this sense is simply the fraction of the classical phase space in which Einstein's Type B motion occurs.

HISTORICAL AND PHILOSOPHICAL IMPLICATIONS

Einstein's 1917 paper may provide further insight into his struggle with finding an acceptable quantum theory. As already noted, Einstein had immediately returned to quantum questions after completing General Relativity and made historic contributions to the quantum theory of radiation in 1916. Often discussions of Einstein and quantum theory skip the next eight years until his last major work in the area, the theory of the quantum ideal (Bose) gas published in 1924 and 1925. We see from his 1917 paper that the lack of a general set of equations of motion in the old quantum theory worried him, and the fact that the Bohr-Sommerfeld prescription was not coordinate independent was to him a symptom of a deeper problem. The old quantum theory was a flimsy logical framework at best. Microscopic particles were assumed to obey the laws of classical mechanics, except that an infinite number of classical states of motion were forbidden to occur, this restriction being expressed by the Bohr-Sommerfeld quantization condition. In addition transitions could occur discontinuously between the different allowed states of motion, for example, when atoms absorb energy from light. In the 1917 paper herein described, Einstein uncovered another fundamental problem with the Bohr-Sommerfeld condition: it is simply meaningless for most bound classical mechanical systems. It provides no prescription for finding the energy levels at all. This is because the condition, even when elegantly generalized by Einstein himself, only works for integrable (non-chaotic) systems, which are not generic.

Surprisingly, the misconception that Bohr-Sommerfeld (or WKB) quantization is a general method to calculate quantum energy levels persists to the present day in most quantum mechanics textbooks and among many practitioners. This misimpression arises from the tendency to focus on highly symmetric and hence integrable example problems in teaching quantum mechanics. More complex real-world problems are solved either by perturbation theory based on integrable cases or numerically on a computer. Einstein was the first person to find the flaw in the old quantum theory that suggested that it was in principle incapable of determining the energy levels of generic systems. One can only speculate on the effect this conclusion had on his subsequent thinking about quantum theory. It seems possible that it would have deepened his conviction that a completely different theory was required, and that the old quantum theory was not on the right track (although his letter to Paul Ehrenfest in June 1917 does not indicate this[12]). Interestingly, when a logically consistent framework was found in 1926 by Schrödinger, he credited Einstein's 1917 paper as the precursor of the quantization condition that emerges from the Schrödinger equation: "The framing of the quantum conditions [in Einstein's 1917 paper] is the most akin, out of all the earlier attempts, to the present one."[13]

Nonetheless, unlike the EBK equations, Schrödinger's equation does in principle determine the bound state energy levels of a generic system, even though they may be difficult to calculate in practice. It is well known that Einstein was initially supportive of Schrödinger's wave mechanics, perhaps because he saw it as, to some extent, a generalization of his 1917 work in a framework that was now applicable to generic systems.

REDISCOVERING EINSTEIN'S INSIGHT[14]

Einstein's 1917 paper was cited during the twenties by such luminaries as de Broglie[15] and Schrödinger (as just noted). In October 1926, Leon Brillouin presented the general theory of what is now known as the WKB method for solving Schrödinger's wave equation.[16] He points out that the Bohr-Sommerfeld quantization condition and its generalization to conserved actions (basically Einstein's condition, Eq. 18.6) follow from the fact that the wavefunction must be single-valued. However he doesn't reference Einstein's 1917 paper, a surprising omission given its recognition by De Broglie and Schrödinger. References to the paper were made in early texts on quantum physics, although there is no indication that anyone understood the difficulty of quantizing chaos that Einstein identified. Therefore the distinction between chaotic wavefunctions of the type shown in Plate 22 and regular wavefunctions familiar (e.g., from quantum analysis of the hydrogen atom) was unknown in the literature of quantum mechanics. The paper then appears to have been forgotten for roughly twenty-five years.

Keller brought the work back into the citation record in 1958 by referencing it in connection with his own independent derivation of the EBK quantization rule.[17] Keller relates that he heard of the Einstein paper a few years earlier (but after his original work) from Professor Fritz Reiche, then a colleague at NYU, who was a physics instructor at Berlin from 1911 to 1921, and who wrote a well-known early text on quantum physics.[18] However Keller's works in the late 1950s and early 1960s do not mention the potential problem with Einstein's Type B motion and give the impression that the method is of general applicability.[19]

Other than the Keller citations, references to Einstein's paper (now numbering several hundred) only begin to appear in the late 1960s with a resurgence of interest in methods for relating quantum properties to the properties of classical motion. Ian Percival reintroduced the community to the paper in his 1973 paper emphasizing that, for a generic system, there should be regular and irregular ("chaotic") levels.[20] Percival coined the term Einstein-Brillouin-Keller (EBK) quantization for the corrected form of Einstein's condition, Equation 18.9. By 1971, although he was unaware of Einstein's work, Gutzwiller had understood that it is not possible to use conventional Bohr-Sommerfeld–EBK quantization to deal with chaotic systems and introduced an entirely new semiclassical approach.[21] The Gutzwiller approach abandons the attempt to find individual chaotic states, but instead derives an equation to calculate the density of states of a chaotic system from knowledge of the (unstable) periodic orbits of the systems. This prescription is known as the Gutzwiller Trace Formula, and it is challenging to implement. For example, the number of periodic orbits of a chaotic system grows exponentially with the length of the orbits considered, so the number of terms that must be summed is formidable. Much beautiful work has gone into refinements of this approach, and it has been successfully applied to a number of systems in atomic and condensed matter physics.

Gutzwiller's work showed that, despite the absence of a Bohr-Sommerfeld–EBK quantization scheme for chaotic systems, there are strong classical-quantum correspondences even in chaotic systems. These correspondences have led to a new interdisciplinary field of quantum physics, unhappily known as quantum chaos theory. The name is unfortunate since there is no exponential sensitivity to initial conditions (the hallmark of classical chaos) in linear quantum mechanics. Nonetheless the search for semiclassical approximations and classical insights into chaotic quantum systems is a lively and useful endeavor, which has flourished now for over three decades. The literature guide and collection of papers edited by Gutzwiller provides a good selection of the relevant work;[22] fittingly, the first paper in this collection is a translation of Einstein's 1917 paper.

Whereas Einstein's antipathy to certain aspects of modern quantum theory is well known, in this centennial year, there appears to be a renewed appreciation of his seminal contributions to the development of quantum physics. With his introduction of the photon concept in 1905, his clear identification of wave-particle duality in 1909, his founding of the quantum theory of radiation in 1916, and his treatment of the Bose gas and its condensation in 1925, Einstein laid much of the foundation of the theory. We should add to this list of illustrious achievements another advance, modest on the scale of Einstein's genius, but brilliant by any other standards: the first identification of the problem of quantizing chaotic motion.

ACKNOWLEDGMENTS

I gratefully acknowledge the assistance of Hakan Tureci and Harald Schwefel in preparing the graphics for this article and for helpful discussions of the EBK method. I thank Martin Gutzwiller for sharing his many insights into the history and physics discussed in this article. I thank Joseph Keller for recounting to me the history related to his rediscovery of the Einstein paper, and Walter Isaacson for making me aware of Einstein's letters to Paul Ehrenfest on this subject.

CHAPTER EIGHTEEN

19

EINSTEIN AND THE QUEST FOR A UNIFIED THEORY

David Gross

MY TOPIC IS AT THE HEART OF Einstein's scientific life, the search for a unified theory of nature. This was Einstein's main pursuit for more than half of his scientific career. Most contemporaries viewed his attempts as a waste of time, a total failure, or at best, premature, but today we look with some admiration at his foresight. Having understood by the mid-1970s, to a large extent, the four forces of nature in the remarkable successful standard model, the scientific community has returned its attention to Einstein's dream of unifying all the forces with gravity. The goal of unification has been at the forefront of fundamental physics for the last three decades.

In this chapter, fully aware of the ease of hindsight, I shall discuss Einstein's goals: his attempts to unify general relativity and electromagnetism, and to include matter. I shall discuss his mistakes, ask why he went wrong, and wonder what might have happened if he had followed a slightly different route. I shall then discuss, very briefly, where we stand today in realizing Einstein's goals.

For many physicists—certainly for me—Einstein is both a hero and a model. He stated the goals of fundamental physics, that small subspecialty that probes the frontiers of physics in a search for the underlying laws and principles of nature. Einstein was a superb epigramist who could capture in a single sentence many deep thoughts.

Here is his definition of the goal of the physicist: "The supreme test of the physicist is to arrive at those universal laws of nature from which the cosmos can be built up by pure deduction."[1] I love this sentence. In this one sentence, Einstein asserts the strong reductionist view of nature: There exist universal, mathematical laws which can be deduced and from which all the workings of the cosmos can (in principle) be deduced, starting from the most elementary laws and building up to the most complex.

More than any other physicist, Einstein was untroubled by either quantum uncertainty or classical complexity and believed in the possibility of a complete, perhaps final, theory of everything. He also believed that the fundamental laws and principles that would embody such a theory would be simple, powerful, and beautiful. The "old one," to whom Einstein often referred, has exquisite taste.

This exciting goal, which I first learned of when I was thirteen by reading popular science books, seemed to me so exciting that I vowed to become a theoretical physicist. Although I certainly had no idea what that meant, I did know that I wanted to spend my life tackling the most fundamental questions of physics. This goal led me to elementary particle physics in the 1960s and to string theory in the 1980s. This goal motivated Einstein to spend the last thirty years of his life in a futile search for a unified theory of physics.

Physicists are an ambitious lot, but Einstein was the most ambitious of all. His demands of a fundamental theory were extremely strong. If a theory contained any arbitrary features or undetermined parameters, then it was deficient, and the deficiency pointed the way to a deeper, more profound, and more predictive theory. There should be no free parameters, no arbitrariness.

"Nature," he stated with confidence, "is constituted so that it is possible to lay down such strong determined laws that within these laws only rationally, completely determined constants occur, not constants therefore that could be changed without completely destroying the theory."[2] This is a lofty goal, under threat nowadays from those who propose the anthropic principle, whereby many of the fundamental constants of nature, even some of the laws, are environmental and therefore might be different in different parts of the universe. For me and for many others, however, the unified and completely predictive theory remains the ultimate goal of physics, and a guiding principle. A theory that contains arbitrary parameters or, worst of all, arbitrarily fine-tuned parameters, is deficient.

After his enormous success at reconciling gravity with relativity, Einstein was troubled by the remaining arbitrariness of the theoretical scheme. First, the separate existence of gravitation and electromagnetism was unacceptable. According to his philosophy, electromagnetism must be unified with general relativity, so that one could not simply imagine that it did not exist. Furthermore, the existence of matter, the mass and the charge of the electron and the proton (the only elementary particles recognized back in the 1920s), were arbitrary features. One of the main goals of a unified theory should be to explain the existence and calculate the properties of matter.

Before passing to a discussion of Einstein's attempts at unification, I wish to make a remark concerning his work on special relativity in 1905, whose centenary we celebrate this year. One of the most important aspects of this work was to revolutionize how we view symmetry. Principles of symmetry have dominated fundamental physics in the 20th century, starting with Einstein in 1905.

Until the 20th century, physicists did not often consciously consider principles of symmetry; those principles were in play, but not consciously considered. The Greeks and others were fascinated by the symmetries of physical objects and believed that these would be mirrored in the structure of nature. Kepler attempted to impose his notions of symmetry on the motion of the planets. The laws of mechanics embodied symmetry principles, notably the principle of equivalence of inertial frames, or Galilean invariance.

The symmetries implied conservation laws. Although these conservation laws, especially those of momentum and energy, were regarded to be of fundamental importance, they were regarded as consequences of the dynamical laws of nature rather than as consequences of the symmetries that underlay these laws. Maxwell's equations, formulated in 1865, embodied both Lorentz invariance and gauge invariance. But these symmetries of electrodynamics were not fully appreciated for over forty years.

This situation changed dramatically in the 20th century, beginning with Einstein. Einstein's great advance in 1905 was to put symmetry first, to regard the symmetry principle as the primary feature of nature that constrains the allowable dynamical laws. Thus the transformation properties of the electromagnetic field were not to be derived from Maxwell's equations, as Lorentz did, but rather were consequences of relativistic invariance, and indeed largely dictate the form of Maxwell's equations. This is a profound change of attitude. Lorentz, who had derived the relativistic transformation laws from Maxwell's equations, must have felt that Einstein cheated. Einstein recognized the symmetry implicit in Maxwell's equations and elevated it to symmetry of space-time itself. This was the first instance of the *geometrization* of symmetry and the beginning of the realization that symmetry is a primary feature of nature that constrains the allowed dynamical laws.

The traditional symmetries discovered in nature were global symmetries, transformations of a physical system in a way that is the same everywhere in space. Global symmetries are regularities of the laws of motion, but are formulated in terms of physical events; the application of the symmetry transformation yields a different physical situation, but all observations are invariant under the transformation. Thus global rotations rotate the laboratory, including the observer and the physical apparatus, and all observations remain unchanged.

Gauge or local symmetry is of a totally different nature. Gauge symmetries are formulated only in terms of the laws of nature; the application of the symmetry transformation merely changes our description of the same physical situation, but does not lead to a different physical situation. Today we realize that local symmetry principles are very powerful—they dictate the form of the laws of nature.

In 1912–17, this point of view scored a spectacular success with Einstein's construction of general relativity. The principle of equivalence, a principle of local symmetry—the invariance of the laws of nature under local changes of the space-time coordinates—dictated the dynamics of gravity, of space-time itself. Fifty years later, gauge theories—invariant under local symmetry transformations, not of space-time but of an internal space of particle labels—assumed a central position in the fundamental theories of nature. They provide the basis for the extremely successful standard model, a theory of the fundamental, nongravitational forces of nature—the electromagnetic, weak and strong interactions.

Surprisingly, Einstein did not follow the symmetry route. He did not, in his attempts to unify physics, search for extensions of the symmetries that he had promulgated. If he had, he might very well have discovered non-Abelian gauge theory or perhaps even supersymmetry. Why not follow this route that has dominated theoretical speculation in the latter half of the 20th century? I think the reason was that Einstein was unaware of the phenomenon of symmetry breaking. All of the new symmetries discovered in the latter half of the 20th century, which are at the heart of the standard model of particle physics and attempts at unification, are approximate, or are broken spontaneously, or hidden by confinement. It was only in the 1960s and early 1970s that these mechanisms of breaking symmetry were elucidated, and the possibility of imagining new symmetries—not directly manifest in the world, but still dictating its dynamics—was possible.

For Einstein, the existence, mass, and charge of the electron and the proton, the only elementary particles recognized back in the 1920s, were arbitrary features. One of the main goals of a unified theory should be to explain the existence and calculate the properties of matter. When he contemplated his equation, he distinguished between the left-hand side of the equation, which was a beautiful consequence of the profound symmetry of general coordinate transformations and which captures the curvature of space-time, and the right-hand side, which was the source of curvature (mass) but had to be arbitrarily put in with no principle to determine the properties of mass. As in politics, Einstein greatly preferred the left to the right. To quote Einstein: "What appears certain to me, however, is that, in the foundations of any consistent field theory the particle concept must not appear in addition to the field concept. The whole theory must by based solely on partial differential equations and their singularity-free solutions."[3]

So Einstein's three goals were:

1. Generalize general relativity to include electromagnetism.

2. Eliminate the right-hand side of his equations, and deduce the existence of matter by constructing singularity-free solutions that would describe stable lumps of energy.

3. And finally, since he abhorred the arbitrary nature of the quantum rules and their probabilistic interpretation, deduce quantum rules from these nonsingular solutions.

He imagined that the prohibition of singularities in the solutions that would describe matter would lead to over-determined equations, whose solutions would exist only for some quantized values of physical parameters, say the radii of electron orbits. Thus he could imagine reproducing the Bohr model of the atom. The core of this program was to include electromagnetism and derive the existence of matter in the form of what we call today solitons. As Einstein understood, nonlinear equations can possess regular solutions that describe lumps of energy that do not dissipate. Thus, one could start with the nonlinear field equations of general relativity and find localized particles. This was his hope:

> If one had the field equation of the total field, one would be compelled to demand that the particles themselves would everywhere be describable as singularity free solutions of the completed field equations. Only then would the general theory of relativity be a complete theory.

As far as I can tell, Einstein knew of no example of solitons or any toy model that exhibited his hopes. Nonetheless, flushed with the success of general relativity, with the faith that electromagnetism had to be unified, that matter needed a reason for its existence, he studied the equations and tried to modify them as well, with the hope of finding such solutions and with the dream that quantization of mass and charge, and even the quantum rules, would emerge from over-determination.

Among all of the extensions of general relativity considered and pursued by Einstein, the idea that the other forces of nature could be reflections of gravity in higher dimensions was the most innovative and enduring. It was not Einstein's idea, but rather that of Theodor Kaluza in 1922, significantly developed by Oscar Klein in 1926. Kaluza and Klein showed that if one assumed general relativity in five dimensions, one of which was curled up, the resulting theory would look like a four-dimensional theory of electromagnetism and gravity. Electromagnetism emerged as a consequence of gravity in five dimensions.

Einstein was immediately attracted to this idea and wrote to Kaluza: "The idea of achieving [a unified field theory] by means of a five-dimensional cylinder world never occurred to me and may well be altogether new. At first glance your idea has great appeal for me."[4] He held this paper for two years before submitting it to be published, probably because he was confused, as was Kaluza, about whether or not the fifth dimension was real. Einstein returned again and again to this idea for over thirty years.

Finally in 1938, Einstein and Peter Bergmann gave the best reasoning for taking the fifth dimension seriously, and argued that it is consistent with observation if it is sufficiently small. Klein had identified the momentum of particles moving around the fifth dimension as electric charge, which is quantized if one assumes the quantum mechanical rules of momentum quantization on circle. In modern versions of Kaluza-Klein, as they appear in string theory, this scenario is greatly amplified. In string theory, there are six or seven extra spatial dimensions. One can imagine that these are curled up to form a small manifold, and remarkably such six-dimension compactifications (achieved by solving the generalization of Einstein's equations in ten dimensions) can produce a world remarkably like our own, in which the shape of the extra dimensions determines the complete matter content and all the forces of nature, as seen by a four-dimensional observer.

Why did Einstein not consider higher dimensional spaces? Much later he did, for a while, play with an eight-dimensional universe, a kind of complexification of Minkowski space—an approach severely criticized by Wolfgang Pauli and rapidly dropped by Einstein. But why didn't he search systematically for higher-dimension theories? If he had done so, he might have discovered non-Abelian gauge theories, much as Oscar Klein almost did in 1938. We cannot know, but I suspect that part of the reason was that Einstein by and large ignored the nuclear forces. His goal was to incorporate electromagnetism together with gravity—for this, one extra dimension sufficed.

Einstein never thought much of this quantization of electric charge. Perhaps he thought, as indeed Klein tried, to turn this around and derive the quantum rules from the quantization of charge. But in any case, Einstein's main goal was to find particles as nonsingular solutions of his equations, so he turned immediately to trying to find nonsingular solutions of Kaluza-Klein theory.

Over the years Einstein came back again and again to this problem and tried to find nonsingular solutions of Kaluza-Klein theory. He published at least three papers in which he proved that such solutions do not exist, with ever increasing generality. The last of these was a paper published with Pauli, who spent part of World War II in Princeton. The remark made in this paper— "When one tries to find a unified theory of the gravitational and electromagnetic fields, he cannot help feeling that there is some truth in Kaluza's five-dimensional theory"[5]—expressed how much Einstein was attracted to this approach. He must have been incredibly disappointed that he could not find matter as solitons in this theory.

But Einstein was wrong. There do exist solitons, nonsingular solutions of his equations in Kaluza-Klein theory, that behave as particles—magnetic monopoles with quantized magnetic charge. These were discovered in the early 1980s by Malcolm Perry and me, and independently by Rafael Sorkin, when Kaluza-Klein theory was revived. In our paper we added the footnote pointing out that these solutions contradicted Einstein. The referees suggested that we remove the footnote because it was disrespectful. We, of course, refused—how could we resist?

I have wondered what would have happened if these solutions had been discovered back in the 1920s. They could have been. It would have given an enormous boost to Einstein's program, even though the solitons were magnetic and not electric, and very massive. But this did not happen, and Einstein's attempts to find nonsingular solutions failed, as did his attempts to construct satisfactory unified theories.

Beginning sometime in the late 1920s, Einstein became more and more isolated from the mainstream of fundamental physics. To a large extent, this resulted from his attitude toward quantum mechanics, the field to which he had made so many revolutionary contributions. Einstein, who understood better than most the implications of the emerging interpretations of quantum mechanics, could never accept it as a final theory of physics. He had no doubt that it worked, that it was a successful as an interim theory of physics, but he was convinced that it would eventually be replaced by a deeper, deterministic theory. His main hope in this regard seems to have been that, by demanding singularity-free solutions of the nonlinear equations of general relativity, one would get an over-determined system of equations that would lead to quantization conditions.

Because of his opposition to quantum mechanics, he allowed himself to ignore most of the important developments in fundamental physics for over twenty-five years, as he himself admitted in 1954: "I must seem like an ostrich who buries its head in the relativistic sand in order not to face the evil quanta."[6] If there is one thing that I fault Einstein for, it is his lack of interest in the development of quantum field theory. To be sure, many of the inventors of quantum field theory were soon to abandon it when faced with ultraviolet divergences, but it is hard to understand how Einstein could not have been impressed with the successes of the marriage of his children, quantum mechanics and special relativity. The Dirac equation and quantum electrodynamics had remarkable successes, especially the prediction of antiparticles. How could Einstein not have been impressed?

The only way to understand this is if general relativity was so important to him that it eclipsed everything else. As Pauli remarked, "If we would have presented Einstein with a synthesis of his general relativity and the quantum theory—then the discussion with him would have been considerably easier."[7] But since general relativity and quantum mechanics seemed so incompatible, a situation that continued until quite recently, he felt free to ignore the exciting advances that were made in special relativistic quantum mechanics.

I turn now to the situation today, or more precisely thirty years ago, after the completion of the standard model of elementary particle physics, where we now have direct evidence of Einstein's dream, the unification of all forces.

One the most important implications of asymptotic freedom is the insight it gave into the unification of all the forces of nature. Almost immediately after the discovery of asymptotic freedom and the proposal of quantum chromodynamics, the first attempts were made to unify all the forces. This was natural, given that one was using very similar theories to describe all the known interactions. The apparently insurmountable barrier to unification—namely, the large difference in the strength between the strong and the electro-weak force—was seen to be a low-energy phenomenon. Since the strong force decreases with increasing energy, all forces could have a common origin at very high energy. Indeed, the couplings run in such a way as to merge about 10^{14} to 10^{16} Gev, close to the point at which gravity becomes equally strong. This is our most direct clue to where the next threshold of fundamental physics lies, and hints that at this immense energy, all the forces of nature, including gravity, are unified.

In more recent times this extrapolation has greatly improved by the beautiful measurements of many experimenters and the hard work done by many theorists. Now the forces can all meet only if we hypothesize a new space-time symmetry-supersymmetry—and if this new symmetry is broken at reasonably low energy, thereby increasing hopes that a new super-world will be revealed at the Large Hadron Collider, soon to be completed at CERN. Supersymmetry is a beautiful, natural, and unique extension of the relativistic and general relativistic symmetries of nature. Einstein would have loved it. It can be thought of as the space-time symmetries of super-space, a space-time with extra dimensions. But the extra dimensions, here denoted collectively by θ, are measured with anticommuting numbers. These are generalizations of ordinary real numbers, much as imaginary or complex numbers are, but because they anticommute, multiplication depends on the order: thus $\theta_1\theta_2 = -\theta_2\theta_1$. If it is hard to imagine a space of four or more dimensions, super-space is even weirder, but mathematically, totally consistent. A theory formulated in super-space, and invariant under transformations or rotations of super-space, has many beautiful and appealing features. Supersymmetric extensions of the standard model can solve many important problems—such as: Why is there such enormous disparity, at low energy, between the strength of the gravitational force and the other forces of nature? The discovery of supersymmetry, which we all hope to see in a few years at the Large Hadron Collider, would be tantamount to the discovery of quantum dimensions of space-time.

Perhaps the most important feature of the extrapolation of the standard model forces is that the energy at which they appear to unify is very close, if not identical, to the point at which gravity becomes equally strong. This indicates that the next stage of unification should include, as Einstein expected, unification of the nongravitational forces and gravity. This is an important clue to that unification because it is not easy to quantize general relativity. A straightforward quantization of Einstein's theory does not work; at the characteristic distance scale of gravity, where the force becomes strong, the quantum fluctuations of the metric are too violent and uncontrollable. It seems inescapable that Einstein's theory is only an effective theory, adequate at long distances, but to be replaced by a more fundamental theory at the Planck scale of 10^{-33} cm.

Luckily such an extension of general relativity is available—string theory. String theory was not invented to describe gravity; it originated in an attempt to describe the strong interactions, wherein mesons can be thought of as open strings with quarks at their ends. The fact that the theory automatically described closed strings as well, and that closed strings invariably produced gravitons and gravity, and that the resulting quantum theory of gravity was finite and consistent, is one of the most appealing aspects of this theory. String theory is a theory in development; we have learned much about it in the last decades, but much more remains. What has been achieved so far?

First, string theory is a consistent logical extension of the conceptual framework of fundamental physics. Such an extension is not easy, and it is rare. Second, string theory provides us, for the first time, with a consistent and finite quantum theory of gravity. This not only proves that quantum mechanics and general relativity are mutually compatible, it also provides us with the tools to explore many of the paradoxical issues that arise when the metric of space-time is quantized. Already, string theory has clarified many of the mysteries of black holes. Thus the suspicion raised by Hawking in fundamental physics—Do black holes indicate a loss of information?—has been dispelled, even to the point where Hawking himself has agreed that information is not lost in the process of the formation and evaporation of black holes.

Finally, string theory has a rich structure that could yield a theory that unifies all of the forces of nature and explains all the constituents of matter. It automatically contains gravity as well as the gauge theories of the standard model. Certain of its four-dimensional compactions give rise to low-energy dynamics remarkably similar to the standard model.

But string theory is a work in progress, and although it has produced many surprises and lessons, it still has not broken dramatically with the conceptual framework of relativistic quantum field theory. Many of us believe that ultimately string theory will give rise to a revolution in physics, as important as the two revolutions of the 20th century, relativity and quantum mechanics. These revolutions are associated with two of the three fundamental dimensionful parameters of nature, the velocity of light and Planck's constant. The revolution in string theory presumably has to do with Newton's constant, which defines a length, the Planck length of 10^{-33} cm. String theory, I believe, will ultimately modify in a fundamental way our concepts at these distances.

Where will the revolution take place? I believe that it will involve our understanding of the nature of space-time, a subject dear to Einstein's heart. To quote some leading string theorists:[8]

Space and time may be doomed. —Edward Witten

I am almost certain that space and time are illusions.—Nathan Seiberg

The notion of space-time is clearly something we're going to have to give up. —Andrew Strominger

The real change that's around the corner is in the way we think about space and time. We haven't come to grips with what Einstein taught us. But that's coming. And that will make the world around us stranger than any of us can imagine.—David Gross

Why is space-time doomed? Among the many reasons, this: In string theory we can change the dimension of space-time by changing the strength of the string force. Thus, the so-called II-A string theory, which semiclassically describes closed strings moving in ten-dimensional flat space for very weak coupling, is equivalent, or dual to, to another theory, M-theory, for strong coupling, which at low energies is described by eleven-dimensional supergravity. By increasing the string coupling, we can grow an extra dimension. How can the spatial continuum be fundamental if the number of spatial dimensions can be so changed?

We can continuously tear the fabric of space. Thus a string theory solution that describes strings moving on a background wherein some of the spatial dimensions are compactified on a manifold M_1 can be continuously deformed, by varying some of the parameters of the solution, to one that describes the strings moving on a background M_2 of different topology. In between, there is no such simple description of the solution as strings moving on a geometric background, but the deformation is continuous and the strings do not mind at all that the fabric of space has been torn and the topology modified. Again this suggests that the spatial continuum cannot be fundamental if its topology can be changed in so smooth a fashion.

On the other hand, in string theory we cannot probe arbitrarily small distances. In string theory, we can ask, what is the smallest distance that can operationally be explored, analyzing (as Heisenberg did in the case of quantum mechanics) how a microscope works. In string theory, the light rays of a microscope are really strings. Consequently, as we increase the energy of the light, to overcome the quantum mechanical uncertainty in the measurement of distance, the strings expand and prevent us from resolving arbitrarily small distances. The minimum distance that we can explore is, not surprisingly, on the order of the Planck length.

We also cannot squeeze spatial volumes to zero size. If one of the spatial dimensions is compactified to form a circle of radius R, it turns out that string theory in this background is identical to string theory in a background where the radius of this circle is 1/R (in Planckian units). Thus if we try to squeeze this dimension and reduce R to zero, we find that the more natural description is in terms of the dual theory, and the minimal size of the compact circle is finite and on the order of the Planck length.

These phenomena suggest that there is no operational meaning to distances smaller than the Planck length, and that the spatial continuum should be replaced by something else. I believe that space, for sure—and presumably time as well—will be emergent. We already have many hints and examples where space is an emergent concept. These include the famous AdS/CFT duality, wherein string theory in ten dimensions with a background geometry of five-dimensional Anti–de Sitter space times a five sphere is dual to a supersymmetric gauge in flat four-dimensional space-time. Six spatial dimensions emerge from the gauge theory description, together with gravity. We have no understanding, however, what it would mean for time itself to be an emergent concept.

I like to depict our confusion in poetic form. Democritus expressed, some 2500 years ago, the atomic hypothesis:

> By convention there is color,
> by convention sweetness,
> by convention bitterness,
> But in reality there are atoms and space.

I say: We are convinced that

> By convention there is space,
> by convention there is time,
> but in reality there is . . .

The problem is that I do not know how to finish the verse.

So did Einstein go wrong in the latter part of his life? The answer is both yes and no. Yes—he refused to accept quantum mechanics, and he ignored the developments in nuclear and particle physics. These mistakes ensured his failure, but they are quite understandable and forgivable. No—he knew that gravity must be unified with the other forces. And this we, too, know is the central issue in fundamental physics today.

And for those of us faced with the fact that we cannot yet directly probe the Planck scale, he believed in the possibility of successful speculative theory. As Einstein stated: "The successful attempt to derive delicate laws of nature, along a purely mental path, by following a belief in the formal unity of the structure of reality, encourages continuation in this speculative direction, the dangers of which everyone vividly must keep in sight who dares follow it."[9]

To all physicists, but especially to those working in speculative areas, Einstein remains an inspiration for his foresight and his unyielding determination and courage.

20

ENERGY IN EINSTEIN'S UNIVERSE

Lisa Randall

IN HIS ANALYSIS OF A 1950 interview with Einstein, Gerald Holton described how "Einstein . . . advises the historian of science to leap across the unavoidable gap between the necessarily too limited 'facts' and the mental construct that must be formed to handle the facts." This is fortunate for me, since I'm a physicist and not a historian of science. But I will attempt this leap to consider how our understanding of energy evolved under Einstein's tutelage.

The notion of energy is critical to all of physics. Yet the categories of energy have continually evolved over time. What we really know is that energy is carried by whatever carries energy.

Einstein expanded the known forms of energy, showing that mass is a form of energy—a logical consequence and inevitable conclusion, but even Einstein found it radical and startling. And after his theory of general relativity, Einstein recognized that energy can exist in a vacuum—the state of the universe with no particles or matter present. It's a logical possibility, and we'll see that, although Einstein didn't quite get it right, such energy exists. But the universe coyly twisted Einstein's logic.

Let's consider the first of these two insights: that mass is a form of energy. Einstein explained to his friend Conrad Habicht in 1905 (a few months after developing special relativity), that "One more consequence of the electrodynamical paper has also occurred to me. The principle of relativity, together with [James] Maxwell's equations, requires that mass be a direct measure of the energy contained in a body; light transfers mass." However, Einstein went on to say, "The argument is amusing and attractive; but I can't tell whether the Lord isn't laughing about it and playing a trick on me."

It wasn't a coincidence that Einstein's paper on relativity was overtly about electrodynamics. Maxwell and Heinrich Hertz in the mid to late 1800s had developed electromagnetic theory. Subsequently, the question of mass as electromagnetic energy had been investigated, even before 1905. Hendrik Lorentz derived the Lorentz contraction and related effects from a microscopic theory of charge interacting with the ether. From this, he derived a velocity-dependent mass. However, the ether and a special frame of reference was critical to his way of thinking.

The physicist Max Abraham was among the principal proponents of the fashionable pursuit of deriving mass as a consequence of electromagnetic energy. Abraham had an electromagnetic world picture in which he tried to derive mechanical properties from electromagnetic ones, for example, derive the mass of the electron from the energy stored in the electromagnetic field.

Einstein's starting point was quite different. His goal was a consistent theory of the "electrodynamics of moving bodies" and relied on very different assumptions to get a theoretically satisfying theory that agreed with experimental data. He assumed two fundamental principles, the constancy of the speed of light and the principle of relativity (that physics is the same for all inertial observers).

Einstein's result implied that mass was equivalent to all forms of energy and did not assume a special role for electromagnetism. This was very different in spirit from the speculations of his contemporaries. And for Einstein, mass was related to energy through c^2, the square of the speed of light.

Einstein was (rightly) proud of his idea. When Johannes Stark mistakenly attributed the relation between mass and energy to Max Planck, Einstein remarked, "I was rather disturbed that you do not acknowledge my priority with regard to the connection between inertial mass and energy." However, later on Einstein relented somewhat and said, "People . . . should not allow pleasure in their common work to be clouded by such matters."

Einstein's first indication of the mass-energy equivalence came from studying the motion of an electron in an electromagnetic field. He could calculate the change in an electron's kinetic energy through the work done by the field to accelerate it from rest to velocity (speed and direction) v, which was $K = m_0c^2(1/\sqrt{(1 - v^2/c^2)} - 1)$. This tells us that $E = m_0c^2(1/\sqrt{(1 - v^2/c^2)})$, which seems to argue for the equivalence of mass and energy. Although this last step, which was key to mass-energy equivalence, is true, it's not clear to me that Einstein saw this as inevitable from only this observation. After all, kinetic energy (energy of motion) is the energy with the mass subtracted away.

But of course Einstein did deduce the equivalence soon afterward. Let's consider his proof: Start with an object of energy E_0 that emits plane waves of energy $L/2$ at an angle Θ with respect to the x-axis and an equal amount in the opposite direction. Now look at the same process in a reference frame that moves with velocity v with respect to the above (at rest) system. Use the relativistic relation $L^* = L\,(1 - v/c\cos\Theta)/\sqrt{(1 - (v/c)^2)}$ to derive the energy of the waves in the new frame of reference.

In the initial frame of reference, conservation of energy gives us $E_0 = E_1 + L/2 + L/2$, whereas in the moving frame of reference, we get $H_0 = H_1 + L/\sqrt{(1 - (v/c)^2)}$. The differences $E_0 - E_1$ and $H_0 - H_1$ measure kinetic energy. So the difference in kinetic energy in the system and rest versus the moving system is $K_0 - K_1 = L(1/\sqrt{(1 - (v/c)^2)} - 1)$.

Einstein's first observation was that his difference in kinetic energy has only to do with the light. His second observation came from doing a power series expansion for small velocity to see the difference, which he noticed had the form of kinetic energy of the electron: $K_0 - K_1 = L/c^2(v^2/2)$. This means that the body releases energy L as radiation, but its mass has decreased by L/c^2!

Of course, the result is really more general. The form of the emitted energy is irrelevant.

All forms of energy are related to mass. Einstein was as startled as anyone by his remarkable result, but nonetheless said, "If the theory agrees with the facts, then radiation transmits inertia between emitting and absorbing bodies."

You might wonder how so basic a fact as the equivalence between mass and energy had been missed for so long. The reason is that the speed of light factor tells us that a large amount of energy corresponds to a very tiny mass. For example, all the energy we use in a year is equivalent to only a few tons' worth of matter. The changes in mass due to kinetic energy are very difficult to measure.

However, that's not true for highly accelerated particles, where Einstein's observation is critical. Even in a TV receiver, electrons are accelerated sufficiently to increase their mass by a few percent.

Though difficult to measure, it was nonetheless important to test the consequences of Einstein's theory. In the early 1900s, the Bonn physicist Walter Kaufmann had already done measurements that in principle should have yielded such a test. He had measured the paths of electrons in an electric and magnetic field. By studying the bending of the path, he could determine the electron's apparent mass.

At first Kaufmann didn't trust his assumptions. But Abraham had already developed a theory in which the electron mass was determined electromagnetically. And Abraham's theory was in agreement with Kaufmann's results. However, they were not in agreement with Einstein's theory. This meant that if Kaufmann was right, Einstein was wrong!

What were the consequences of this measurement for Einstein and his contemporaries? Einstein's work gave the same velocity-dependent mass as Lorentz' theory. Lorentz had developed an electromagnetic theory that explained electromagnetic data in detail. In fact, Einstein's theory was originally (mistakenly) considered a generalization of Lorentz' work. Lorentz' reaction to Kaufmann's data was disappointment: "Unfortunately my hypothesis of the flattening of electrons is in contradiction with Kaufmann's results, and I must abandon it."

Einstein's reaction was very different. In his 1907 review of relativity, Einstein discusses Kaufmann's result and is skeptical. He's also extremely dubious about the theory it would support. In his words, "In my opinion, however, a rather small probability should be ascribed to these theories, since their fundamental assumptions about the mass of a moving electron are not supported by theoretical systems that embrace wider complexes of phenomena."(!)

I find this a fascinating episode because it so clearly demonstrates Einstein's attitudes toward scientific advances. As Einstein says, "The reciprocal relationship of epistemology and science is of a noteworthy kind. They are dependent upon each other."

"The first point is obvious; the theory must not contradict experimental facts. However evident this demand may in the first place appear, its application turns out to be quite delicate. . . . The second point of view [is concerned with] the premises of the theory itself, with what may briefly but vaguely be characterized as the 'naturalness' or 'logical simplicity' of the premises."

Max Planck had looked at Kaufmann's results and concluded the data were not sufficiently accurate to decide which theory, Einstein's or Abraham's, was correct. Einstein came to the same conclusion. Einstein evaluated theories as critically as he did experiments. He didn't believe a nonsensical theory, even if the data seemed to support it, so he was happy to dismiss Abraham's theory. I admire his confidence.

Sure enough, when Kaufmann's Bonn colleague, Albert Bucherer, measured the velocity-dependent electron mass correctly in 1908, he drew the following conclusion: "by means of carefully experiments, I have elevated the validity of the principle of relativity beyond any doubt."

Actually, that wasn't the end of the story.

Neumann in 1914 addressed an objection, but Charles Thomas Zahn and A. H. Spees in 1938 showed Bucherer's velocity measurement wasn't really any better than Kaufmann's!

However, Einstein's results were accepted by this point, so this observation didn't have any influence.

Just to finish the story: Einstein said $E = mc^2$ was experimentally demonstrated in 1932 when Ernest Walton and John Cockcroft split the atom. They accelerated protons and shot them at lithium metal. The proton merged with lithium and created two helium nuclei. They measured the energy gained (in motion of helium nuclei) and found that it agreed with the mass that was lost.

Of course, we now know $E = mc^2$ is correct, and that it has spectacular consequences. For example, nuclear processes (fueled by mass converting to energy) sustain the burning of the sun. In fact, Charles Darwin realized life was hundreds of millions of years old, even when Kelvin, the physicist, thought it could be no more than thirty million years old. The conflict was resolved only when the sun's true source of energy from nuclear conversion of mass to energy became clear.

$E = mc^2$ also explains the creation of chemical elements (energy added to nuclei during supernovae). And we can understand radioactivity as the reverse process. Furthermore, in the Big Bang picture of the universe, which describes a universe that is initially very hot, particles and antiparticles were constantly created and destroyed as mass and energy were interconverted. Today this happens at particle accelerators, where mass is converted to energy when particles are annihilated and then transformed back into the mass of heavy particles. Moreover, energy accounts for most of the mass of known matter. The mass of protons and neutrons actually arises primarily from the energy stored in the strong nuclear interactions. And of course, we have nuclear energy and nuclear bombs.

But there are further scientific implications. This understanding of the connection of mass and energy is critical to another of Einstein's major accomplishments, the theory of general relativity. At the time of Einstein's work, physicists were trying to show the relationship between mechanics and electromagnetism. Einstein not only showed that such an endeavor would be fruitless, but he developed relativity, which now underlies any theory of either. With relativistic principles, Einstein showed how energy and momentum create a gravitational field (playing the role previously occupied by mass alone).

He arduously developed general relativity, the culmination of this understanding, between 1907 and 1915. In his theory, he related the *metric*, describing the geometry of space-time, to the gravitational field arising from any energy and momentum present. He showed that gravity could be encoded in the geometry of space-time. Mass and energy distort space-time, and the distorted space-time in turn accounts for gravitational effects on matter moving through it.

In 1917, with general relativity and knowledge about matter and energy in the universe, Einstein could derive the equations that tell the metric of the space-time of the universe as a whole. The solution he found was time-dependent, which Einstein didn't like: "The most important fact we draw from experience is that the relative velocity of the stars are very small as compared with the velocity of light." According to well-known lore, Einstein then added the cosmological (which he called universal) constant, to find a static solution.

In fact Einstein had identified a new form of energy not carried by matter and not the result of a force. It was a bold assumption: the existence of energy, but nothing carrying it. That is, there was energy and momentum, but *no* quanta responsible for them. The cosmological (universal) constant was a property of space-time, a constant energy density that didn't clump or disperse. And it was and is theoretically entirely legitimate (and expected—albeit with hindsight). However, Einstein's use of this constant was incorrect.

Before explaining that further, let's consider the nature of the cosmological constant, cc. Actually calling the cc a form of energy is a bit of a cheat. It's in fact a form of energy *and* pressure consistent with relativistic principles that relate the two. The key aspect of the cc is that it has positive energy but negative pressure. The net total gravitational force is sometimes said to look like antigravity, since the net contribution to the universe's expansion is opposite in sign to other forms of energy we know about.

However, $E = mc^2$ was just the beginning. It said that both mass and energy exert gravitational forces. But, also according to Einstein, both energy and pressure contribute to the gravitational field in the combination $\rho + 3p$, where ρ is the energy density and p is the pressure. This combination is important because it means that negative pressure has the opposite gravitational effect to positive energy, and in fact negative pressure can accelerate the expansion of the universe. And that is precisely what happens when there is a cc, or energy stored in the vacuum.

Einstein tried to use the fact that the quantity $\rho + 3p$ was relevant to get a static universe. He said that if the pressure was correct and the cc took a particular value, the net gravitational mass would vanish, leading to a static universe. The particular solution he found corresponded to a closed universe: a big sphere with finite radius that stayed constant over time.

However, we now know this is not what the universe looks like; Edwin Hubble's 1929 measurements established the expansion of the universe, and additional measurements since have confirmed it. Hubble discovered that the expansion rate is proportional to distance and could observe the red-shifts of galaxies consistent with this relationship.

However, even before Hubble's measurements, there were problems with Einstein's static universe. Alexandre Friedmann, a Russian physicist, had discovered that the static solution was unstable. He observed that if the universe were slightly bigger, the vacuum energy would be the same, but the matter density would be lower. So the universe would accelerate outward and grow. Conversely, a smaller universe would give rise to a bigger mass density, and that would make the universe accelerate inward and shrink.

There were other misconceptions as well about the cc. For example: Einstein thought that with the cc, there was no cosmological solution without matter, which he favored because it seemed to be consistent with Mach's principle in which matter determined the inertial frame. Einstein thought he had found the unique solution with a cc, and that the solution required matter. But that was only early on. Soon afterward (1917), the Dutch astronomer Willem de Sitter found a static solution with cc and no matter. And in 1923, it was discovered that de Sitter's universe isn't actually static either. An empty universe seems static, but small particles move apart under its gravitational field!

On top of that, it was soon realized that Einstein's equations permitted even more cosmological solutions. Friedmann also demonstrated expanding solutions of the form $ds^2 = dt^2 - R(t)^2 dr^2/(1 - kr^2) + r^2 d\Omega^2$. Notice that these solutions involve a function $R(t)$; that is, the solutions are time-dependent. This is the cosmological form of Einstein's solution. It really is an expanding universe.

Einstein originally disagreed with Friedmann's results. But eight months later, he acknowledged a calculation error: "I am convinced that Mr. Friedmann's results are both correct and clarifying. They show that in addition to the static solutions to the field equations there are time varying solutions with a spatially symmetric structure." Even so, Einstein still held that Friedmann's solutions, though mathematically legitimate, were irrelevant to the universe. (His true feelings were perhaps revealed by the words he crossed out when commenting on the work: "A physical significance can hardly be ascribed.")

The resolution to all these questions was a matter of measurement. Already between 1910 and the early 1920s, Vesto Slipher had observed red-shifts of about 6 percent in wavelength (analogous to the shift in wavelength of the sound from an ambulance as it recedes). In 1923, Hermann Weyl realized the de Sitter solutions exhibited red-shifts. But Friedmann's 1922 solutions had shown there can be red-shift with no cc at all. And in a 1923 letter, Einstein wrote to Weyl, "If there is no quasi-static world, then away with the cosmological term!" And indeed, with Hubble's data, it became clear there was no quasi-static world. This cosmology further pursued by Georges Lemaitre, whose original French work (of 1927) had been ignored. Sadly, Friedmann died in 1925 just as his theory was being vindicated.

But the story of the cosmological term did not end there. According to current observations, we live in universe in which the rate of expansion is accelerating, indicating that 70 percent of the universe's energy is of the form of a cc. A cc explains current observations, including the rate of dimming of distant supernovae. Furthermore, observations of the cosmic microwave background radiation show that the universe contains critical energy density corresponding to a flat universe. Since we know dark matter is only about 30 percent of the energy density, the remaining energy should take the form of a cc.

It's worth noting that the amount of cosmological energy is still incredibly tiny compared to what one would estimate in quantum field theory. The energy has to be small enough for the universe to have lasted thirteen billion years. That energy is 120 orders of magnitude smaller than what field theory leads you to believe. This is a major embarrassment to physicists today.

But to return to Einstein's story. It is now commonly said that Einstein referred to the cc as his "biggest blunder," which are the words John Wheeler quotes. It certainly sounds enticing to repeat this. The words "Einstein," "biggest," and "blunder" sound interesting together. Even if an accurate quote, I doubt it was the idea of the cc that Einstein viewed as his biggest blunder. The mistake was not believing that the universe was expanding, which was a remarkable prediction that followed from Einstein's theory of gravity. The cc was actually another brilliant insight, and Einstein was still referring to it in the 1930s. Einstein, who thinks about the big picture, likely thought of the static universe as his big mistake.

We now know that vacuum energy is expected to be present. One source is quantum fluctuations even in the absence of particles. The surprise is that the cc is so small. That is the reason it didn't seem to be there (at first). The mistake wasn't thinking the cc existed, but thinking it had to be there in a specific amount to guarantee a unique static solution (for a given amount of matter). It wasn't such a bad mistake given the observations of the time, which were consistent with a static universe.

We've already mentioned that the cc leads to accelerated expansion of the universe today. Let's now consider a couple of its other consequences. First, cosmological inflation, which results when the cc dominates the energy density. During inflation, there is an exponential expansion of the universe. By making causally connected regions of the universe much bigger, inflation solves the horizon, flatness, and monopole problems. Cosmological inflation furthermore provides observable and measurable perturbations in the cosmic microwave background radiation.

The negative pressure that I mentioned earlier is a prerequisite for inflation, so that when the universe expands, energy must be added to maintain the constant energy density in a larger universe. If you imagine universe as a piston, you need to pull on it and work against pressure that would pull inward. So you need negative pressure and positive energy to get inflation. Right now, there is a gathering array of experimental evidence supporting an inflationary era in the early universe.

The cosmological vacuum energy has other consequences that go well beyond anything Einstein could have anticipated. We have recently found some rather surprising results when a cc is considered in the presence of an additional dimension of space.

Einstein's equations work for any number of spatial dimensions. In 1919, Kaluza suggested an additional dimension of space beyond the three that were known. Einstein refereed the paper. He didn't like the fact that the size of the extra dimension was undetermined. Oskar Klein suggested one way for dimensions to be hidden, but the question of the size is still an important issue and one that we have recently investigated using general relativity and "branes."

According to relativity, matter and energy determine geometry, and the curved geometry gives rise to gravity. Matter warps space-time, and the distorted space yields gravitational force. This would be true in any number of dimensions.

In higher-dimensional space, there can be a cc that exists throughout the entire (bulk) space. But there can also be cc terms (vacuum energy terms) that exist only on a subspace of the full higher-dimensional geometry. This is because of the existence of objects called branes. Branes are objects that distinguish some of the dimensions of space. There are directions along the branes and directions perpendicular to them. Branes can carry energy in the form of a cc that resides on its surface. However, a large flat brane with cosmological vacuum energy mandates an additional cc in the bulk. This is necessary if we are to have a large flat brane; otherwise the brane energy (according to Einstein's theory) would curve the brane, and it wouldn't look like the large flat dimensions that we experience.

The end result with the two compensating energy terms, the brane and the bulk, is that there is flat brane but a warped higher-dimensional bulk. According to Einstein's theory, the bulk energy bends space-time, and space-time geometry is dramatically warped.

The consequences of this warped space-time can be quite significant. One of the most important consequences is that space-time can be so warped that gravity can be highly concentrated in the vicinity of a brane, making an infinite extra dimension possible. Because of warping, an extra dimension can be infinite, yet hidden. The graviton would be so strongly peaked near a brane that gravity wouldn't leak away. The graviton—the particle that communicates the force of gravity—is essentially bound near the brane. The force of gravity appears as if there were only three dimensions of space because the force spreads very asymmetrically, and the force wouldn't spread far into the bulk. Gravity essentially hugs the brane, and spreads out only in the three infinite dimensions of the brane—not the direction that is perpendicular to it.

Even more dramatic effects are possible. In the scenario I just described, gravity is localized near a brane. But no matter where you are in five-dimensional space-time, you would observe gravity to appear to be four-dimensional. Even in regions far from the brane, localized gravity would mimic four space-time dimensions.

We also found something even more startling: four space-time dimensions can be a local phenomenon. That is, we might be living in a pocket of four space-time dimensions inside a higher-dimensional space-time. The Copernican revolution continues as we realize the rest of the universe might be very different than what is obvious from our perspective.

These results are exciting, but for a physicist, they are also frustrating. There can be something as dramatic as an infinite fifth dimension or higher-dimensional regions of the universe where gravity is very different from what we see, yet we wouldn't know about it directly through experiments. This leads to the obvious question: Can the warped space-time geometry be experimentally tested? The generic answer is no. But the answer is yes if warped geometry has consequences for particle properties we can observe.

In particular, a space-time geometry, similar to the warped geometry just described, but with two branes instead of one, can explain the weakness of gravity relative to the other three forces we know of: the strong, weak, and electromagnetic forces. Essentially, if we live anywhere aside from the Gravity-brane, where gravity is highly concentrated, gravity will appear to be weak. The graviton probability near a second brane, the Weak-brane, which is slightly displaced from the Gravity-brane, will be feeble—exponentially suppressed in the warped space-time geometry. This idea, though quite exotic, has direct experimental consequences that will be explored in the next few years.

In this essay, we've covered quite a bit of ground. I hope I have made it clear that Einstein's work on energy was extraordinarily influential. His work changed all of physics and has implications for nuclear physics, particle physics, general relativity, string theory, cosmology, and much more. We have yet to understand all the connections and all the implications.

Furthermore, the cc remains one of the outstanding problems facing string theorists, particle physicists, and cosmologists today. No one understands it. Is it Einstein's biggest blunder, or theoretical physicists' biggest embarrassment?

CHAPTER TWENTY

But the most important lesson for this essay is that Einstein's legacy is everywhere. And it will certainly continue to stay with us throughout the 21st century.

Sources

All quotations from Einstein's letters and publications are taken from:

The Collected Papers of Albert Einstein (Princeton, NJ: Princeton University Press, 1987ff.).

Arthur I. Miller, *Albert Einstein's Special Theory of Relativity: Emergence and Early Interpretation* (Reading, MA: Addison Wesley, 1981).

One of the most comprehensible presentations of the special and general theory of relativity remains Einstein's own:

Einstein, Albert. *Relativity: The Special and the General Theory*, 2nd ed., trans. Robert W. Lawson (New York: Crown Trade Paperbacks, 1995).

The modern classic exposition is:

Weinberg, Steven. *Gravitation and Cosmology: Principles and Applications of the General Theory of Relativity* (New York: Wiley, 1972).

A critical assessment of the general relativity is given in:

Will, Clifford M. *Theory and Experiment in Gravitational Physics*, rev. ed. (Cambridge and New York: Cambridge University Press, 1993).

Will, Clifford M. *Was Einstein Right? Putting General Relativity to the Test*, 2nd ed. (New York: Basic Books, 1993).

A historical perspective on the development of modern cosmology can be obtained from:

Kragh, Helge. *Cosmology and Controversy: The Historical Development of Two Theories of the Universe* (Princeton, NJ: Princeton University Press, 1996).

Kragh, Helge. *Matter and Spirit in the Universe: Scientific and Religious Preludes to Modern Cosmology* (London: Imperial College Press; Singapore and Hackensack, NJ: World Scientific, 2004).

For readily accessible expositions of recent developments:

Greene, B. Brian. *The Elegant Universe: Superstrings, Hidden Dimensions, and the Quest for the Ultimate Theory* (New York: W.W. Norton, 1999).

Greene, B. Brian. *The Fabric of the Cosmos: Space, Time, and the Texture of Reality* (New York: A. A. Knopf, 2004).

Randall, Lisa. *Warped Passages: Unravelling the Mysteries of the Universe's Hidden Dimensions* (New York: Ecco, 2005).

NOTES

Note to Chapter 1

[1]But let us at least put into some time capsule the opinion of the then-reigning Dean of theoretical physics, P.A.M. Dirac. At the Centenary Symposium in Jerusalem in 1979, Dirac said: "It might very well be that [a] new quantum mechanics will have determinism in the way that Einstein wanted. . . . I think it is very likely, or at any rate quite possible, that in the long run Einstein will turn out to be correct, even though for the time being physicists have to accept the Bohr probability interpretation, especially if they have examinations in front of them."

Notes to Chapter 2

[1]Albert Einstein, "Autobiographical Notes," in Paul Arthur Schilpp, ed., *Albert Einstein: Philosopher Scientist*, vol. 1 (1949; La Salle, IL: Open Court, 1982), pp. 4f. Einstein's German original is as follows: "Es ist mir klar, daß das so verlorene religiöse Paradies der Jugend ein erster Versuch war, mich aus den Fesseln des 'Nur-Persönlichen' zu befreien, aus einem Dasein, das durch Wünsche, Hoffnungen und primitive Gefühle beherrscht ist. Da gab es draußen diese große Welt, die unabhängig von uns Menschen da ist und vor uns steht wie ein großes, ewiges Rätsel, wenigstens teilweise zugänglich unserem Schauen und Denken. Ihre Betrachtung winkte als eine Befreiung, und ich merkte bald, daß so Mancher, den ich schätzen und bewundern gelernt hatte, in der hingebenden Beschäftigung mit ihr innere Freiheit und Sicherheit gefunden hatte. Das gedankliche Erfassen dieser außerpersönlichen Welt im Rahmen der uns gebotenen Möglichkeiten, schwebte mir halb bewußt, halb unbewußt als höchstes Ziel vor. Ähnlich eingestellte Menschen der Gegenwart und Vergangenheit sowie die von ihnen erlangten Einsichten waren die unverlierbaren Freunde. Der Weg zu diesem Paradies war nicht so bequem und lockend wie der Weg zum religiösen Paradies; aber er hat sich als zuverlässig erwiesen, und ich habe es nie bedauert, ihn gewählt zu haben."

[2]Einstein, "Autobiographical Notes," pp. 6f.: Bei einem Menschen meiner Art liegt der Wendepunkt der Entwicklung darin, dass das Hauptinteresse sich allmählich weitgehend loslöst vom Momentanen und Nur-Persönlichen und sich dem Streben nach gedanklicher Erfassung der Dinge zuwendet.

[3]On the history of the scientific persona, see Christopher Lawrence and Steven Shapin, eds., *Science Incarnate: Historical Embodiments of Natural Knowledge* (Chicago and London: University of Chicago Press, 1998); Lorraine Daston and H. Otto Sibum, eds., "Scientific Personae," in *Science in Context* 16, nos. 1–2 (2003).

[4]Albert Einstein, "Isaac Newton," in his *Out of My Later Years* (New York: Philosophical Library, 1950), p. 219.

[5]Thomas Nagel, *The View from Nowhere* (New York, Oxford: Oxford University Press, 1986), p. 5.

[6]Charles Sanders Peirce, "Three Logical Sentiments," in *Collected Papers of Charles Sanders*

Peirce, vol. 2: *Elements of Logic* (Cambridge, MA: Harvard University Press, 1932), p. 398; Peirce, "Grounds of Validity," in *Collected Papers*, vol. 5: *Pragmatism and Pragmaticism* (1934), p. 221.

[7]Michael Polanyi, *The Logic of Liberty: Reflections and Rejoinders* (Chicago: University of Chicago Press, 1951), p. 26. I thank Mary Jo Nye, who allowed me to read the text of her History of Science Society Distinguished Lecture, "Cultural and Political Sources of Science as Social Practice," delivered Vancouver, BC, November 4, 2000, and Karl Hall for an enlightening conversation on Polanyi.

[8]Albert Einstein, "Paul Langevin in Memoriam," in his *Out of My Later Years*, p. 231.

[9]I. Bernard Cohen, "Einstein and Newton," in A. P. French, ed., *Einstein: A Centenary Volume* (Cambridge, MA: Harvard University, 1979), pp. 40–42.

[10]E. Schwartz, "Diogenes (40)," in G. Wissowa and W. Kroll, eds., *Paulys Realencylopädie der classischen Altertumswissenschaft* (Stuttgart: Metzler, 1962–95), vol. 5, cols. 738–63.

[11]Diogenes Laertius, *Lives of Eminent Philosophers*, R. D. Hicks, trans. (London: W. Heinemann, 1925), vol. 1, pp. 163–67.

[12]Samuel Smiles, *Character* (London: John Murray, 1871), p. 114.

[13]Philip Gilbert Hamerton, *The Intellectual Life* (London: Macmillan & Co., 1873), p. 16.

[14]Hamerton, *The Intellectual Life*, p. 311.

[15]Janet Malcolm, *In the Freud Archives* (1983; New York: New York Review of Books, 2002).

[16]The literature on this subject is vast, but see the remarkable essay by Pierre Hadot, "The Figure of Socrates," in his *Philosophy as a Way of Life: Spiritual Exercises from Socrates to Foucault*, Arnold I. Davidson, ed., Michael Chase, trans. (Oxford: Blackwell, 1995), pp. 147–78.

[17]Plato, *Symposium*, Walter Hamilton, trans. (1951; Harmondsworth: Penguin, 1988), pp. 108, 221a.

[18]Hadot, "The Figure of Socrates," p. 158.

[19]Seneca, *Naturales quaestiones*, T. H. Corcoran, trans. (London: W. Heinemann, 1972), vol. 2, pp. 138–41, VI. 2. 8.

[20]Seneca, *Naturales Quaestiones*, vol. 1, pp. 7f.; I. Pref. 7.

[21]Pierre Hadot, "Forms of Life and Forms of Discourse in Ancient Philosophy," in his *Philosophy as a Way of Life*, p. 59.

[22]Hadot, "Forms of Life," p. 60.

[23]Jean d'Alembert, "Essai sur la société des gens de lettres et des grands, sur la réputation, sur les mécènes et les récompenses littéraires," in his *Mélanges de littérature, d'histoire et de philosophie*, nouv. ed. (Amsterdam: Z. Chatelain, 1770), vol. 1, p. 362.

[24]Thomas Babington Macaulay, "Francis Bacon" [1837], in *Literary Essays Contributed to the Edinburgh Review* (London: Oxford University Press, 1932), p. 291.

[25]Abelard and Heloise, *The Story of His Misfortunes and the Personal Letters*, Betty Radice, trans. (London: The Folio Society, 1977), p. 28.

[26]Gadi Algazi, "Scholars in Households: Refiguring the Learned Habitus, 1480–1550," in *Science in Context* 16 (2003), pp. 9–42.

[27]Jean de La Bruyère, "Of Fashion," in his *Characters* [1688], Jean Stewart, trans. (Harmondsworth: Penguin, 1970), p. 254; *Les caractères de Théophraste traduit du grec avec les caractères ou les moeurs de ce siècle*, Robert Pignarre, ed. (Paris: Garnier-Flammarion, 1965), p. 338.

[28]Jimena Canales, "Photogenic Venus: 'The Cinematographic Turn' and Its Alternatives in Nineteenth-Century France," in *Isis* 93 (2002), pp. 585–613; F. Gonnessiat, *Recherches sur l'équation personnelle dans les observations astronomiques de passage* (Paris: G. Masson, 1892).

[29]Wilhelm Wundt, *Grundzüge der physiologischen Psychologie* (Leipzig: Wilhelm Engelmann, 1874), p. 709.

[30]See, for example, Ewald Hering, "Über individuelle Verschiedenheiten des Farbensinnes," in *Lotos* 6 (1885), p. 156; and, more generally, R. Steven Turner, *In the Eye's Mind: Vision and the Helmholtz-Hering Controversy* (Princeton, NJ: Princeton University Press, 1994).

[31]Gottlob Frege, *Die Grundlagen der Arithmetik. Eine logisch-mathematische Untersuchung über den Begriff der Zahl* (1884; Breslau: M. & H. Marcus, 1934), p. 37n.

[32]Gottlob Frege, "Thoughts," in his *Logical Investigations*, Peter Geach, ed. (New Haven, CT: Yale University Press, 1977), pp. 8f.; quoted in Thomas G. Ricketts, "Objectivity and Objecthood: Frege's Metaphysics of Judgment," in Leila Haaparanta and Jaakko Hintikka, eds., *Frege Synthesized: Essays on the Philosophical and Foundational Work of Gottlob Frege* (Dordrecht, Boston: D. Reidel, 1986), pp. 65–95.

[33]Henri Poincaré, *La Valeur de la science* (1905; Paris: Flammarion, 1970), p. 179.

[34]Henri Poincaré, *Science et l'hypothèse* (Paris: Flammarion, 1902), p. 281.

[35]Ernst Mach, *Die Geschichte und die Wurzel des Satzes von der Erhaltung der Arbeit*, 2nd ed. (1872; Leipzig: Johann Ambrosius Barth, 1909), p. 1.

[36]Charles Sanders Peirce, "A Critical Review of Berkeley's Idealism" [1871], in Philip Wiener, ed., *Values in a Universe of Chance: Selected Writings of Charles Peirce (1839–1914)* (New York: Double-day, 1958), p. 82.

[37]Peirce, "A Critical Review," p. 83; compare Peirce, "Three Logical Sentiments," p. 398.

[38]Max Planck, *Acht Vorlesungen über theoretische Physik, gehalten an der Columbia University in the City of New York im Frühjahr 1909* (Leipzig: S. Hirzel, 1910), p. 6.

[39]Henri Poincaré, "La Morale et la science," in his *Dernières pensées* (Paris: Flammarion, 1913), pp. 232f.

[40]Robert Nozick, *Invariances: The Structure of the Objective World* (Cambridge, MA: Harvard University Press, 2001), p. 96.

[41]Hermann Weyl, *Philosophy of Mathematics and Natural Science* (Princeton, NJ: Princeton University Press, 1949), p. 123.

[42]Friedrich Nietzsche, *Von Nutzen und Nachteil der Historie für das Leben* [1874], in his *Unzeitgemäße Betrachtungen*, Peter Pütz, ed. (Munich: Goldmann, 1992), p. 113.

NOTES TO CHAPTER 3

The main sources of the quotations in this paper are: Albert Einstein, *About Zionism: Speeches and Letters*, trans. and ed. Leon Simon (London: Soncino Press, 1930); Albert Einstein, *Ideas and Opinions* [based on *Mein Weltbild*] (New York: Three Rivers Press, 1982 [originally published in 1954]); and documents in the Albert Einstein Archives, The Hebrew University of Jerusalem. Related references include John Stachel, "Einstein's Jewish Identity," in *Einstein from B to Z* (Basel: Birkhauser, 2001); Uriel Tal, "Jewish and Universal Social Ethics in the Life and Thought of Albert Einstein," in *Albert Einstein, Historical and Cultural Perspectives*, ed. G. Holton and Y. Elkana (Princeton, NJ: Princeton University Press, 1982); Max Jammer, *Einstein and Religion: Physics and Theology* (Princeton, NJ: Princeton University Press, 1999).

[1]Einstein to Abba Eban, Israel's Ambassador in the United States, in response to Ben-Gurion's offer of presidency, 18 November 1952, AEA doc. 28 943, Albert Einstein Archives, The Hebrew University of Jerusalem (italics introduced by author).

[2]"Autobiographical Notes," in *Albert Einstein: Philosopher-Scientist*, 3rd ed., ed. Paul A. Schilpp (La Salle, IL: Open Court, 1970).

[3]Albert Einstein, "Assimilation and Nationalism,"in *About Zionism: Speeches and Letters*, trans. and ed. Leon Simon (London: Soncino Press, 1930), p. 27.

[4]Published in the *Vossische Zeitung*, 8 October 1929, in response to an article by Prof. W. Hellpach, the minister of education of the State of Baden, Germany. The letter is reprinted in Albert Einstein, *Ideas and Opinions* (New York: Three Rivers Press, 1982), p. 171.

[5]"Why Do They Hate Jews?" *Colliers Weekly Magazine*, 26 November 1938, p. 102; reprinted in Einstein, *Ideas and Opinions*, p. 191.

[6]Einstein, "Assimilation and Nationalism," p. 23.

[7]"Wie ich Zionist wurde," *Judische Rundschau*, 21 June 1921; reprinted in *The Collected Papers of Albert Einstein*, vol. 7 (Princeton, NJ: Princeton University Press).

[8]Einstein, "Assimilation and Nationalism," p. 29

[9]Ibid., p. 30.

[10]Ibid., p. 27.

[11]Einstein to Judge Jerome Frank, AEA doc. 35 071.

[12]Einstein, "Assimilation and Nationalism," p. 34.

[13]Einstein's travel diary to Japan, Palestine, Spain, AEA doc. 29 129, entry on 10 November 1922. Until the early 1930s Einstein occasionally used "the race" when he referred to the Jewish people. Before the Nazis came to power, this was an innocent and accepted term.

[14]Ibid., entry on 3 February 1923.

[15]Ibid., entry on 9 February 1923.

[16]*Palestine Weekly*, 9 February 1923, at a reception in the Lemel School in Jerusalem.

[17]"Jewish Ideals," published in *Temoinage de nos temps*, 2 September 1933; reprinted in Einstein, *Ideas and Opinions*, p. 185.

[18]"Addresses on Reconstruction in Palestine, V," from an address during Einstein's first visit in the United States, April–May, 1921; reprinted in Einstein, *Ideas and Opinions*, p. 181.

[19]"Jewish Recovery," written in 1931; reprinted in Einstein, *Ideas and Opinions*, p. 181.

[20]Draft for a radio broadcast for the United Jewish Appeal, 27 November 1949, AEA doc. 28 862.

[21]Einstein to Chaim Weizmann, 25 November 1929, AEA doc. 33 411.

[22]"Addresses on Reconstruction in Palestine, I," from an address during Einstein's second or third visit to the United States, in the winter of either 1930 or 1931; reprinted in Einstein, *Ideas and Opinions*, p. 176.

[23]Einstein to Zvi Lurie, 4 January 1955, the Zionist Archives, doc. 12/50/K.

[24]Einstein's letter to Hendrik Lorentz, 22 February 1921, AEA doc. 16 537.

[25]Einstein's letter to Fritz Haber, 21 September 1921, AEA doc. 36 217.

[26]From an address delivered at a reception in Singapore, 2 November 1922; published in *Singapore Times*, 3 November 1922.

[27]"The Mission of Our University," message to the Hebrew University upon its opening; published in *New Palestine*, 27 March 1925.

[28]Einstein's testament, 18 March 1950.

[29]Isaiah Berlin, "Einstein and Israel," in Albert Einstein, *Historical and Cultural Perspectives*, ed. G. Holton and Y. Elkana (Princeton, NJ: Princeton University Press, 1982), p. 281.

Notes to Chapter 4

[1]Max Jammer, *Einstein and Religion: Physics and Theology* (Princeton, NJ: Princeton University Press, 1999).

[2]Letter of August 1927, quoted by Alice Calaprice: *The Expanded Quotable Einstein* (Princeton, NJ: Princeton University Press, 2000), p. 204; hereafter, AC.

[3]Written in 1954, AC, p. 219.

[4]Written in 1930, AC, p. 241.

[5]Written in 1936, AC, p. 278.

[6]Written in 1955, AC, p. 281.

[7]Written in 1952, AC, p. 80.

[8]Written in 1923, AC, p. 202.

[9]Written in 1952, AC, p. 217.

[10]Written ca. 1920, AC, p. 204.

[11]Written in 1929, AC, p. 204.

[12]Written in the late 1930s, AC, p. 258.

[13]In Carl Seelig, *Helle Zeit, dunkle Zeit. In Memoriam Albert Einstein* (Zürich: Europa-Verlag, 1956), p. 72.

[14]Max Born and Albert Einstein, *The Born-Einstein Letters: Correspondence between Albert Einstein and Max and Hedwig Born from 1916 to 1955*, Irene Born, trans. (New York: Walker, 1971), p. 91.

[15]Originally said to Thorsten Veblen in Princeton in 1921; quoted by Abraham Pais, who called his book (quoting Einstein) *Subtle Is the Lord: The Science and the Life of Albert Einstein* (New York: Oxford University Press, 1982). Later, as quoted in AC, p. 241: "I have second thoughts. Maybe God is malicious."

[16]AC, p. 241.

[17]Written in 1942, AC, p. 250.

[18]AC, p. 252.

[19]Cited in A. P. French, ed., *Einstein: A Centenary Volume* (Cambridge, MA: Harvard University Press, 1979), p. 6.

[20]To Elsa Loewenthal [his future wife] in February 1914, in *Collected Papers of Albert Einstein,*

vol. 5, Martin J. Klein, A. J. Kox, and Robert Schulmann, eds. (Princeton, NJ: Princeton University Press, 1993), Doc. 509.

[21] Gerald Holton and Stephen G. Brush, *Physics, the Human Adventure: From Copernicus to Einstein and Beyond*, 3rd ed. (Princeton, NJ: Princeton University Press, 2001), p. 514.

[22] Written in 1930, AC, p. 205.

[23] Steven Weinberg, *Facing Up: Science and Its Cultural Adversaries* (Cambridge, MA: Harvard University Press, 2001), p. 20.

[24] Weinberg, *Facing Up*, p. 68.

[25] AC, p. 254.

[26] Ibid.

[27] Written to one of his most intimate friends and correspondents, Maurice Solovine, in 1951, by then a fifty-year-old friendship. In Albert Einstein, *Letters to Solovine, 1906–1955* (New York: Philosophical Library, 1987), p. 123.

[28] Compare with the fascinating interview with one of the great string theorists, Brian Greene, in *Focus* 52 (2004), pp. 85f.

[29] Einstein, *Letters to Solovine*, p. 258.

[30] Albert Einstein, "Physics and Reality," in *Journal of the Franklin Institute* 221 (1936), pp. 349–82.

[31] From "Principles of Research," a speech given in Berlin in honor of Max Planck's sixtieth birthday in 1918. Quoted from Albert Einstein, *Ideas and Opinions. Based on Mein Weltbild, and Other Sources*, Carl Seelig, ed. (New York: Three Rivers Press, 1982), p. 226.

[32] Gerald Holton, *Victory and Vexation in Science* (Cambridge, MA: Harvard University Press, 2005), ch. 1.

[33] Eric H. Erikson, "Psychoanalytic Reflections on Einstein's Centenary," in Gerald Holton and Yehuda Elkana, eds., *Albert Einstein: Historical and Cultural Perspectives. The Centennial Symposium in Jerusalem* (Princeton, NJ: Princeton University Press, 1982), pp. 151–73. Roman Jakobson, "Einstein and the Science of Language," ibid., pp. 139–50.

[34] From Albert Einstein, "Autobiographical Notes," in Paul Arthur Schilpp, ed., *Albert Einstein: Philosopher-Scientist*, vol. 1 (1949; La Salle, IL: Open Court, 1982), p. 9; quoted by Max Wertheimer, who interviewed Einstein at length for his psychological studies in thinking and published it in his *Productive Thinking* (New York: Harper & Row, 1959), pp. 213–28. Especially on this nonverbal thinking, see Roman Jakobson, who also correlates this claim with the Sapir-Whorf hypothesis, and with Einstein's famous exchange with the mathematician Jacques Hadamard, who appended Einstein's answer to his essay. See Jacques Hadamard, *Essay on the Psychology of Invention in the Mathematical Field* (Princeton, NJ: Princeton University Press, 1945).

[35] Einstein, "Autobiographical Notes." Schilpp translates *Gedankenwelt* as "mental world," which I believe diverges from the real meaning, in much the same way as translating Sigmund Freud's *Ich* to "ego."

[36] Einstein, "Physics and Reality," pass.

[37] Ibid.

[38] Klaus Meyer-Abich, "Bohr's Complementarity and Goldstein's Holism," in *Mind and Matter* 2 (2004), p. 92.

[39] Said on the occasion of Newton's 300th birthday, AC, p. 95.

[40] Einstein, *Ideas and Opinions*, p. 270; Einstein, "Physics and Reality," pass.

[41] Ibid.

[42] Ibid.

[43] Ibid.

[44] Ibid.

[45] Ibid.

[46] Ibid.

[47] Ibid.

[48] Ibid.

[49] From "Principles of Research"; quoted in Einstein, *Ideas and Opinions*, pp. 224f.

[50]Ibid.

[51]Ibid.

[52]AC, p. 317.

Notes to Chapter 5

[1]Louis Dumont, *Essays on Individualism: Modern Ideology in Anthropological Perspective* (Chicago: University of Chicago Press, 1992).

[2]Cited in Yaron Ezrahi, *The Descent of Icarus: Science and the Transformation of Contemporary Democracy* (Cambridge, MA: Harvard University Press, 1990), p. 106.

[3]Albert Einstein, "Remarks on Bertrand Russell's Theory of Knowledge," in Paul Arthur Schilpp, ed. and trans., *The Philosophy of Bertrand Russell* (Evanston, IL: Library of Living Philosophers, 1946), p. 281.

[4]W.V.O. Quine, *From a Logical Point of View*, 2nd ed. (New York: Harper Torchbooks, 1961), p. 49.

[5]Cited in Robert H. Kargon, *The Rise of Robert Millikan* (Ithaca, NY: Cornell University Press, 1982), p. 32.

[6]On virtual witnessing, see Steven Shapin and Simon Schaffer, *Leviathan and the Air-Pump: Hobbes, Boyle, and the Experimental Life* (Princeton, NJ: Princeton University Press, 1985).

[7]In this connection, see also Garry Wills, *Reagan's America: Innocents at Home* (New York: Doubleday Company, 1987); and for further discussion, Ezrahi, *The Descent of Icarus*, pp. 254–62.

[8]Thomas Hobbes, *Leviathan*, C. B. Macpherson, ed. (Harmondsworth, UK: Penguin Books, 1976), pp. 131f.

[9]For the citation and the discussion, see Yaron Ezrahi, "Einstein and the Light of Reason," in Gerald Holton and Yehuda Elkana, eds. *Albert Einstein: Historical and Cultural Perspectives* (Princeton, NJ: Princeton University Press, 1982), p. 257.

[10]Albert Einstein, *Out of My Later Years* (New York: Philosophical Library, 1950), p. 59.

[11]John Hartley, *The Politics of Pictures: The Creation of the Public in the Age of Popular Media* (London: Routledge, 1992), p. 147.

[12]W.J.T. Mitchell, *Picture Theory* (Chicago: University of Chicago Press, 1994), p. 367.

[13]Mitchell, *Picture Theory*, p. 367.

[14]Bas C. van Fraassen, "The Manifest Image and the Scientific Image," in D. Aerts, J. Broekaert, and E. Mathijs, eds., *Einstein Meets Magritte: An Interdisciplinary Reflection* (Dordrecht: Kluwer, 1999), pp. 29–52.

[15]Yaron Ezrahi, "Science and the Political Imagination in Contemporary Democracies," in Sheila Jasanoff, ed., *States of Knowledge: The Co-Production of Science and the Social Order* (London: Routledge, 2004), pp. 268–73.

[16]Rafael Sanchez, "Channel-Surfing: Media, Mediumship, and State Authority. Maria Lionza Possession Cult (Venezuela)," in Hent De Vries and Samuel Weber, eds., *Religion and Media* (Stanford, CA: Stanford University Press, 2001), p. 423.

[17]Hartley, *The Politics of Pictures*, p. 40.

[18]See Yaron Ezrahi, "Nature as Dogma, a Review of Bruno Latour's Politics of Nature: How to Bring the Sciences into Democracy," in *The American Scientist* 93 (2005), pp. 89f.; and Yaron Ezrahi, *Necessary Fictions: Imagining Democracy Between Modernity and Post-Modernity* (forthcoming, 2008).

Notes to Chapter 6

[1]Frederic Golden, "Albert Einstein: Person of the Century," *Time Magazine*, December 12, 1999, p. 34.

[2]Die Zeit (Einstein Ausgabe, Anfang 2005) Ulrich Schnabel: Und Action, Albert . . . , DieZeit, Dez. 16th, 2004, No. 52.

[3]Denis Brian, *Einstein: A Life* (New York: Wiley, 1995), p. 235.

[4]Albrecht Fölsing, *Albert Einstein: A Biography*, trans. Ewald Osers (New York: Penguin, 1997), p. 641.

[5]Fölsing, *Albert Einstein*, p. 458.

[6]Fred Jerome, *The Einstein File: J. Edgar Hoover's Secret War Against the World's Most Famous Scientist*, 1st ed. (New York: St. Martin's, 2002), p. 111.

[7]Fölsing, *Albert Einstein*, p. 349.

[8]Jerome, *The Einstein File*, p. 7.

[9]Ibid., p. 11.

[10]Abraham Pais, *Einstein Lived Here* (New York: Oxford University Press, 1994), p. 196.

[11]Ze'ev Rosenkranz, *Albert Through the Looking-Glass: The Personal Papers of Albert Einstein* (Jerusalem: The Albert Einstein Archives, The Jewish National and University Library, The Hebrew University of Jerusalem, 1998), p. 72.

[12]Jerome, *The Einstein File*, p. 110.

[13]Albert Einstein, *Out of My Later Years: The Scientist, Philosopher, and Man Portrayed Through His Own Words*, rev. ed. (New York: Wings Books, 1993), p. 128.

[14]Albert Einstein, "Remarks on the Essays Appearing in this Collective Volume," in *Albert Einstein: Philosopher-Scientist*, 3rd ed., ed. Paul A. Schilpp (La Salle, IL: Open Court, 1970), p. 680.

[15]Fölsing, *Albert Einstein*, p. 623.

[16]Otto Nathan and Heinz Norden, eds., *Einstein on Peace* (New York: Simon and Schuster, 1960), p. 393.

[17]See http://nobelprize.org/nobel_prizes/literature/laureates/1949/faulkner-speech.html (accessed June 26, 2007).

[18]Immanuel Kant, *The Critique of Practical Reason*, ed. and trans. Lewis White Beck (New York: Open Court, 1969), p. 162.

Notes to Chapter 7

[1]A. Stern, "Interview with Einstein," in *Contemporary Jewish Record* (1945), pp. 245–49.

[2]See, for example, his fairly extensive correspondence with Henri Barbusse during the 1920s. Barbusse tried to enlist Einstein in many left-wing causes. Einstein agreed to be associated with some of them; in others he forthrightly refused to participate.

[3]O. Nathan and H. Norden, eds., *Einstein on Peace* (New York: Schocken Books, 1968), pp. 347–51; hereafter EoP.

[4]It had been Freud, who, in their correspondence of 1932, had succinctly pointed out the necessity of an international court: "There is but one sure way of ending war and that is the establishment, by common consent, of a central control which shall have the last word in every conflict of interests. For this two things are needed: first the creation of such a supreme court of jurisdiction; secondly, its investment with adequate executive force. Unless this second requirement is fulfilled, the first is unavailing. Obviously the League of Nations, acting as a Supreme Court, fulfills the first condition; it does not fulfill the second. It has no force at its disposal and can only get it if the members of the new body, its constituent members, furnish it. And, as things are, this is a forlorn hope." EoP, pp. 195f.

[5]In 1932, while still a convinced pacifist, he had stated his views as follows:

greatest obstacle to disarmament has been the inability of most people to appreciate the enormity of the problem. Most objectives are accomplished in small steps. Think, for example, of the transition from absolute monarchy to democracy! But we are here concerned with an objective that cannot be attained step by step. So long as the possibility of war exists, nations will continue to insist on their being as perfectly militarily prepared as they can in order to emerge triumphant from the next war. They will also find it unavoidable to educate the young people in warlike traditions and cultivating narrow nationalistic vanities. The glorification of the war spirit will proceed as long as there is reason to believe that situations will arise where that spirit will need to be invoked for the purpose of waging war. To arm means simply to approve and prepare for war, not for peace. Hence, disarmament cannot come in small steps, it must come about in one stroke or not at all. To accomplish so profound a change in the life of nations, a mighty moral effort and a deliberate departure from deeply ingrained traditions are required. Anyone who is not prepared to let the fate of his country, in the event of conflict, depend without qualification upon the decisions of an international court of arbitration, and who is not prepared to see his country enter into treaties that provide for such a procedure without any reservation, is not really resolved to avoid war. This is a case of all or nothing.

See EoP, pp. 163f. See also Albert Einstein, *Ideas and Opinions*, S. Bargmann, trans. (New York: Crown Publishers, 1954), pp. 102f.; hereafter I&O.

[6]A word of caution. Nathan and Norden (EoP) aimed at translating Einstein's German texts into smooth and literate English. At times certain liberties were taken to do so, and the translation does not correspond to the literal meaning of the original text. They themselves in some of their footnotes indicate that they have made certain "revisions." The translations of Sonja Bargmann, found in I&O, are more faithful to the original texts, but these likewise at times leave something to be desired.

[7]G. Holton and Y. Elkana, eds., *Albert Einstein, Historical and Cultural Perspectives: The Centennial Symposium in Jerusalem* (Princeton, NJ: Princeton University Press, 1982).

[8]See, for example, http://www.dannen.com/szilard.html (accessed May 2007).

[9]H. G. Wells, *The World Set Free: A Story of Mankind* (London: Macmillan, 1914). P. Doty, "Einstein and International Security," in Holton and Elkana, *Albert Einstein*, pp. 347–68. B. T. Feld, "Einstein and the Politics of Nuclear Weapons," in Holton and Elkana, *Albert Einstein*, pp. 369–96. See also http://www.online-literature.com/wellshg/worldsetfree/ (accessed May 2007).

[10]Whether the chain reaction could be controlled or would produce an explosion depended on the energy of the produced neutrons, on the probability of the two uranium isotopes that make up natural uranium to capture a neutron and fission, and on the concentration and geometrical arrangement of the uranium isotopes. In any case, the amount of energy released per fission was orders of magnitude greater than in a chemical reaction.

[11]It was Szilard's knowledge of engineering and materials properties that was responsible for the use of pure (boron-free) graphite as a moderator in the Fermi-Szilard pile. This turned out to be a decisive factor in the successful construction of a critical assembly. The Germans—who after 1941 concentrated all their efforts in building a reactor—were never able to build one because the graphite they used was impure and absorbed too many neutrons. They thus had to rely on heavy water as a moderator. But the Allies could prevent their acquisition of heavy water by sabotaging and bombing the Norwegian hydroelectric plant that produced it.

[12]In the 1932 presidential campaign, Sachs had written speeches for Roosevelt on the economy, and from 1933 until 1936, he held a fairly high position in the National Recovery Administration.

[13]Sachs saw Roosevelt on October 11, 1939, at which time he presented an oral interview and submitted Einstein's letter dated August 2, the memorandum by Szilard dated August 15, copies of articles in scientific journals on uranium research, and a memorandum by Sachs dated March 10 entitled "Notes on Imminent World War in Perspective Accrued Errors and Cultural Crisis of the Inter-War Decades." Albert Einstein Archives, National Library, Jerusalem, Israel, 39-4888.1; hereafter AEA.

[14]Oliphant as quoted in A. P. Brown, *The Neutron and the Bomb: A Biography of Sir James Chadwick* (Oxford: Oxford University Press, 1997).

[15]Sachs in AEA, 39-488.i.7.

[16]AEA, 39-488.8.

[17]Einstein to Sachs, March 7, 1935; AEA, 39-488.12.

[18]Roosevelt to Sachs April 5, 1940; AEA, 39-488-16, Exhibit 8a.

[19]Watson to Sachs, April 8, 1939; AEA, 29-488.17.

[20]Sachs to Watson, April 19, 1939; AEA, 29-488.18.

[21]Sachs; AEA, 39-488.1, p. 15.

[22]Einstein to Biggs, April 25, 1939; AEA, 29-488.24.

[23]See B. T. Feld, "Einstein and the Politics of Nuclear Weapons," in *Bulletin of the Atomic Scientists* 35 (March 1979), pp. 5–16.

[24]In *Bohr Political Papers*, The Niels Bohr Archive, Copenhagen. I thank Finn Aaserud for pointing me to this entry.

[25]Wallace Akers was the Director of the Tube Alloys, the British Atomic Bomb project. M. W. Perrin was an important manager at Imperial Chemical Industries, who early on became associated with Tube Alloys and became Akers' deputy. See M. Gowing, *Britain and Atomic Energy 1939–1945* (London: Macmillan, 1964). I thank Finn Aaserud for pointing me to this letter.

[26]Although Otto Stern continued to live in Pittsburgh—he was a member of the faculty at Carnegie Tech—he traveled every six weeks to Chicago for the information meetings held there. It seems that he knew a fair amount about the state of atomic bomb development.

²⁷See A. K. Smith, *A Peril and a Hope. The Scientist Movement in America: 1945–1947* (Berkeley: University of California Press, 1965) for details.

²⁸In his biography of Einstein, Clark gives the dates of the meeting and letter as December 11 and 12, respectively. The letter to Bohr in the Einstein Archive is not dated. Clark gives as his reference for both the letter and the dates a "diplomatic source." R. W. Clark, *Einstein: The Life and Times* (London: Hodder and Stoughton, 1973), p. 645.

²⁹Einstein to Bohr, mid-December 1945; AEA, 08-094-1.

³⁰Einstein to Bohr, mid-December 1945; AEA, 08-094-1. The relevant paragraph in the original German reads as follows: "Gestern war nun Stern wieder da, und es schien uns, daß es doch einen Weg gibt, der—wenn auch geringe—Aussichten auf Erfolg gibt. Es gibt in den hauptsächlichen Ländern Wissenschaftler, die wirklich einflußreich sind und bei den politischen Leitern Gehör finden können. Da sind *Sie* mit Ihren internationalen Beziehungen, A. Compton hier in U.S.A., Lindeman in England, Kapitza und Joffe in Russland etc. Die Idee ist, diese zu gemeinsamer Aktion auf die Leiter der Politik in ihren Ländern zu bringen, um eine Internationalisierung der Militärmacht zu erreichen–ein Weg, der als zu abenteuerlich schon geraume Zeit fallen gelassen worden ist. Aber dieser radikale Schritt mit all seinen weitgehenden politischen Voraussetzungen betreffend übernationale Regierung scheint die einzige Alternative gegen das technische Geheimwettrüsten zu sein."

³¹For an insightful exposition of Bohr's intentions, his efforts, his meetings with Roosevelt and Churchill, and the influential figures around them, see Gowing, *Britain and Atomic Energy*, pp. 346–66.

³²Clark, *Einstein*, p. 576.

³³I am indebted to Finn Aaserud for this information.

³⁴Clark, *Einstein*, p. 577.

³⁵Einstein to Stern, December 26, 1945; AEA, 22-240.

³⁶Szilard's letter can be found in W. Lanouette, *Genius in the Shadows* (New York: Scribner, 1992), pp. 261f.

³⁷The U.S. Department of Energy Archive offers this note on "MAUD": Although many people assume MAUD is an acronym, it actually stems from a simple misunderstanding. Early in the war, while Bohr was still trapped in German-occupied Denmark, he sent a telegram to his old colleague Frisch. Bohr ended the telegram with instructions to pass his words along to "Cockroft and Maud Ray Kent." "Maud," mistakenly thought to be a cryptic reference for something atomic, was chosen as a codename for the committee. Not until after the war was Maud Ray Kent identified as the former governess of Bohr's children. Available at http://www.mbe.doe.gov/me70/manhattan/maud.htm (accessed May 2007).

³⁸See, for example, EoP, pp. 350, 519. In September 1952, the editor of *Kaizo* asked Einstein: "Why did you co-operate in the production of the atomic bomb although you were well aware of its tremendous destructive power?" His answer: "My participation in the production of the atomic bomb consisted of one single act: I signed a letter to President Roosevelt, in which I emphasized the necessity of conducting large-scale experimentation with regard to the feasibility of producing an atomic bomb. I was well aware of the dreadful danger which would threaten mankind were the experiments to prove successful."

³⁹January 23, 1950; EoP, p. 519.

⁴⁰*Linus and Ava Helen Pauling Papers*, Oregon State University Library, Box 104, E: individual correspondence. Available at http://osulibrary.oregonstate.edu/specialcollections/coll/pauling/calendar/1954/11/16.html (accessed May 2007).

⁴¹Einstein to Hutchins, September 10, 1945; EoP, p. 337.

⁴²The Reveses amassed an extensive art collection, consisting of some 1,400 pieces and valued at some $40 million, which is now housed in the Dallas Museum of Art. See: M. Gilbert, ed., *Winston Churchill and Emery Reves: Correspondence, 1937–1964* (Austin: University of Texas Press, 1997). M. Soames, ed., *Speaking for Themselves: The Personal Letters of Winston and Clementine Churchill* (New York: Doubleday, 1998). R. Jenkins, *Churchill* (London: Macmillan 2001). See also the entries under Emery Reves in Google, and in particular William Van Dusen Wishard's inaugural Emery Reves Memorial Lecture delivered at the Reves Center for International Studies, College of William and Mary on November 13, 2000: http://www.tay.fi/FAST/AK11/EXTRACTS/wishard.html (accessed January 2005, May 2007).

⁴³In October 1931, Einstein wrote to Russell that Dr. Imre Revesz would soon be in England and asked Russell to grant him a brief interview. AEA, 33.157.

⁴⁴See, for example, Orville Prescott's review of it on June 13, 1945, in the *New York Times*, p. 21.

[45]Reves to Einstein, August 24, 1945; AEA, 57-290.00.

[46]Reves to Einstein, August 24, 1945; AEA, 57-290.00. Reves to Einstein, October 17, 1945; AEA, 57–299. Reves to Einstein, November 27, 1945; AEA, 57-300.

In fact, the reaction continued well into 1946. The December issue of *Reader's Digest*, of some 10 million copies, carried the first of three installments of Reves' book, *The Anatomy of Peace* (New York: Harper and Brothers, 1945). The same three extracts were published in the British, Spanish, Portuguese, Swedish, and Arabic edition of *The Reader's Digest*. In addition, during the first half of 1946, Reader's Digest organized discussion of the *Anatomy of Peace* in some 15,000 American clubs and discussion groups.

[47]Reves to Einstein, August 24, 1945; AEA, 57-290.

[48]Einstein to Reves, August 28, 1945; AEA, 57-292. In the fall of 1945 during his interview with Raymond Swing, Einstein commended Reves "whose book *The Anatomy of Peace* is intelligent, brief, clear, and if I may use the abused term, dynamic on the topic of war and the need for world government." Before Einstein's endorsement, some 8,000 copies of the book had been sold; during the six months after Einstein's endorsement, some 180,000 copies were sold in the United States. It was subsequently translated into twenty-five languages, and eventually some 800,000 copies of the book were sold worldwide.

[49]EoP, pp. 340f.

[50]Reves to Churchill, December 19, 1945, quoted in Gilbert, *Winston Churchill and Emery Reves*, p. 257.

[51]Shortly after the death of Roosevelt, Truman, on his way back from the San Francisco U.N. Charter meeting, addressed a crowd in Kansas City: "It will be just as easy for nations to get along in a republic of the world as it is for you to get along in the republic of the United States. Now when Kansas and Colorado have a quarrel over water in the Arkansas River, they don't call out the National Guard in each state and go to war over it. They bring suit in the Supreme Court of the United States and abide by the decision. There isn't a reason in the world why we cannot do that internationally." Truman's statement was quoted in the *Open Letter to the American People* initiated by Justice Roberts and signed by Einstein. The entire open letter was reprinted on the cover of Emery Reves' *The Anatomy of Peace*.

[52]Reves to Einstein, September 27, 1945; AEA, 57-293.00; EoP, pp. 337f.

[53]At the time, William Golden was attached to the office of Lewis Strauss at the Atomic Energy Commission. He recorded his interview with Einstein in AEA document 32-447-1.

[54]*Foreign Relations of the United States, 1947*, vol. I (Washington, DC: U.S. Government Printing Office), pp. 487–89.

[55]See H. York, *The Advisors: Oppenheimer, Teller, and the Superbomb. With a Historical Essay by Hans A. Bethe* (Stanford, CA: Stanford University Press, 1989).

[56]For the biography of this remarkable man, see http://www.ajmuste.org/ajmbio.htm (accessed May 2007).

[57]Muste to Einstein, January 19, 1950; AEA, 60-630-1,2.

[58]Einstein to Muste, January 23, 1950; AEA, 60-6311; EoP, pp. 519f.

[59]Muste to Einstein, January 30, 1950; AEA, 60-635-1,2.

[60]Einstein to Muste, January 30, 1950; AEA, 60-636; EoP, pp. 520f.

[61]Einstein to Muste, April 6, 1954; AEA, 60-651. The reference to the voice of angels is to be understood as slightly mocking: Muste always had ministers joining him in his appeals—in this case Paul Scherer, "the country's foremost Lutheran preacher," and Howard Thurman, "negro poet and chaplain of Boston University."

[62]AEA, 28-925-1.

[63]AEA, 28-920-1.

[64]EoP, pp. 455f.

[65]Fred Jerome, *The Einstein File* (New York: St. Martin Press, 2002).

[66]Let me give just one example. On June 12, 1953, Einstein made headlines in the *New York Times* for the advice he had given to William Frauenglass, a high-school English teacher, who had been called before the Senate Internal Security Sub-committee and had refused to be a cooperative witness even though it meant the loss of his job. Frauenglass had written Einstein for a letter of support. Einstein complied and agreed to have the letter sent to the *New York Times*. In it he advised civil disobedience: "What ought the minority of intellectuals do against this evil? Frankly, I can only see the revolutionary way of non-cooperation in the sense of Gandhi's. Every intellectual who is called before one of the

committees ought to refuse to testify, i.e., he must be prepared for jail and economic ruin, in short for the sacrifice of his personal welfare in the interests of the cultural welfare of his country." In its June 13 editorial, the *New York Times* chastised Einstein by interpreting his statement to mean that "Congressional committees have no right to question teachers and scientists or to seek out subversives wherever they can find them." Congressmen have that right, the *Times* asserted. "What is wrong is the way some of them are exercising it. . . . To employ the unnatural and illegal forces of civil disobedience, as Professor Einstein advises, is in this case to attack one evil with another." But of course, Einstein only suggested civil disobedience when Congressional committees were abusing their powers; he wasn't questioning the right of Congress to investigate under specified circumstances and specified guidelines. Moreover, Einstein was prepared to go to jail, face economic ruin, and sacrifice his personal welfare in the interest of the cultural welfare of his adopted country.

[67] B. Russell, *Autobiography*, vol. 3 (London: George Allen and Unwin Ltd., 1969), p. 18.

[68] Russell, *Autobiography*, p. 18. The above biographical material on Muste is based on http://www.san.beck.org/GPJ24-Russell,Muste.html#3 (accessed May 2007).

[69] Bertrand Russell, "Man's Peril from the Hydrogen Bomb," BBC radio address, December 23, 1954.

[70] Bertrand Russell, Letter to Albert Einstein, February 11, 1955.

[71] EoP, pp. 625f.

[72] EoP, p. 631.

[73] A copy of the Manifesto, known also as the Einstein-Russell Manifesto, can be found on the web at http://www.pugwash.org/about/manifesto.htm (accessed May 2007).

[74] Consult the site of the Pugwash Organization for further details about its conferences, workshops, and publications: http://www.pugwash.org/site_index.htm.

[75] Joseph Rotblat was one of the very few scientists to leave Los Alamos; he did so in December 1944 when it had become clear that Germany had been defeated. He was a Polish Jew who had emigrated to England before the war, and was one the sixteen British physicists to join the Los Alamos project in late 1943. At a private dinner in March of 1944 at the home of James Chadwick, the head of the British Mission to Los Alamos, Groves had stated, "You realize, of course, that the whole purpose of this project is to subdue the Russians." Groves's statement shocked Rotblat. He had thought that the purpose of Los Alamos had been to cope with the German threat, but evidently its mission had already changed. Groves's declaration initiated Rotblat's doubts about the enterprise. Toward the end of 1944, he was informed by Chadwick, who was close to British intelligence, that Germany had stopped work on the bomb. He accordingly decided to end his participation in the project. Rotblat had believed "that if the U.S. and Great Britain had developed the bomb, then even if Germany had it, we could have made the Germans give up using it. It was the idea of deterrence." J. Rotblat, *Scientists in the Quest for Peace: A History of the Pugwash Conferences* (Cambridge, MA: MIT Press, 1995). Groves testified at the Oppenheimer hearings in 1954 that the views of the aim of the Los Alamos project expressed at the Chadwick dinner had shaped his policy regarding security matters at Los Alamos: "There was never from about 2 weeks from the time I took charge [of the atomic bomb] project any illusion on my part but that Russia was our enemy and the project was conducted on that basis." Philip M. Stern, Foreword, in *In the Matter of J. Robert Oppenheimer. Transcript of Hearing before Personnel Security Board and Texts of Principal Documents and Letters*, U.S. AEC (Cambridge, MA: MIT Press, 1971).

[76] First explicitly stated in Einstein's article in the *The London Times*, November 28, 1919.

[77] I&O, p. 228.

NOTES TO CHAPTER 8

[1] See *Time* 154 (December 31, 1999). On Einstein's general cultural import, see, e.g., Gerald Holton, "Einstein's Influence on Our Culture," in his *Einstein, History, and Other Passions* (Cambridge, MA: Harvard University Press, 1996), ch. 6. The fullest overview of 19th- and 20th-century art and science to date is Lynn Gamwell's beautifully illustrated volume *Exploring the Invisible: Art, Science, and the Spiritual* (Princeton, NJ: Princeton University Press, 2002). Drawing on my discussions of Einstein and art in *The Fourth Dimension and Non-Euclidean Geometry in Modern Art* (1983; 2nd ed., Cambridge, MA: MIT Press, 2008) and adding new material, especially on astronomy and cosmology, Gamwell's book provides a useful introduction to this topic; see, e.g., pp. 195–205, 224–37, 252–57). With its 200-year scope, however, her book must necessarily treat artists rather summarily.

²See Philip Courtenay, "Einstein and Art," in *Einstein: The First Hundred Years*, ed. Maurice Goldsmith, Alan Mackay, and James Woudhuysen (Oxford: Pergamon Press, 1980), p. 156. Courtenay extends the error to Marcel Duchamp, as well, asserting that his *Nude Descending a Staircase, No. 2* "gives us an artistic expression of Einstein's space-time continuum" (p. 149).

³Mary Acton, *Learning to Look at Modern Art* (London: Routledge, 2004), p. 15.

⁴See Sigfried Giedion, *Space, Time and Architecture: The Growth of a New Tradition* (Cambridge, MA: Harvard University Press, 1941); see also, e.g., Paul Laporte, "Cubism and Science," in *The Journal of Aesthetics and Art Criticism* 7 (1949), pp. 243–56. For a discussion of this phenomenon, including Giedion and Laporte's several articles on this subject, see Linda Dalrymple Henderson, "Four-Dimensional Space or Space-Time? The Emergence of the Cubism-Relativity Myth in New York in the 1940s," in *The Visual Mind II*, ed. Michele Emmer (Cambridge, MA: MIT Press, 2005), pp. 349–97. On Einstein's comments rejecting both Giedion's and Laporte's assertions, see pp. 362f., 374–77, as well as Holton's discussion of Laporte in *Einstein, History, and Other Passions*, p. 131.

⁵Holton, *Einstein, History, and Other Passions*, ch. 6, n. 10. On the problems with the most recent attempt to argue the Cubism-Relativity connection, *Inside Modernism: Relativity Theory, Cubism, Narrative* by Delo Mook and Thomas Vargish (New Haven, CT: Yale University Press, 1999), see Henderson, "Four-Dimensional Space or Space-Time," p. 384, n. 16. For a survey of the writing on art and science that first emerged in earnest in the 1940s, see Linda Dalrymple Henderson, "Editor's Introduction: I. Writing Modern Art and Science—An Overview; II. Cubism, Futurism, and Ether Physics in the Early Twentieth Century," in *Science in Context* 17 (2004), pp. 423–66.

⁶A collage-like "essay" of more than eighty pages, "Einstein and Cubism: Science and Art" was assembled by scholar Joseph Masheck after Schapiro's death from his notes. On this text, see Henderson, "Four-Dimensional Space or Space-Time," p. 385, n. 16.

⁷See Henderson, *Fourth Dimension*, ch. 2, app. A. This argument initially appeared in Henderson, "A New Facet of Cubism: 'The Fourth Dimension' and 'Non-Euclidean Geometry' Reinterpreted," in *The Art Quarterly* 34 (Winter 1971), pp. 410–33.

⁸For a history of the fourth dimension's popularization, including Abbott's *Flatland*, see Henderson, *Fourth Dimension*, ch. 1 and bibliography; see also Henry P. Manning, *The Fourth Dimension Simply Explained: A Collection of Essays Submitted in the Scientific American's Prize Competition* (New York: Munn & Co., 1910; rpt. ed., New York: Dover Publications, 1960). In addition to Manning, the best single introduction to the popular fourth dimension in this period is Claude Bragdon's *A Primer of Higher Space: The Fourth Dimension* (Rochester, NY: The Manas Press, 1913), which is also available in reprint editions.

⁹Pablo Picasso, as quoted in Ramón Gómez de la Serna, "Completa y verídica historia de Picasso y el cubismo," *Revista de Occidente* 25 (July 1929), p. 100. On the importance of Henri Bergson's philosophy for Cubism as well, see Mark Antliff and Patricia Leighten, *Cubism and Culture* (New York: Thames & Hudson, 2001), ch. 2.

¹⁰Henri Poincaré, *La Science et l'hypothèse* (Paris: Flammarion, 1902), pp. 72f., 89f. On Cubist theory and painting in relation to Jouffret and Poincaré, see Henderson, *Fourth Dimension* (1983), pp. 57–58, 71–73, 81–85, 93–99. Despite its title, Arthur I. Miller's book *Einstein, Picasso: Space-Time and the Beauty That Causes Havoc* (New York: Basic Books, 2001) does not suggest a direct connection between Picasso and Einstein, but rather examines their "parallel biographies" and finds commonalities in sources such as four-dimensional geometry and the ideas of Poincaré. For an art historian's comments on Miller's argument, see Henderson, "Four-Dimensional Space or Space-Time," p. 386, n. 16; and Henderson, "Reintroduction," *Fourth Dimension* (2008).

¹¹On this subject, see, e.g., Henderson, "Editor's Introduction: II. Cubism, Futurism, and Ether Physics in the Early Twentieth-Century," in *Science in Context*, pp. 445–58; Henderson, "X Rays and the Quest for Invisible Reality in the Art of Kupka, Duchamp, and the Cubists," in *Art Journal* 47 (1988), pp. 323–40; and Henderson, "Vibratory Modernism: Boccioni, Kupka, and the Ether of Space," in *From Energy to Information: Representation in Science and Technology, Art, and Literature*, ed. Henderson and Bruce Clarke (Stanford, CA: Stanford University Press, 2002), pp. 126–49.

¹²Observations made during a solar eclipse in 1919 confirmed predictions of the General Theory of Relativity (1916), and only then did Einstein gain international acclaim. On Special and General Relativity—and the reception of Einstein's theories—see, e.g., Helge Kragh, *Quantum Generations: A History of Physics in the Twentieth Century* (Princeton, NJ: Princeton University Press, 1999),

pp. 90–104. On the continued allegiance to the ether, see Henderson, "Editor's Introduction II," *Science in Context*, pp. 450–53, as well as n. 95 below.

[13]On Le Bon, see Henderson, "Editor's Introduction II," *Science in Context*, pp. 449–51, as well as Gustave Le Bon, *L'Evolution de la matière* (Paris: Flammarion, 1905). See also Robert Kennedy Duncan, *The New Knowledge* (New York: A. S. Barnes, 1905), p. 5.

[14]See Linda Dalrymple Henderson, *Duchamp in Context: Science and Technology in the Large Glass and Related Works* (Princeton, NJ: Princeton University Press, 1998); for a summary of Duchamp's scientific interests, see Henderson, "The *Large Glass* Seen Anew: Reflections of Science and Technology in Marcel Duchamp's 'Hilarious Picture,'" *Leonardo* 32, no. 2 (1999), pp. 113–26.

[15]See Henderson, Reintroduction, *Fourth Dimension* (2008).

[16]For Duchamp's notes from the *White Box*, see *Salt Seller: The Writings of Marcel Duchamp*, ed. Michel Sanouillet and Elmer Peterson (New York: Oxford University Press, 1973). This collection also includes Duchamp's notes from his *Box of 1914* and 1934 *Green Box*. On the production of these boxes, see Francis M. Naumann, *Marcel Duchamp: Making Art in the Age of Mechanical Reproduction* (New York: Harry N. Abrams, 1999), pp. 54–56, ch. 4.

[17]See, e.g., the advertisement published in *New York Times Book Review*, Mar. 19, 2000, p. 5.

[18]On Höch, see Maud Lavin, *Cut with the Kitchen Knife: The Weimar Photomontages of Hannah Höch* (New Haven: Yale University Press, 1993), pp. 19–24.

[19]See Kathleen James, "Expressionism, Relativity, and the Einstein Tower," in *Journal of the Society of Architectural Historians* 53 (December 1994), pp. 392–413.

[20]For Mendelsohn's quote, see James, "Expressionism," p. 397; for Mendelsohn's interest in energy and high-speed motion in Relativity Theory, as well as his organic interpretation of technology, see pp. 397f., 406f.

[21]Salvador Dalí, "The Rotting Donkey" (1930), in *The Collected Writings of Salvador Dalí*, ed. and trans. Haim Finkelstein (Cambridge, UK: Cambridge University Press, 1998), p. 223. Gavin Parkinson's forthcoming *Surrealism, Art and Modern Physics: Relativity, Quantum Mechanics, Epistemology* (New Haven, CT: Yale University Press, 2007) is an exemplary study of Surrealism and both Relativity Theory and quantum physics, including Dalí. For Parkinson's discussion of *The Persistence of Memory* (pp. 148–51) as well as other entries related to Dalí and science, see Dawn Ades and Michael Taylor, *Dalí*, exhibit cat., Palazzo Grassi, Venice, and Philadelphia Museum of Art (New York: Rizzoli, 2004).

[22]Dalí, *The Conquest of the Irrational* (1935), in his *Collected Writings*, p. 272. For the artist's 1930 comments on "dilation," see Dalí, "The Sanitary Goat," ibid., p. 229.

[23]Portraits of Einstein by artists such as Max Lieberman, Jacob Epstein, Ben Shahn, and Wolfgang Paalen would also be included in this category, although they are not discussed here.

[24]See David Cassidy, *Einstein and Our World* (Atlantic Highlands, NJ: Humanities Press International, 1995), in which a portrait drawing of Einstein is compressed laterally.

[25]D'Alembert in the 18th century had suggested that time be could considered a fourth dimension, but apart from H. G. Wells' *The Time Machine* (1895), the later 19th and early 20th century was dominated by the interpretation of the fourth dimension as a spatial phenomenon (see Henderson, *Fourth Dimension* [1983], pp. 9, 17–41). For Minkowski's original paper, see H. Minkowski, "Space and Time," in *The Principle of Relativity: A Collection of Original Memoirs on the Special and General Theory of Relativity by H. A. Lorentz, A. Einstein, H. Minkowski and H. Weyl*, trans. W. Perrett and G. B. Jeffery ([1923]; New York: Dover Publications, 1952), pp. 73–91.

[26]Writing of this period in Berlin, Alexander Moskowski declared, "during that time no name was quoted so often as that of this man. Everything sank away in the face of this universal theme which had taken possession of humanity. . . . Newspapers entered on a chase for contributors who could furnish them with short or long, technical or non-technical, notices about Einstein's theory. All contemporary questions had gained a fixed center from which threads could be spun to each. Relativity had become a sovereign password." See Alexander Moszkowski, *Conversations with Einstein* (1921; New York: Horizon Press, 1971), pp. 13f.

[27]On the response to Relativity in Germany and Russia, see Thomas Glick, ed., *The Comparative Reception of Relativity* (Dordrecht: D. Reidel Publishing, 1987), chs. 3, 8. See also Henderson, *Fourth Dimension* (1983), pp. 243f., for a sampling of accessible Russian sources during the 1910s, including Alexander Vasiliev's 1913 *New Ideas in Mathematics*.

[28]On Gabo's studies and recollection of hearing of Einstein, see Martin Hammer and Christina Lodder, *Constructing Modernity: The Art & Career of Naum Gabo* (New Haven, CT: Yale University Press, 2000), pp. 20f. For Lissitzky's training, see, e.g., Henk Puts, "El Lissitzky (1890–1941): His Life and Work," in *El Lissitzky 1890–1941: Architect, Painter, Typographer, Photographer* (Eindhoven, Netherlands: Stedelijk Van Abbemuseum, 1991), p. 14.

[29]For Gabo's *Kinetic Construction*, see Hammer and Lodder, *Constructing Modernity*, p. 69, as well as Gabo's 1969 account of making the work in *Gabo on Gabo: Texts and Interviews*, ed. Martin Hammer and Christina Lodder (East Sussex, UK: Artists Bookworks, 2000), pp. 259–62.

[30]Naum Gabo and Anton Pevsner, "Realistic Manifesto," in *Gabo on Gabo*, p. 26f. For Gabo's Berlin title, see Hammer and Lodder, *Constructing Modernity*, p. 110. Hammer and Lodder are circumspect about associating Gabo's and Pevsner's manifesto directly with Relativity Theory, suggesting instead parallels with Kant's philosophical discussion of space and time (pp. 65f.). Yet, with Gabo having heard about Relativity Theory in Germany as early as 1911 or 1912 and with the general enthusiasm for Einstein's revolutionary views by 1920, it is unlikely that any avant-garde figure at this moment would have found Kant a more compelling figure than Einstein or Minkowski with whom to associate a new artistic style. Even with his engineering training, however, Gabo's grasp of the new theories would have been at the level of popularization, and he later commented on their limited usefulness to him in a 1929 diary entry (pp. 393f.). Nonetheless, Hammer and Lodder (pp. 399–401) do connect Gabo's later "Spheric Theme" series to Einstein's model of the universe as curved back upon itself, as discussed by Arthur Eddington in his chapter on "Spherical Space" in *The Expanding Universe* (1933). As in this instance, the authors do an excellent job of establishing Gabo's primary scientific and mathematical sources during his career—from late 19th-century physics and the popular books of James Jeans and Eddington to D'Arcy Wentworth Thompson and mathematical models (ch. 14).

[31]Gabo, as quoted in a 1957 interview with Abram Lassaw and Ilya Bolotowsky, in *Gabo on Gabo*, p. 213. For the subsequent development of kinetic art, see, e.g., Frank Popper, *Origins and Development of Kinetic Art* (Greenwich, CT: New York Graphic Society, 1968); and Guy Brett, *Force Fields: Phases of the Kinetic* (Barcelona: Museu d'Art Contemporani de Barcelona, 2000). On the rhetoric of space-time that developed around kinetic art in Paris in the 1950s and elsewhere, see n. 74 below as well as Henderson, Reintroduction, *Fourth Dimension* (2008). The Russian sculptor Alexander Archipenko, in Berlin from 1921 to 1923, made a unique contribution to the genre after he relocated to the United States in 1923. His *Archipentura* or "Apparatus for Displaying Changeable Pictures" used moving slats on the side of a vertical box-like structure to present one image that gradually turned into another. Archipenko dedicated the *Archipentura* to "T. Edison and A. Einstein." For a reproduction of the work, as well as Archipenko's additional commentary on Einstein, see Katherine Jánszky Michaelson and Nehama Guralnik, *Alexander Archipenko: A Centennial Tribute* (Washington, DC: National Gallery of Art, 1986), pp. 64–66.

[32]On Moholy-Nagy's art, including the *Light Prop*, see, e.g., Krisztina Passuth, *Moholy-Nagy* (London: Thames and Hudson, 1985). For an overview of Moholy-Nagy's response to Einstein and Relativity (as well as the altered title question), see Henderson, "Four-Dimensional Space or Space-Time," pp. 370–72, 392f., n. 67. *Vision in Motion* (Chicago: Paul Theobold, 1947), which appeared the year after his death, is the fullest statement of Moholy-Nagy's theory and practice; see, e.g., the chapter on "Space-Time Problems" (pp. 244–69).

[33]Moholy-Nagy recounted his meeting with Einstein in an August 24, 1924, letter to Theo van Doesburg (Theo van Doesburg, *Grondbegrippen van de nieuewe beeldende kunst* [Nijmegan: Sun, 1983], p. 109), to which Oliver Botar kindly alerted me. Einstein was recruited as one of the "Friends of the Bauhaus" in its struggle with the Weimar government in 1924; see "An Invitation to Join the Circle of Friends," in Hans M. Wingler, *The Bauhaus: Weimar, Dessau, Berlin, Chicago* (1969; Cambridge, MA: MIT Press, 1978), p. 78. Letters in the Bauhaus Archive document the continued correspondence between Einstein the Bauhaus's founding director, Walter Gropius (Gamwell, *Exploring the Invisible*, p. 320, n. 38).

[34]For van Doesburg's extensive engagement with the fourth dimension and space-time in the 1920s, summarized here, see Henderson, *Fourth Dimension* (1983), pp. 321–36. See also Henderson, "Theo van Doesburg, 'Die Vierte Dimension,' und die Relativitätstheorie in den zwanziger Jahren," in *Zeit: Die vierte Dimension in der Kunst*, ed. Michel Baudson (Mannheim: Kunsthalle, 1985; Weinheim: VCH, 1985), pp. 195–205.

³⁵Theo van Doesburg, "The Will to Style: The Reconstruction of Life, Art and Technology," quoted in Henderson, *Fourth Dimension* (1983), p. 322. van Doesburg also linked Dada's liberating iconoclasm to the fourth dimension and declared Einstein a Dadaist, along with Bergson and Charlie Chaplin, in his 1923 text, "What is Dada?????????" (Henderson, *Fourth Dimension*, pp. 328f.).

³⁶For this history, including van Doesburg's assertion in *De Stijl* in 1921 that the two filmmakers were "assisted by the best talents . . . (for example, by Einstein)," see Standish Lawder, *The Cubist Cinema* (New York: New York University Press, 1975), pp. 46–49. See also Justin Hoffman, "Hans Richter: Constructivist Filmmaker," in Hans Richter, *Activism, Modernism, and the Avant-Garde*, ed. Stephen C. Foster (Cambridge, MA: MIT Press, 1998), pp. 72–91.

³⁷See Henderson, *Fourth Dimension* (1983), pp. 323–26, figs. 102, 103.

³⁸In contrast to Giedion's comments on Picasso and Einstein, noted at this essay's beginning, van Doesburg's arguments on the subject of architecture were legitimately grounded in avant-garde theory and practice of the 1920s.

³⁹See Henderson, *Fourth Dimension* (1983), pp. 332f., fig. 109. See again n. 32 above on the question of Moholy's title. Russian filmmaker Sergei Eisenstein associated his technique of "overtonal montage" with the fourth dimension and Einstein in his 1929 essay "The Filmic Fourth Dimension"; see Henderson, *Fourth Dimension* (1983), pp. 298f., 338, n. 120. In *Vision in Motion*, Moholy-Nagy illustrated Van Doesburg's "filmcontinuum" hypercube drawing and noted the potential of Eisenstein's montage techniques for future development "to conquer psychological and physical space-time" (p. 279).

Relativity Theory and Moholy-Nagy's space-time theories were vital sources for American painter and experimental filmmaker Jim Davis, who sought to capture abstract patterns of light on film. Davis taught modern art at Princeton and photographed Einstein as well. On Davis's films of light in motion and his writings, see *Jim Davis: The Flow of Energy*, ed. Robert Haller (New York: Anthology Film Archives, 1992). The Einstein photographs are reproduced in Haller, *Galaxy: Avant-Garde Film-Makers Look Across Space and Time* (New York: Anthology Film Archives, 2001), pp. 16f.

⁴⁰On Pereira's art and theory, including her grounding in the spatial fourth dimension, see Karen A. Bearor, *Irene Rice Pereira: Her Paintings and Philosophy* (Austin: University of Texas Press, 1993); see also Henderson, Reintroduction, *Fourth Dimension* (2008).

⁴¹Pereira, "An Abstract Painter on Abstract Art," in *American Contemporary Art* 1 (1944), p. 4; quoted in Bearor, *Irene Rice Pereira*, p. 135.

⁴²Athena Tacha, "Sculptured Light," *Art International* 11 (Christmas 1967), pp. 29, 49. For the "time-spatial energy" reference, see Tacha, "Sculptured Light," p. 33; and László Moholy-Nagy, "*The New Vision*" and "*Abstract of an Artist*," trans. Daphne M. Hoffman (New York: Wittenborn, Schultz, 1946), p. 50.

⁴³For a fuller version of the discussion of Lissitzky here, see Henderson, *Fourth Dimension* (1983), pp. 294–98. For Lissitzky's training in engineering and architecture, see again n. 28; for his art and career, see Peter Nisbet, "An Introduction to El Lissitzky," in Nisbet, *El Lissitzky 1890–1941* (Cambridge, MA: Busch-Reisinger Museum, Harvard University, 1987). As Nisbet has noted, the initial Prouns have a pronounced architectural orientation and often bear specific references to buildings in their titles (pp. 18–21).

⁴⁴El Lissitzky, "Proun," unpublished manuscript, trans. John Bowlt, in *El Lissitzky* (Cologne: Galerie Gmurzynska, 1976), p. 60. On Malevich's Suprematism and the fourth dimension, see Henderson, *Fourth Dimension* (1983), pp. 274–94.

⁴⁵Yves-Alain Bois has emphasized the "radical reversibility" in Lissitzky's Prouns that reflects the artist's reconception of the canvas surface as more akin to "geographical maps" viewed from above and at a variety of angles. See, e.g., Bois, "El Lissitzky: Radical Reversibility," in *Art in America* 76 (1988), pp. 170f.

⁴⁶Lissitzky, "A. and Pangeometry" (originally published 1925), in Sophie Lissitzky-Küppers, *El Lissitzky: Life, Letters, Texts* (London: Thames and Hudson, 1967), p. 351.

⁴⁷For the "interchange station" quote, see Lissitzky, "The Film of El's Life," in Lissitzky-Küppers, *El Lissitzky*, p. 329. Lissitzky also used reversible axonometric perspective to render these rooms; see, e.g., *El Lissitzky* (Eindhoven), pp. 130–33, 182–85.

⁴⁸Lissitzky, "A. and Pangeometry," pp. 351–53.

[49]Lissitzky, "A. and Pangeometry," p. 351. Although the irregular curvature of General Relativity's model of space-time was Riemannian, Nikolai Lobachevsky, the Russian pioneer of a negatively curved non-Euclidean geometry, was a particular hero for the Russian avant-garde. See, e.g., Henderson, *Fourth Dimension* (1983), pp. 241f., 254f., as well as pp. 3–6 and 11–17 on non-Euclidean geometry and its philosophical implications.

[50]El Lissitzky letter to J.J.P. Oud, June 30, 1924, in Galerie Gmurzynska, *El Lissitzky*, p. 73. See also, however, Lissitzky's equivocation on Relativity Theory's implications for understanding time in "A. and Pangeometry," p. 351.

[51]André Breton, *Le Surréalisme et la peinture* (New York: Brentano's, 1945), p. 152. See also Henderson, *Fourth Dimension* (1983), pp. 346–49, as well as the expanded discussion of Matta and Onslow Ford in Henderson, Reintroduction, *Fourth Dimension* (2008). For the fullest treatment of this subject, see Parkinson, *Surrealism and Science*.

[52]By this time, Dalí had been excommunicated from Surrealism by Breton, whose focus was now on the renewed interest in psychic automatism among the young painters he names here. In his 1936 essay "Le Surrationalisme," Gaston Bachelard touted non-Euclidean geometry as a Surrealist weapon in the fight against logic and reason; see Bachelard, "Le Surrationalism," in Julien Levy, *Surrealism* (New York: Black Sun Press, 1936), p. 186.

[53]On this subject, see Henderson, Reintroduction, *Fourth Dimension* (2008).

[54]On Einstein's visits and the publicity surrounding them, see, e.g., Ronald W. Clark, *Einstein: The Life and Times* (New York: World Publishing Co., 1971), pp. 426–35, 444f. The *Reader's Guide to Periodical Literature* lists over fifty articles under Einstein's name for the period 1929–32, along with seventeen entries under "Einstein theory" (and "Einstein unified field theory"), compared to the total of twenty-three articles under "Einstein, Albert" and "Einstein theory" for 1925–28.

[55]See Herman Hupfeld, "As Time Goes By" (HARMS, Inc., 1931). Although usually sung with the phrase "fourth dimension," the lyrics as printed in the sheet music issued by Warner Brother Publications after the 1942 film *Casablanca* refer to "third dimension" instead of "fourth dimension." It is conceivable that Hupfeld was making a sly joke about the fourth dimension, emphasizing instead the new *third* dimension being explored in early 3-D films during the 1920s. See Hal Morgan and Dan Symmes, *Amazing 3-D* (Boston: Little, Brown and Co., 1982), pp. 14–17.

[56]On Davis's art and theory, see, e.g., John R. Lane, *Stuart Davis: Art and Theory* (New York: Brooklyn Museum, 1978) as well as the catalogs of the major exhibitions of his work in 1991 and 1997. Bearor documents Pereira and Davis' friendship in *Irene Rice Pereira*, p. 18.

[57]Stuart Davis, "Autobiography" (1945), in *Stuart Davis*, ed. Diane Kelder (New York: Praeger, 1971), p. 30.

[58]See Davis, "Daybook," p. 55. Jeans explained of his diagram, "Thus by welding together a length . . . and a time . . . , we have obtained an area having one dimension of space and one of time" (Sir James Jeans, *The Mysterious Universe* [New York: The Macmillan Co., 1931], p. 109).

[59]Davis, "Daybook," p. 55. For more of Davis's ideas on "angular contrast" as a design technique, see Davis, "Daybook," p. 56f., and Lane, *Stuart Davis*, pp. 23–25.

[60]Lane, *Stuart Davis*, p. 58.

[61]Davis, January 2, 1947, quoted in Lane, *Stuart Davis*, p. 66; for Davis's usage of "color-space," see Lane, *Stuart Davis*, ch. 4. For a further discussion of Davis and Relativity Theory, see Henderson, Reintroduction, *Fourth Dimension* (2008).

[62]This category could also include artists who themselves invented symbolic means to "diagram" aspects of Relativity. The primary example of that approach may be the painting-collage by Duchamp's friend, American artist John Covert's *Time* (1919). On this witty, diagrammatic image, dominated by its rows of upholstery tacks (tic-toc), see Henderson, *Fourth Dimension* (1983), fig. 68 and pp. 228–30, as well as Judith Zilczer, "Decoding John Covert's *Time*," *Art Bulletin* 75 (1993): 713–22.

[63]For this series, see Henry Geldzahler and Robert Rosenblum, *Andy Warhol: Portraits of the Seventies and Eighties* (London: Anthony d'Offay Gallery and Thames and Hudson, 1993), p. 152.

[64]See Halsman's description of his session with Einstein and one of his images in *Einstein: A Centenary Volume*, ed. A. P. French (Cambridge, MA: Harvard University Press, 1979), pp. 26–28. Halsman, whose name Einstein had succeeded in getting onto the list for emergency visas after the fall of France, was also a collaborator of Dalí. See, e.g., Philippe Halsman, *Dali's Mustache: A Photographic Portrait* (New York: Simon and Schuster, 1954).

[65]See H. Minkowski, "Space and Time" (1908), in *The Principle of Relativity* (1923, 1952), fig. 1, pp. 78–83. Similar diagrams were readily available, for example, in Sir Arthur Eddington's *Space, Time and Gravitation: An Outline of the General Theory of Relativity* (1920; New York: Harper & Row, 1959), ch. 3 ("The World of Four Dimensions"), fig. 7, pp. 53–56. Although Warhol might have relied in part on the Dover cover photograph, his drawing is closer to Halsman's photographs of a slightly older Einstein.

[66]Another instance of this approach may be present in the 1963 painting by Spanish artist Remedios Varo, which appeared on the dust jacket of Peter G. Bergmann's 1968 book, *The Riddle of Gravitation* (New York: Charles Scribner's Sons, 1968). Varo, who died in 1963, was closely involved in the Surrealist communities in Paris in the 1930s and in Mexico in the 1940s. Once in Mexico, Varo drew increasingly on science as well as fantasy to express her unique re-visioning of "reality." In her painting, usually translated *Phenomenon of Anti-Gravity* or *Phenomenon of Weightlessness*, she creates a magical world in which an astronomer's globe has broken free of his hands and is pulled toward an extraterrestrial object in what has been described as another dimension of space interpenetrating his own (see Bergmann, *Riddle of Gravitation*, dust jacket; and Janet A. Kaplan, *Unexpected Journeys: The Art and Life of Remedios Varo* [New York: Abbeville, 1988], pp. 175–77). However, the overlapping of the vertical window by another window at an angle to it suggests that Varo might have conceived of her interpenetrating dimensions as diagrammatic frames of reference as well.

[67]In his 1955 essay "The Brain of Einstein," Roland Barthes observed the way in which not only Einstein's person but his brain had come to function as a metonymical sign for his genius (see Barthes, *Mythologies*, trans. Annette Lavers [New York: Hill and Wang, 1972], pp. 68–70). For the relevant still from *The Day the Earth Stood Still* (with Sam Jaffe in the Einstein-like role), see *The Universe: A Convergence of Art, Music, and Science*, ed. Jay Belloli (Pasadena, CA: Armory Center for the Arts, 2001), p. 111; compare the well-known photograph of Einstein at a blackboard during one of his Cal Tech visits in the early 1930s (p. 92).

[68]See "Albert Einstein Action Figure," marketed by Accoutrements: Seattle, 2003.

[69]The designer for the dust jacket for Martin Gardner's *Relativity for the Millions* (New York: The Macmillan Co., 1962) added only a small segment of curved grid under the book's title to signify its content. Akin to the curved lettering above the Einstein action figure's head, curved movie titles floating in space became popular in the 1950s (see, e.g., *Forbidden Planet* [1955]).

[70]For Wilson's talk of Einstein and sailing as well as the emphasis on mathematical, clockwork-like counting in the opera, see the 1976 interview with Renate Klett at www.robertwilson.com. For photographic documentation of *Einstein on the Beach*, see Franco Quadri, Franco Bertoni, and Robert Stearns, *Robert Wilson* (New York: Rizzoli, 1998), pp. 86–93.

[71]For Wilson's comment, see Lawrence Shyer, *Robert Wilson and His Collaborators* (New York: Theatre Communications Group, 1989), p. 216. Shyer suggests the "fellow dreamer" comparison (p. 215) and quotes Glass's description of his own preparation, including Ronald Clark's biography (p. 216), in this chapter titled "Philip Glass" (pp. 213–32).

[72]Wilson discussed Einstein and trains as well as his interest in time and space in an e-mail communication to the author, June 8, 2001.

[73]Wilson to Henderson, June 8, 2001.

[74]See again n. 31 above and the related text; see also Henderson, Reintroduction, *Fourth Dimension* (2008).

[75]See Denise René Gallery, New York, *Le Mouvement/The Movement Paris 1955* (Paris: Editions Denise René, 1975).

[76]Although interest in Einstein and space-time was flagging at this point, young artists were increasingly interested in time itself as an element in new art forms that went beyond kinetic art. That interest is documented in several essays on contemporary art, including video, in Michel Baudson, *L'Art et le temps: Régards sur la quatrième dimension* (Brussels: Société du Palais des Beaux-Arts, 1984); see again n. 34 for the German edition of the catalog. See also the essays addressing later 20th-century art in *Tempus Fugit: Time Flies*, ed. Jan Schall (Kansas City: The Nelson-Atkins Museum of Art, 2000).

[77]See Robert Smithson, "The Iconography of Desolation," in *Robert Smithson: The Collected Writings*, ed. Jack Flam (Berkeley: University of California Press, 1996), pp. 321f.

[78]See David Bourdon, "E = MC² à Go-Go," in *Art News* 64 (1966), pp. 22–25, 57–59.

⁷⁹Di Suvero, as quoted in Bourdon, "E = MC² à Go-Go", p. 25. See also the essay by Irving Sandler in *Mark di Suvero at Storm King Art Center* (Mountainville, NY: Storm King Art Center, 1996), pp. 19–46; see p. 24 for di Suvero's historical models. The Park Place sculptors were also interested in José de Rivera (Bourdon, "E = MC² à Go-Go," p. 59); de Rivera had produced his revolving, curvilinear *Construction #1, Homage to Minkowski* (Metropolitan Museum of Art, New York) in 1955.

⁸⁰Robert Smithson, "Entropy and the New Monuments," *Artforum* 5 (June 1966), pp. 26–31. On the interest of both Smithson and the Park Place group in the fourth dimension, see Henderson, Reintroduction, *Fourth Dimension* (2008). Smithson responded specifically to Duchamp's *White Box* notes on the fourth dimension, which he may have seen in an exhibition in January 1965, before their publication in February 1967 (Henderson, *Fourth Dimension*, 2008).

⁸¹For the information on Forakis, Fleming, and Park Place here and following, see Henderson, Reintroduction, *Fourth Dimension* (2008); and Henderson, "Dean Fleming, Ed Ruda, and the Park Place Gallery: Spatial Complexity and the 'Fourth Dimension' in 1960s New York," in *Blanton Museum of Art: American Art Since 1900*, ed. Annette DiMeo Carlozzi and Kelly Baum (Austin, TX: Blanton Museum of Art, The University of Texas at Austin, 2006).

⁸²P[eter] D[emianovich] Ouspensky, *Tertium Organum, The Third Canon of Thought: A Key to the Enigmas of the World* (1911), trans. Nicholas Bessaraboff and Claude Bragdon, 2nd. Amer. ed. rev. (New York: Knopf, 1922), p. 257. Buckminster Fuller announced in a lecture of 1964 his renewed interest in the fourth dimension as space (versus time), as he came to focus more on his developing "synergetic geometry." His first publication, *4D Time Lock* in 1928, had grafted Einstein and time onto his roots in the spatial fourth dimension. In the intervening decades, however, Fuller sought to link his ideas to Einstein as an icon of radically new thinking. See Henderson, *Fourth Dimension*, Reintroduction, (2008).

⁸³Lawrence Alloway, "Peter Forakis Since 1960," *Artforum* 6 (1968), p. 28.

⁸⁴For a sampling of these developments, see Max Jammer, *Concepts of Space: The History of Theories of Space in Physics*, 3rd enl. ed. (New York: Dover Publications, 1993), ch. 6. See also Clifford M. Will, *Was Einstein Right? Putting General Relativity to the Test* (New York: Basic Books, 1986), ch. 1.

⁸⁵Among the books given to Robbin were H. P. Manning's 1911 *Geometry of Four Dimensions*, Duncan M. Y. Sommerville's 1929 *An Introduction to the Geometry of N Dimensions*, and Robert Marks's 1964 *Space, Time, and the New Mathematics* (Robbin's e-mail communication to the author, October 12, 2003).

⁸⁶See Tony Robbin, *Fourfield: Computers, Art & the Fourth Dimension* (Boston: Bullfinch Press, 1992).

⁸⁷See, e.g., Robbin, *Fourfield*, colorplates, pp. 4–13.

⁸⁸In Robbin, *Fourfield*, see pp. 104f. on *Lobofour*, and ch. 3 on Robbin and Banchoff.

⁸⁹See Tony Robbin, *Shadows of Reality: The Fourth Dimension in Cubism, Relativity, and Modern Thought* (New Haven, CT: Yale University Press, 2006). On his most recent work, see also Tony Robbin, "Four-Dimensional Projection: Art and Reality," in *The Visual Mind II*, ed. Emmer, pp. 433–49.

⁹⁰See, e,g., Brian Greene, *The Elegant Universe: Superstrings, Hidden Dimensions, and the Quest for the Ultimate Theory* (New York: W. W. Norton & Co., 1999); and Paul Halpern, *The Great Beyond: Higher Dimensions, Parallel Universes, and the Extraordinary Search for a Theory of Everything* (Hoboken, NJ: John Wiley & Sons, 2004), ch. 10.

⁹¹James Rosenquist, as quoted in Walter Hopps and Sarah Bancroft (Houston: The Menil Collection and The Museum of Fine Arts, Houston, 2003), p. 230.

⁹²See also William Jeffett, "James Rosenquist: 'what you see is what you don't get,'" in *James Rosenquist: The Stowaway Peers Out at the Speed of Light* (New York: Gagosian Gallery, 2001), pp. 9–13.

⁹³See, for example, the celebration of Einstein and space-time by the Venezuelan kinetic artist Jésus Raphael Soto, who was in Paris in the 1950s, in "The Role of Scientific Concepts in Art," in *Leonardo* 27, no. 3 (1994), pp. 227–30.

⁹⁴Marcel Duchamp, as quoted in Pierre Cabanne, *Dialogues with Marcel Duchamp*, trans. Ron Padgett (New York: Viking Press, 1971), p. 26.

⁹⁵See again n. 12; see also, e.g., the numerous books and articles by Lodge on the ether and studies such as Leo Corry, "From Mie's Electromagnetic Theory of Matter to Hilbert's Unified Foundations of

Physics," in *Studies in the History and Philosophy of Modern Physics* 30 (1999), pp. 159–83. On ether physics, see also Russel McCormmach, "H. A. Lorentz and the Electromagnetic View of Nature," *Isis* 61 (Winter 1970), pp. 459–97.

[96]Peter Galison, Introduction to "Focus: The Elusive Icon: Einstein, 1905–2005," in *Isis* 95 (2004), p. 611.

[97]On Hammer and Lodder, see again n. 30 above. See also Holly Henry, *Virginia Woolf and the Discourse of Science: The Aesthetics of Astronomy* (Cambridge, UK: Cambridge University Press, 2003), which is an exemplary model for other literature scholars examining writers' responses to science.

[98]On the prominence of energy in the writings of modern artists and critics (before the popularization of Relativity Theory), see Henderson and Clarke, eds., *From Energy to Information*, Introduction. In her chapter on "The Atomic Sublime" (*Exploring the Invisible*, pp. 264–79), Gamwell addresses the impact of the atomic bomb and the resultant discourse of energy that was at times attributed more directly to Einstein than was historically accurate.

[99]Athena Tacha, "Rhythm as Form," in *American Artists on Art from 1940 to 1980*, ed. Ellen H. Johnson (New York: Harper & Row, 1982), p. 218. Tacha's works have continued to evolve in response to contemporary scientific developments, including her remarkable recent series of *Dark Energy* and *Singularity* drawings.

[100]Agnes Denes, "Evolution and the Creative Mind," in *American Artists on Art from 1940 to 1980*, p. 139.

[101]See Lynn M. Herbert, *Matthew Ritchie: Proposition Player* (Houston, TX: Contemporary Arts Museum, 2003), p. 24.

NOTES TO CHAPTER 9

[1]See in particular the pathbreaking essay by Erwin Panofsky, *Perspective as Symbolic Form* [1927], translated and published with an introduction by Christopher Wood (Cambridge: Zone and MIT Press, 1991).

[2]Here I am thinking of the Gospel from the Echternach scriptorium, called the "Codex Aureus," ca. 1030, now in the Germanisches Nationalmuseum, Nürnberg, Hs. 156142.

[3]Underscoring this point are narratives such as the Christian Bible's parable of the vineyard, chosen for one extraordinary layout in the Codex Aureus. In this parable, a landowner pays the men he hires at the end of the day the same amount he paid those he hired at the beginning of the day. Questioned about this time-defying pay scale, the landowner defended his actions with a statement that became a proverb: "The last shall be first, and the first, last" (Matthew 20). Secular time has no meaning in God's larger universe.

[4]Martin Heidegger, "The Age of the World Picture," [1938] in *The Question Concerning Technology and other Essays*, trans. William Lovitt, (New York: Harper Torchbook, 1977), pp. 115–154.

[5]See http://www.search.com/reference/Et_in_Arcadia_ego for images and a brief discussion.

[6]See these well-known quotations from the community of Impressionists: "To paint after nature is not a matter of copying the objective world, it's giving shape to your sensations" (Paul Cezanne), or "Art is a corner of nature viewed through a temperament" (Emile Zola).

[7]See Jan Goldstein, *The Post-revolutionary Self: Politics and Psyche in France, 1750–1850* (Cambridge, MA: Harvard University Press, 2005); see also Goldstein's "Saying 'I': Victor Cousin, Caroline Angebert, and the politics of selfhood in 19th-century France," in Michael S. Roth, ed., *Rediscovering History: Culture, Politics, and the Psyche* (Stanford, CA: Stanford University Press, 1994).

[8]Attributed to Cézanne by Ambroise Vollard in *Paul Cézanne*, cited by Erle Loran in *Cézanne's Composition* (Berkeley: University of California Press, 1946), p. 46.

[9]It seems clear that Monet was preoccupied with churning out paintings to meet his substantial international market by the time Bergson was just finishing his doctorate. See Paul Hayes Tucker, *Monet in the '90s: The Series Paintings* (Boston and New Haven: Museum of Fine Arts and Yale University Press, 1989), p. 93. As one art historian put it: "Bergson doesn't explain Monet's exquisite sensibility regarding the depiction of time in space." See Eugenia Parry Janis, review of "Monet in the 90s. London, Royal Academy" in *The Burlington Magazine*, vol. 132, no. 1053 (December 1990), p. 888. For a broad perspective, see Mark Antliff, *Inventing Bergson: Cultural Politics and the Parisian*

Avant-Garde (Princeton, NJ: Princeton University Press, 1993), as well as his later "Creative Time: Bergson and European Modernism," *Tempus Fugit* (2000), pp. 35–65.

[10]From two articles published in the *Revue des deux mondes*, February 1 and 15, 1912, later included in E. Le Roy, *A New Philosophy: Henri Bergson*, trans. Vincent Benson (New York: Holt, 1913); available online at "The World Wide School," http://www.worldwideschool.org/library/books/phil/specificphilosophies/AnewPhilosophyHenriBergson/chap5.html (accessed May 2007).

[11]Wolfgang Schivelbusch, *The Railway Journey: The Industrialization and Perception of Time and Space* (Berkeley: University of California Press, 1986), p. 33.

[12]Peter Galison, *Einstein's Clocks, Poincaré's Maps* (New York: Norton, 2003).

[13]This argument draws on Meyer Schapiro's similar comment about brushstrokes in American Abstract Expressionism during the 1950s, which the art historian and critic saw as uniquely resistant to the regimentation of industrial labor. See his "Recent Abstract Painting," [1957], anthologized in his *Modern Art: Nineteenth and Twentieth Centuries, Selected Papers* (New York: George Braziller, 1978), pp. 213–26.

[14]For a hint of the complexity of Bergson's relation to Impressionism, I turn again to his important apologist Le Roy: "Is Mr. Bergson only a poet, and does his work amount to nothing but the introduction of impressionism in metaphysics? It is an old objection. If the truth be told, Mr. Bergson's immense scientific knowledge should be sufficient refutation. . . . [but f]or absolute intuition, in the full sense of the word, we must have integral experience; that is to say, a living application of rational theory no less than of working technique." See note 10, *supra*.

[15]"In late January or early February 1892, Monet rented rooms across from Rouen cathedral. He remained until spring, painting its looming facade many times. . . . The next winter he returned to paint the cathedral again, making in all more than thirty views of it. But it was not so much the deeply carved Gothic facade that was Monet's subject as it was the atmosphere—the *enveloppe*—that surrounded the building. 'To me the motif itself is an insignificant factor,' Monet said. 'What I want to reproduce is what exists between the motif and me.'" From the National Gallery of Art Web site on the Monet cathedral painting illustrated in this essay, http://www.nga.gov/collection/gallery/gg85/gg85-46371.0.html (accessed February 2007).

[16]Quotes are from the liner notes published with the CD of *Einstein on the Beach* (New York: Elektra Nonesuch, 1993).

[17]Some of these thoughts on Ritchie have been published as "Fields of Intuition," in Elisabeth Sussman, ed., *Remote Viewing: Invented Worlds in Recent Painting and Drawing* (New York: Whitney Museum of American Art, 2005).

[18]Einstein introduced (and then abandoned) the Cosmological Constant, which causes the universe to expand, in order to counteract what he took to be the universe's tendency to contract. His mathematical tool made it possible to introduce quantities that would show expansion, and in fact recent observations have shown an accelerating form of expansion that seems to make use of Einstein's original calculation. The existence of dark energy (the Cosmological Constant) and its expansionary forces must be balanced by enough dark matter to keep galaxies gravitationally bound. See "New Evidence for Dark Energy in the Universe," in *Jodrell Bank Observatory*, Press Release, November 7, 2002 (Manchester, UK: University of Manchester), p. 11, online at http://www.jb.man.ac.uk/news/darkenergy/ (accessed May 2007).

[19]The Wilkinson Microwave Anisotropy Probe (WMAP) is a project managed by NASA to measure "the temperature of the cosmic background radiation over the full sky with unprecedented accuracy." From http://map.gsfc.nasa.gov/ (accessed May 2007). More about time (the origins of the universe) than about space, it nonetheless has implications for the distribution of matter and the organization of space, viz.: "The shape of the universe (flat, saddle, or spherical) is not specified by the Big Bang theory so all shapes are permitted. . . . So far, measurements indicate that space is very nearly flat. This is, of course, the easiest shape for us to mentally picture." From WMAP Web site, "Frequently Asked Questions," http://map.gsfc.nasa.gov/m_help/h_faq.html (accessed May 2007).

[20]From "WMAP Mission": "The Wilkinson Microwave Anisotropy Probe (WMAP) team has made the first detailed full-sky map of the oldest light in the universe. It is a 'baby picture' of the universe. . . . The microwave light captured in this picture is from 379,000 years after the Big Bang, over 13 billion years ago: the equivalent of taking a picture of an 80-year-old person on the day of their birth." From http://map.gsfc.nasa.gov/m_mm.html (accessed May 2007).

[21]The concept of the "meat machine" was generated by William Gibson and other cyberpunk authors of the 1980s to contrast the human body with the conceptual freedom of the mind in cyberspace.

Notably, the "meat" was usually female, the virtual cyberspace operator male. See in particular Gibson, *Neuromancer* (New York: Ace Books, 1984).

[22]Compare with that lovely "baby picture" (as they describe it) at the WMAP Web site, http://map.gsfc.nasa.gov/m_mm.html (accessed May 2007).

[23]"Expansion mods" are bits of software, often user-generated, that add a new dimension to a computer game. Note the "Hello Kitty Mod" for the game "Doom 3," crafted and distributed by users to provide a humorous projection of the Japanese pop icon as "flashlight" illumination for the dark scenography of a single-shooter game. Referenced on "Boing Boing" site, http://www.boingboing.net/2004/08/09/hello_kitty_flashlig.html, accessed June 7, 2007.

[24]Heraclitus, fragment 41, quoted by Plato in *Cratylus*: "You could not step twice into the same river; for other waters are ever flowing on to you." From http://en.wikiquote.org/wiki/Heraclitus (accessed May 2007).

[25]Rosalind Krauss, "Grids," *October* 9 (Summer 1979), pp. 50–64. See also her *Originality of the Avant-Garde and Other Modernist Myths* (Cambridge, MA: MIT Press, 1985), pp. 8–22.

[26]Greenberg was laying out "The Crisis of the Easel Picture" in *Partisan Review* (April 1948), anthologized in Greenberg, *Collected Essays and Criticism*, vol. 2 (Chicago: University of Chicago Press, 1986), p. 224.

NOTES TO CHAPTER 10

[1]Albert Einstein, *Ideas and Opinions* (Pinebrook, NJ: Dell, 1954).

[2]George Steiner, *Grammars of Creation* (London: Faber and Faber, 2001), p. 275.

[3]Krzysztof Pomian, "Vision and Cognition," Caroline A. Jones and Peter Galison, eds., in *Picturing Science, Producing Art* (Oxford: Routledge, 1998), p. 218.

NOTES TO CHAPTER 11

[1]See Leon Botstein, *Judentum und Modernität: Essays zur Rolle der Juden in der Deutschen und Österreichischen Kultur 1848–1938* (Vienna: Böhlau, 1991), pp. 126–48.

[2]See Albrecht Fölsing, *Albert Einstein: A Biography*, Ewald Osers, trans. (New York: Penguin, 1993), p. 26; and Arthur I. Miller, *Einstein, Picasso: Space, Time, and the Beauty That Causes Havoc* (New York: Basic Books, 2001), p. 44.

[3]All quoted in Fölsing, *Albert Einstein*, p. 26.

[4]Miller, *Einstein, Picasso*, p. 45. See also Ronald W. Clark, *Einstein: The Life and Times* (New York: Avon Books, 1971), pp. 140f.

[5]See J. L. Heilbron, *The Dilemmas of an Upright Man: Max Planck and the Fortunes of German Science* (Cambridge, MA: Harvard University Press, 1996), p. 34. For a description of Einstein and Planck playing Beethoven together, see Lise Meitner's reminiscence in Brian Foster, "CERN, the Violin, and the Music of the Spheres," in *CERN Courier* 45, no. 1 (2005), http://www.cerncourier.com/main/toc/45/1 (accessed May 2007). On the interrelated careers of Einstein and Planck, see Fritz Stern, *Einstein's German World* (Princeton, NJ: Princeton University Press, 1999).

[6]Clark, *Einstein*, p. 746.

[7]See Miller, *Einstein, Picasso*, p. 45; and Fölsing, *Albert Einstein*, p. 53.

[8]See Boris Schwartz, "Musical and Personal Reminiscences of Albert Einstein," in Gerald Holton and Yehuda Elkana, eds., *Albert Einstein: Historical and Cultural Perspectives: The Centennial Symposium in Jerusalem* (Princeton, NJ: Princeton University Press, 1982; Mineola, NY: Dover Publications, 1997), pp. 409–16. Citations are from the Dover edition.

[9]For an account of this meeting, see Robert Mann, "On Playing with Scientists: Remarks at the Einstein Centennial Celebration Concert by the Juilliard Quartet," in Harry Woolf, ed., *Some Strangeness in Proportion: A Centennial Symposium to Celebrate the Achievements of Albert Einstein* (Reading, MA: Addison Wesley, 1980), pp. 526f.

[10]Quoted in Clark, *Einstein*, p. 141.

[11]See John S. Rigden, *Einstein 1905: The Standard of Greatness* (Cambridge, MA: Harvard University Press, 2005), pp. 147–49.

[12]For more on the "back to Mozart" movement at the turn of the 20th century, see Leon Botstein, "Nineteenth-Century Mozart: The Fin de Siècle Mozart Revivial," in James M. Morris, ed., *On Mozart* (Cambridge, UK: Cambridge University Press, 1994), pp. 204–26.

[13]See Clark, *Einstein*, pp. 140f.; and Miller, *Einstein, Picasso*, pp. 186f.

[14]Miller, *Einstein, Picasso*, p. 186.

[15]See Fölsing, *Albert Einstein*, pp. 616f.; Miller, *Einstein, Picasso*, pp. 229f., 233f., 262f.; and Clark, *Einstein*, pp. 51f.

[16]On the sexual politics of domestic music making, see Richard Leppert, "Sexual Identity, Death, and the Family Piano," in *19th-Century Music* 16 (1992), pp. 105–28; and Peter J. Rabinowitz, "'With Our Own Dominant Passions': Gottschalk, Gender, and the Power of Listening," in *19th-Century Music* 16 (1993), pp. 242–52. On the musical construction of gender, see Lucy Green, *Music, Gender, Education* (New York: Cambridge University Press, 1997), esp. ch. 3, pp. 52–81.

[17]See Fölsing, *Albert Einstein*, p. 54.

[18]Miller, *Einstein, Picasso*, p. 186.

[19]See Schwartz, "Musical and Personal Reminiscences," pp. 410, 412.

[20]Jacob Burckhardt, *Die Kultur der Renaissance in Italien*, in his *Gesammelte Werke*, vol. 3 (Basel and Stuttgart: Schwabe, 1978), pp. 20f., 40, 108, 114, 122, 125f., 187, 265, 286, 334, 358. See also Burckhardt's *Der Cicerone. Eine Anleitung zum Genuß der Kunstwerke Italiens* (Leipzig: Alfred Kröner, 1927), in which he contrasts the golden age of early 16th-century Italian art with the "raw and deviant" Mannerist and Baroque forms that followed.

[21]On Einstein's musical tastes, see Ze'ev Rosenkranz, *Albert Through the Looking Glass* (Jerusalem: Albert Einstein Archives, Jewish National and University Library, Hebrew University of Jerusalem, 1998), pp. 102–105; Fölsing, *Albert Einstein*, p. 26; and Miller, *Einstein, Picasso*, p. 186.

[22]It is notable that Thomas Levenson's *Einstein in Berlin* (New York: Bantam Books, 2003) does not confront this anomaly. Despite Einstein's deep attachment to music, the book's discussion about Einstein's Berlin years revolves around architecture and art. The literature on the musical life in Berlin in the 1920s is extensive. The interested reader would be best served by consulting the music criticism of Paul Bekker (1882–1937); see, for example, *Die Weltgeltung der deutschen Musik* (Berlin: Schuster & Loeffler, 1920), and Christopher Hailey, *Franz Schreker, 1878–1934: A Cultural Biography* (New York: Cambridge University Press, 1993).

[23]See David Z. Kushner, *The Ernest Bloch Companion* (Westport, CT: Greenwood Press, 2002), pp. 102f.

[24]See John Covach, "Twelve-Tone Theory," in Thomas Christensen, ed., *The Cambridge History of Western Musical Theory* (New York: Cambridge University Press, 2002), pp. 603–27. On Schreker's life and works, see Hailey, *Franz Schreker*.

[25]See Leon Botstein, "Artur Schnabel and the Ideology of Interpretation," in *The Musical Quarterly* 85 (2001), pp. 587–94.

[26]See Stephen Lehmann and Marion Faber, *Rudolf Serkin: A Life* (New York: Oxford University Press, 2003), esp. ch. 3 and 6.

[27]See William Drabkin, "Heinrich Schenker," in Thomas Christensen, ed., *The Cambridge History of Western Musical Theory* (New York: Cambridge University Press, 2002), pp. 812–43.

[28]See Leon Botstein, "Pfitzner and Musical Politics," in *The Musical Quarterly* 85 (2001), pp. 67–69.

[29]See Eduard Hanslick, *The Beautiful in Music*, Gustav Cohen, trans. (New York: Liberal Arts Press, 1957).

[30]See Carl Dahlhaus, *Die Idee der absoluten Musik* (Basel: Bärenreiter, 1978).

[31]See Ethan Haimo, "Schoenberg and the Origins of Atonality," in Juliane Brand and Christopher Hailey, eds., *Constructive Dissonance: Arnold Schoenberg and the Transformations of Twentieth-Century Culture* (Berkeley: University of California Press, 1997), pp. 71–86.

[32]On Wittgenstein's conservative musical tastes, see Ray Monk, *Ludwig Wittgenstein: The Duty of Genius* (New York: The Free Press, 1990), pp. 13, 78. For a comprehensive collection of Wittgenstein's views about music, consult Martin Alber, Brian McGuinness, and Monika Seekircher, eds., *Wittgenstein und die Musik: Ludwig Wittgenstein–Rudolf Koder: Briefwechsel* (Innsbruck: Haymon, 2000).

[33]On Wittgenstein's view of Wagner, see Alber et al., *Wittgenstein und die Musik*, pp. 141–44, 151f., p. 156, pp. 162f., 171f., p. 175, 179; and Wittgenstein, *Culture and Value*, G. H. von Wright, ed.,

Peter Winch, trans. (Chicago: University of Chicago Press, 1980), p. 41. Wittgenstein's famous quote derives from his *Philosophical Investigations* (New York: Macmillan, 1953).

[34]Robert Musil criticized Mach's positivism in a 1908 doctoral thesis (*Beitrag zur Beurteilung der Lehren Machs*) for the University of Berlin. See the English translation, *On Mach's Theories*, Kevin Mulligan, trans. (Washington, DC: Catholic University of America Press, 1982).

[35]See Ernst Mach, *The Analysis of the Sensations and the Relation of the Physical to the Psychical* (Chicago: Open Court Publishing Company, 1914).

[36]For examples of Mach's prose style, see Rudolf Haller and Friedrich Stadler, eds., *Werk und Wirkung* (Wien: Hölder-Pichler-Tempsky, 1988), and *Popularwissenschaftliche Vorlesungen*, 4th, rev. ed. (Leipzig: Johann Ambrosius Barth, 1910). On clarity in modern scientific language, see Gerald Holton, *Einstein, History, and Other Passions: The Rebellion Against Science at the End of the Twentieth Century* (Cambridge, MA: Harvard University Press, 1996), p. 163; and Holton, *Thematic Origins of Scientific Thought: Kepler to Einstein* (Cambridge, MA: Harvard University Press, 1973), p. 224.

[37]Einstein's theory of relativity is misconstrued in, for example, composer George Rochberg's *The Aesthetics of Survival: A Composer's View of Twentieth-Century Music*, rev. ed. (Ann Arbor: University of Michigan Press, 2004), p. 154.

[38]See Hermann von Helmholtz, *On the Sensations of Tone as a Physiological Basis for the Theory of Music*, 2nd ed. (New York: Dover Publications, 1954).

[39]Dane Rudyar's essay was published in *The Musical Quarterly* 8 (1922), pp. 108–18. On Rudyar, see Carol J. Oja, *Making Music Modern: New York in the 1920s* (New York: Oxford University Press, 2000), pp. 97–110.

[40]Lazare Saminsky, *Physics and Metaphysics of Music and Essays on the Philosophy of Mathematics* (The Hague: Martinus Nijhoff, 1957).

[41]Saminsky, *Physics and Metaphysics of Music*, p. 21. He argued further that "The triton [fourth + major seventh] and its resolution mark a definite position in the tonal universe even as gravitational and electro-magnetic tension in the physical world signify a change in the curvature of the space-time continuum," p. 21.

[42]See Saminsky, *Physics and Metaphysics of Music*, pp. 33–40.

[43]Saminsky, *Physics and Metaphysics of Music*, p. 36.

[44]Quoted in Rosenkranz, *Albert Through the Looking Glass*, p. 24. Einstein's statement appeared in the German journal *Menschen* on January 27, 1921.

[45]On Richard Strauss and realism, see Morten Kristiansen, *Richard Strauss's Feuersnot in its Aesthetic and Cultural Context: A Modernist Critique of Musical Idealism* (Ph.D. diss., Yale University, 2000), pp. 32–172.

[46]Ferruccio Busoni, *Sketch of a New Aesthetic of Music*, Theodore Baker, trans. (New York: G. Schirmer, 1911).

[47]Hans Pfitzner, *Futuristengefahr: Bei Gelegenheit von Busoni's Ästhetik* (Leipzig: Süddeutsche Monatshefte, 1918).

[48]See Botstein, "Pfitzner and Musical Politics," p. 65, pp. 72f.

[49]Heinrich Schenker, *Harmony*, Oswald Jonas, ed., Elisabeth Mann-Borghese, trans. (Chicago: University of Chicago Press, 1954). Schenker's *Harmonielehre* was the first volume of his *Neue musikalische Theorien und Phantasien*. On Schenker, see also Leon Botstein, "Gedanken zu Heinrich Schenkers jüdischer Identität," in Evelyn Fink, ed., *Rebell und Visionär. Heinrich Schenker in Wien* (Vienna: Lafite, 2003), pp. 11–17.

[50]Arnold Schönberg, *Theory of Harmony*, Roy E. Carter, trans. (Berkeley: University of California Press, 1978).

[51]Clark, *Einstein*, p. 414.

Notes to Chapter 13

[1]Friedrich Adler to Victor Adler, 19 July 1908, cited in Julius Braunthal, *Victor und Friedrich Adler* (Vienna: Wiener Volksbuchhandlung, 1965), p. 196. Fritz to Victor, from Friedrich Adler, *Victor Adlers Briefwechsel mit Karl Katusky und August Bebel* (Vienna, 1954), cited in Ronald Florence, *Fritz: The Story of a Political Assassin* (New York: Dial, 1971), p. 44.

[2]Friedrich Adler to the Board of Education the canton of Zurich, 1908, cited in Florence, *Fritz*, pp. 44–45.

[3]Friedrich Adler to Victor, 28 October 1909, cited in Rudolf Ardelt, *Friedrich Adler* (Vienna 1984), p. 166. Also cited in Albrecht Fölsing, *Albert Einstein* (New York: Viking, 1997), p. 258f.

[4]Einstein to Alfred Stern (Zurich), 6 December 1910, cited in John Stachel et al. (eds.), *Einstein Collected Papers*, Vol. 5, doc. 236 (hereafter CPAE; Princeton, NJ: Princeton University Press, 1987ff.). Princeton University Press's English translation of CPAE: Vol. 5 (1995) is by Anna Beck, vol. 6 (1997) by Alfred Engel, vol. 7 (2002) by Alfred Engel, vol. 8 (1998) by Ann M. Hentschel. Hereafter these English translations designated E; this reference, p. E168.

[5]Florence, *Fritz*, p. 45.

[6]Leon Trotsky, "Political Profiles, Victor and Friedrich Adler," in *Novy Mir* No. 903, 5 February 1917, http://www.marxists.org/archive/trotsky/works/1940/profiles/victor-friedalder.htm (accessed 8 December 2004).

[7]Trotsky, "Political Profiles."

[8]Trotsky, "Political Profiles."

[9]J. W. Brügel, ed., *Friedrich Adler vor dem Ausnahmegericht* (Vienna, Frankfurt, Zurich: Europa Verlag, 1967), p. 147; also Florence, *Fritz*, p. 235.

[10]AE to Heinrich Zangger, after 10 March 1917, CPAE, vol. 8, doc. 310, pp. E299f., on p. 299.

[11]AE to FA, 13 April 1917, CPAE, vol. 8, doc. 324, p. E315.

[12]Walther Rathenau to Einstein, 10–11 May 1917, CPAE, vol. 8, doc. 337, pp. E327–29, on p. E. 328.

[13]"Draft for a petition by Albert Einstein to the Austrian Emperor, requesting a pardon for Friedrich Adler," probably spring 1917. Manuscript in the collection of the Max Planck Institute for History of Science, Berlin, Germany. Also in CPAE, vol. 10.

[14]J. W. Brügel, ed., *Ausnahmegericht*, p. 241.

[15]J. W. Brügel, ed., *Ausnahmegericht*, pp. 241, 243f.

[16]J. W. Brügel, ed., *Ausnahmegericht*, p. 64.

[17]Trotsky, "Political Profiles," in *Novy Mir* 903.

[18]Otto Nathan and Heinz Norden, *Einstein on Peace*, preface by Bertrand Russell (New York: Simon and Schuster, 1960).

[19]Einstein to Romain Rolland, 22 March 1915, in *Einstein on Peace*. Insurance quotation, p. xx.

[20]AE to Katja Adler, 20 February 1917, cited in CPAE, vol. 8, doc. 301, p. E288.

[21]Quotation "I am suddenly . . ." from Friedrich Adler to Katja Adler, mid-February 1917, in Rudolf Neck, *Arbeiterschaft und Staat im ersten Weltkrieg 1914–19*, 2 vols. (Vienna: Europa-Verlag, 1968), vol. 1, p. 235, doc. no. 140. Quotation including "everything in a new light" from Neck, *Arbeiterschaft*, vol. 1, p. 244, doc. no. 147. Fritz Adler to Katja, 22 February 1917, parts of letter from Friedrich Adler to Katja, Victor, and Emma Adler cited in Florence, *Fritz*, p. 199.

[22]Philipp Frank, *Einstein: His Life and Times* (New York: Knopf, 1953), pp. 174f.

[23]FA to AE, 9 March 1917, CPAE, vol. 8, doc. 307, p. E294f., on p. 294.

[24]Friedrich Adler, *Ernst Machs Überwindung des mechanischen Materialismus* (Wien: Verlag der Wiener Volksbuchhandlung Ignaz Brand & Co., 1918).

[25]Friedrich Adler, *Ernst Machs*, pp. 133–40.

[26]FA to AE, 9 March 1917, CPAE, vol. 8, doc. 307, p. E294f., on p. 294.

[27]FA to AE, 9 March 1917, CPAE, vol. 8, doc. 307, p. E294f., on p. 295.

[28]"Friedrich Adler als Physiker. Eine Unterredung mit A. Einstein," in *Vossische Zeitung*, 23 May 1917, morning edition, p. 2, cited in Fölsing, *Einstein*, p. 786 (note 38).

[29]All from Kurt Joel, "Friedrich Adler als Physiker," in *Vossische Zeitung*, 23 May 1917, p. 2.

[30]FA to AE, Stein-on-the-Danube, 6 July 1918, CPAE, vol. 8, doc. 582, pp. E607f.

[31]FA to AE, Stein-on-the-Danube, 6 July 1918, CPAE, vol. 8, doc. 582, pp. E607f.

[32]Friedrich Adler, *Ortszeit, Systemzeit, Zonenzeit und das ausgezeichnete Bezugssystem der Elektrodynamik. Eine Untersuchung über die Lorentzsche und die Einsteinsche Kinematik* (Vienna: Wiener Volksbuchhandlung, 1920), p. 70.

[33]AE to FA, 4 August 1918, CPAE, vol. 8, doc. 594, pp. E616–20, on p. 617.

[34]Adler, *Ortszeit*, pp. 75, 79.

[35]AE to FA, 4 August 1918, CPAE, vol. 8, doc. 594, pp. E616–20, on p. 618.

NOTES TO CHAPTER THIRTEEN

³⁶AE to FA (Berlin), 29 September 1918, CPAE, vol. 8, doc. 628, p. E658–60, first emphasis added, second in original.

³⁷AE to FA, 29 September 1918, CPAE vol. 8, doc. 628, p. E658–60, on p. E660.

³⁸Adler, *Ortszeit*, pp. 60f. On train time, city time, master clocks, telegraphic exchange of time signals, and the different zones chosen for train and ship travel, see ibid., chapter 2. For a much more extended discussion of the more-than-metaphors of trains and longitude determination in the development of relativistic clock coordination procedures, see Peter L. Galison, *Einstein's Clocks, Poincaré's Maps* (NewYork: Norton, 2003).

³⁹Adler, *Ortszeit*, p. 70.

⁴⁰Adler, *Ortszeit*, p. 70.

⁴¹Adler, *Ortszeit*, p. 130, figures on pp. 81, 128f.

⁴²AE to FA, 29 September 1918, CPAE, vol. 8, doc. 628, pp. E658–60, on p. 660.

⁴³See AE to FA, 30 September 1918, CPAE, vol. 8, doc. 629, on pp. E660f–61f. See also FA to AE, 12 October 1918, CPAE, vol. 8, doc. 632, on pp. E664–66, where Adler also switches to an electron traveling with a velocity. Einstein replied that this was an improvement, but now Adler would have no paradox: "You will certainly discover, though, that the condition relating to the moment of passing is independent of the choice of coordinates." From AE to FA, 20 October 1918, CPAE, vol. 8, doc. 636, pp. E670f. The article in *Die Naturwissenschaften* was called "Dialog über Einwände gegen die Relativitätstheorie," published on 29 November 1918 in vol. 6, pp. E697–702.

⁴⁴Einstein, "On the Need for a National Assembly," 13 November 1918, CPAE, vol. 7, doc. 14, pp. E76–77, on p. E76.

⁴⁵Suzanne Sturmthal Russin, *Democracy Under Fire: Memoirs of a European Socialist—Adolf Sturmthal* (Duke University Press, 1989), p. 15.

⁴⁶AE to FA, 20 October 1918, CPAE, vol. 8, doc. 636, p. E670f., on p. E671. On the release of Adler, see Florence, *Fritz* (1971), pp. 301–307. On the Adler and the Second and Third International, see ibid., p. 313.

⁴⁷Lenin, "Democracy and Dictatorship," written 23 December 1918, published 3 January 1919, Pravda 2, in Vladimir Ilyich Lenin, *Collected Works*, vol. 28, pp. 368–72, http://www.cddc.vt.edu/marxists/archive/lenin/works/1918/dec/23.html.

⁴⁸Trotsky, "Political Parties," in *Novy Mir* 903.

⁴⁹Adler, *The Witchcraft Trial in Moscow: Commission of Enquiry into the Conditions of Political Prisoners, November 1936* (New York: Published for the Socialist Party of the United States by Pioneer Publishers, 1937), pp. 34–36.

⁵⁰Adler, *The Witchcraft Trial*, pp. 34f.; impending war, fortress, p. 36.

⁵¹Emma Adler to Einstein, 9 July 1930, Albert Einstein Archives, Adler File, doc. 34-408.1. I would like to thank and acknowledge the gracious help of the Einstein Papers Project for helping me to locate this and the other documents mentioned below.

⁵²Einstein Archives, Adler File, doc. 6-040.

⁵³Florence, *Fritz*, p. 315.

⁵⁴FA to AE, 28 June 1945, Einstein Archives, Adler File, doc. 6-019.

NOTES TO CHAPTER 14

¹H. Reichenbach, *Philosophie der Raum-Zeit-Lehre* (Berlin: de Gruyter, 1928), pp. 47f.; translated as *The Philosophy of Space and Time* (New York: Dover, 1958), pp. 35f.

²A. Einstein, *Geometrie und Erfahrung. Erweiterte Fassung des Festvortrages gehalten an der Preußischen Akademie der Wissenschaft zu Berlin am 27. Januar 1921* (Berlin: Springer, 1921), pp. 3f.; translated in G. Jeffrey and W. Perrett, eds., *Sidelights on Relativity* (London: Methuen, 1923), pp. 28f.

³See Einstein, *Geometrie und Erfahrung*, pp. 4f. [Jeffrey and Perrett, pp. 30f.]: "Geometry treats of objects that are designated with the words line, point, etc. No kind of acquaintance or intuition of these objects is presupposed, but only the validity of those axioms which are likewise to be conceived as purely formal, i.e., as separated from every content of intuition and experience. . . . These axioms are free creations of the human spirit. All other geometrical propositions are logical consequences of the (only nominalistically conceived) axioms. The axioms first define the objects of which geometry treats.

Schlick therefore designated the axioms very appropriately as 'implicit definitions' in his book on theory of knowledge."

[4]Einstein, *Geometrie und Erfahrung*, p. 6 [Jeffrey and Perrett, p. 33].

[5]Einstein, *Geometrie und Erfahrung*, p. 7 [Jeffrey and Perrett, p. 34].

[6]Einstein, *Geometrie und Erfahrung*, p. 8 [Jeffrey and Perrett, pp. 35f.].

[7]For a more detailed discussion of Helmholtz and Poincaré in this connection, see my "Geometry as a Branch of Physics: Background and Context for Einstein's 'Geometry and Experience,'" in D. Malament, ed., *Reading Natural Philosophy* (Chicago: Open Court, 2002).

[8]See Einstein, *Geometrie und Erfahrung*, p. 6 [Jeffrey and Perrett, p. 33]: "I attach particular importance to this conception of geometry, because without it I would have found it impossible to establish the theory of relativity. Without it, namely, the following consideration would have been impossible. In a reference system that is rotating relative to an inertial system the situational laws [*Lagerungsgesetze*] of rigid bodies, due to the Lorentz contraction, do not correspond to the rules of Euclidean geometry. Therefore, the admission of non-inertial systems as equally justified must lead to the abandonment of Euclidean geometry. The decisive step in the transition to generally covariant equations would certainly not have taken place, if the above interpretation had not been taken as basis."

[9]See J. Stachel, "Einstein and the Rigidly Rotating Disk," in A. Held, ed., *General Relativity and Gravitation* (New York: Plenum, 1980); reprinted as "The Rigidly Rotating Disk as the 'Missing Link' in the History of General Relativity," in D. Howard and J. Stachel, eds., *Einstein and the History of General Relativity* (Boston: Birkhäuser, 1989), pp. 48–62.

[10]Again, for a more detailed discussion of Schlick, see Friedman, "Geometry as a Branch of Physics."

[11]See note 8, above.

Notes to Chapter 15

Citation style: Books referenced in the text or notes, listed below, are designated by a bracketed letter. When a specific page number is cited, it is shown behind a colon; e.g., [F: 104]. References to the Collected Papers pertain to volume one (unless indicated otherwise) and cite not pages but document numbers; e.g., [C: D127] or [C8: D87]. When more than one item is cited, each is separated by a semicolon; e.g., [HP; JB; R;].

[C] *Collected Papers of Albert Einstein*, John Stachel, et al., eds., vol. 1 (1879–1902): *The Early Years*; vol. 2 (1900–1909): *The Swiss Years* (Princeton, NJ: Princeton University Press, 1987)

[JB] Jeremy Bernstein, *Secrets of the Old One: Einstein, 1905* (New York: Copernicus, 2006)

[DC] David Cassidy, *Einstein and Our World* (Amherst, NY: Humanity Books, 1995)

[E] Albert Einstein, Autobiographical Notes, in Paul A. Schilpp, Ed., *Albert Einstein: Philosopher-Scientist* (Chicago: Open Court, 1970)

[F] Albrecht Fölsing, *Albert Einstein: A Biography* (New York: Viking, 1997)

[G] Peter Galison, *Einstein's Clocks, Poincaré's Maps* (New York: Norton, 2003)

[HA] Gerald Holton, *The Advancement of Science and Its Burdens* (Cambridge, MA: Harvard University Press, 1998)

[HD] Banish Hoffman and Helen Dukas, *Albert Einstein, Creator and Rebel* (New York: Viking, 1972)

[HE] G. Holton and Y. Elkana, eds., *Albert Einstein Historical and Cultural Perspectives: Centennial Symposium in Jerusalem* (Princeton, NJ: Princeton University Press, 1982; Dover, 1997)

[HP] Gerald Holton, *Einstein, History, and Other Passions* (Boston: Addison-Wesley, 1996)

[HT] Gerald Holton, *Thematic Origins of Scientific Thought* (Rev. ed., Cambridge, MA: Harvard, 1988)

[I] Walter Isaacson, *Einstein, His Life and Universe* (New York: Simon and Schuster, 2007)

[P] Abraham Pais, *"Subtle Is the Lord . . ."* (New York: Oxford, 1982)

[R]   John S. Rigden, *Einstein 1905: The Standard of Greatness* (Cambridge, MA: Harvard, 2005)

[C]   Julian Schwinger, *Einstein's Legacy* (Scientific American Library, 1986)

[W]   Harry Woolf, Ed., *Some Strangeness in the Proportion: Centennial Symposium* (March, 1979; Boston: Addison Wesley, 1980)

Remarks: Readers seeking a quick overview will be well served by a new website (http://www/aip.org/history/einstein/) prepared by the Center for History of Physics of the American Institute of Physics, or by the short books by Hoffman and Dukas [HD] and by Cassidy [DC]. For comprehensive biographies, see Fölsing [F] and Isaacson [I]. For scholarly detail, see the editorial commentary included in the collected papers [C]. The scientific biography by Pais [P] is splendid, as are books with less broad focus by Holton, [HP] and [HT], Galison [G], Rigden [R], Bernstein [JB], and Schwinger [S]. Other invaluable sources include Einstein's Autobiographical Notes and exchanges with contemporaries in [E], as well as essays from centennial symposia in [HE] and [W].

[1]See [HD: 46]; the added remark about Newton's portrait was made by Helen Dukas to Gerald Holton (private communication).

[2]Gerald Holton [HA: 197]: "Much of my work has had its origin in the notion that science should treasure its own history, that historical scholarship should treasure science, and that the full understanding of each is deficient without the other."

[3]Specific references to items listed in Tables 15.1–15.5 can readily be found in [C] and [F] or [I], which are also organized chronologically; here for the most part only sources of quotations are given.

[4]The Swiss Polytechnic Institute was later renamed the Eidgenössische Technische Hochschule, usually referred to as ETH, as I do hereafter.

[5]Maja Winteler-Einstein, *Albert Einstein—Beitrag für sein Lebensbild*, manuscript (1924). Excerpt published in [C, V1] with extensive editorial notes, and included in the accompanying paperback English "pony" by Anna Beck.

[6]Maurice Solovine, ed., *Albert Einstein: Lettres à Maurice Solovine, 1906–1966* (Paris: Gauthier-Villars, 1956).

[7]Martin J. Klein, "Thermodynamics in Einstein's Thought," in *Science* 157 (1967), p. 509.

[8]Peter Galison, *How Experiments End* (Chicago: University of Chicago Press, 1987), p. 34.

[9]Solovine, *Lettres* (1956).

[10]Klein, "Thermodynamics," p. 509.

[11]Klein, "Thermodynamics," p. 509; Gerald Holton, "Einstein's Scientific Program: The Formative Years," in [W: 49] and in [HA: 57–76]; Martin J. Klein, "No Firm Foundation: Einstein and the Early Quantum Theory," in [W: 161]; [G; P; R; Sc].

[12]J. S. Rowlinson, *Cohesion: A Scientific History of Intermolecular Forces* (Cambridge: Cambridge University Press, 2002). John N. Murrell and Nicole Grobert, "The Centenary of Einstein's First Scientific Paper," in *Notes Rec. R. Soc.* (London) 56, 1 (2002), pp. 89–94.

[13]Klein, "Thermodynamics," p. 509.

[14]Bretislav Friedrich kindly provided the observations noted here about *Annalen der Physik* (private communication).

[15]Henri Poincaré, *Science and Hypothesis* (New York: Dover, 1952), pp. 179f.

[16]William Lanouette, *Genius in the Shadows* (New York: Scribners, 1992), pp. 60–64.

[17]Albert Einstein, "What I Believe," in *Forum and Century* 84 (1930), p. 193; reprinted as "The World As I See It," in his *Ideas and Opinions*, C. Seelig, ed. (New York: Crown Press, 1954).

[18]Dudley Herschbach, "Changes in the Gardens of Science, Wrought by Women," in *Annals of New York Academy of Sciences* 869 (1999), pp. 66–74

[19]Michael Polanyi, "The Republic of Science. Its Political and Economic Theory," in *Minerva* 54, 1 (1962), pp.54–74.

[20]Story told to me many years ago by Edward Purcell (private communication); I have been unable to find it in print.

[21]Gerald Holton, "Einstein and the Cultural Roots of Modern Science," in *Daedalus*, 127, 1 (1998); and in [HA: xiii–xlix].

[22]Edward O. Wilson, *Naturalist* (Washington, DC: Island Press/Shearwater Books, 1994).

[23]Oliver Sacks, *Uncle Tungsten* (New York: Knopf, 2001).

[24]Remark quoted in a class by George Polya many years ago; I have been unable to find it in print.

[25]Dudley Herschbach, "The Impossible Takes a Little Longer," in *Science Literacy for the Twenty-First Century*, S. P. Marshall, J. A. Scheppler, and M. J. Palmisano, eds. (Amherst, NY: Prometheus Books, 2003), pp. 131–44.

[26]Herschbach, "Changes in the Gardens," pp. 69–72.

[27]Freeman Dyson, "To Teach or Not to Teach," in his *From Eros to Gaia* (New York: Penguin Books, 1992), p. 197.

[28]Steven Weinberg, "Four Golden Lessons," in *Nature* 426 (2003), p. 389.

NOTES TO CHAPTER 16

[1]The author would like to thank Susan Neiman for hosting the stimulating conference at the Einstein Forum in Potsdam at which this paper was presented.

[2]See *The Genesis of General Relativity*, 4 vols., *Boston Studies in the Philosophy of Science*, vol. 250 (Dordrecht: Springer, 2007), as well as "The Relativity Revolution from the Perspective of Historical Epistemology" *Isis* [Special Issue: *The Elusive Icon: Albert Einstein 1905–2005*] (2004), pp. 640–48, on which this outline is based. See also *Auf den Schultern von Riesen und Zwergen: Einsteins unvollendete Revolution* (Berlin: Wiley-VCH, 2005).

[3]See the three volumes resulting from the Einstein exhibition in Berlin in 2005: *Albert Einstein—Chief Engineer of the Universe; Einstein's Life and Work in Context; A Hundred Authors for Einstein; Documents of a Life's Pathway* (Berlin: Wiley-VCH, 2005). The virtual Einstein exhibition can be seen at: http://einstein-virtuell.mpiwg-berlin.mpg.de.

[4]See "Challenges from the Past: Innovative Structures for Science and the Contribution of the History of Science, Ringberg Symposium October 2000," in *Innovative Structures in Basic Research* (München: Max-Planck-Gesellschaft, 2002), pp. 25–36, and "From Research Challenges of the Humanities to the Epistemic Web (Web 3.0)" (with Malcolm Hyman), available at www.sis.pitt.edu/~repwkshop/groups.html.

NOTES TO CHAPTER 17

[1]The original Einstein letter can be found in A. Pais, *Subtle Is the Lord: The Science and the Life of Albert Einstein* (New York: Oxford University Press, 1982), p. 431.

[2]*Collected Papers of Albert Einstein*, Volumes 1–9 (Princeton, NJ: Princeton University Press, 1988–2004).

[3]Pais, *Subtle Is the Lord*.

[4]A. Fölsing, *Albert Einstein* (New York: Viking Press, 1997). For Einstein's miraculous year, see C. Lanczos, *The Einstein Decade, 1905–1915* (London: Elek Science, 1974).

[5]Incidentally, Planck's law plays a key role in recent experimental tests of Big Bang cosmology, in connection with the 3K cosmic microwave radiation.

[6]The gas constant R is introduced through the ideal gas law $pV = nRT$, which is believed to be universal for *all* gases at sufficiently high temperature and low density.

[7]Historically, the situation is of course much more complicated—roughly speaking, one might say that it occurred the other way around. However, much of 19th-century physics dealt with macroscopic phenomena. It was Maxwell and, particularly, Boltzmann, who were the key contributors in trying to answer the question, What is the atomistic basis of the macroscopic laws? See, for example, R. D. Purrington, *Physics in the Nineteenth Century* (New Brunswick, NJ: Rutgers University Press, 1997). I wish to note, in passing, that the passage from a continuum theory of matter to atomism can be understood as a *deformation*, in a mathematically precise sense. But further discussion is beyond the scope of this presentation.

[8]All macroscopic properties of a simple system in thermal equilibrium are determined by $s(u,v)$.

[9]A. Einstein, "Über einen die Erzeugung und Verwandlung des Lichtes betreffenden heuristischen Gesichtspunkt," in *Annalen der Physik* 17 (1905), pp. 132–48.

[10]Millikan also made the first direct determination of Planck's constant h using the photoelectric

effect, and his subsequent studies of Brownian motion in gases were decisive in putting an end to all opposition against the atomistic and kinetic theories of matter.

[11]See Pais, *Subtle Is the Lord*, p. 431.

[12]A. Einstein, "Die Plancksche Theorie der Strahlung und die Theorie der spezifischen Wärme," in *Annalen der Physik* 22 (1907), pp. 180–90.

[13]A. Einstein and J. de Haas, "Experimenteller Nachweis der Ampèrischen Molekularströme," in *Verhandlungen der Deutschen Physikalischen Gesellschaft* 17 (1915), pp. 152–70; A. Einstein and J. de Haas, "Experimental proof of the Existence of Ampère's Molecular Currents. *Verhandlungen der Koninklijke Akademie der Wetenschappen te Amsterdam* 18 (1915–16).

[14]A. Einstein, "Zur Quantentheorie der Strahlung," in *Physikalische Zeitschrift* 18 (1917), pp. 121–28.

[15]A. Einstein, "Quantentheorie des einatomigen idealen Gases," in *Naturwissenschaften* 12 (1924), pp. 601f.

[16]See, for example, A. Einstein, "Der Angst-Traum," in *Berliner Tagblatt*, December 25, 1917.

[17]I thank my colleague and friend Joel L. Lebowitz for setting such a great example, in this and other respects.

NOTES TO CHAPTER 18

[1]A. Einstein, "Zur Quantentheorie der Strahlung," in *Physikalische Zeitschrift* 18 (1917), p. 121.

[2]T. S. Kuhn, *Black-Body Theory and the Quantum Discontinuity, 1894–1912* (Oxford: Clarendon Press, 1978), p. 187: "The concepts of particles of light and resonators restricted to energy nhv had entered physics together in Einstein's paper of 1905 and 1906."

[3]A. Einstein, "Die Plancksche Theorie der Strahlung und die theorie der spezifischen Wärme," in *Annalen der Physik* 22 (1907), p. 180; quotation is taken from Anna Beck's translation in *The Collected Works of Albert Einstein*, vol. 2 (Princeton, NJ: Princeton University Press, 1987), doc. 38, p. 218.

[4]Quoted in A. Pais, *Subtle Is the Lord: The Science and Life of Albert Einstein* (Oxford: Oxford University Press, 1982), p. 9.

[5]A. Einstein, "Zum Quantensatz von Sommerfeld und Epstein [On the Quantization Condition of Sommerfeld and Epstein]," in *Verhandlungen der deutschen Physikalischen Gesellschaft*, vol. 19 (1917), pp. 82–92. All subsequent Einstein quotations not otherwise attributed are taken from Anna Beck's translation in *The Collected Works of Albert Einstein*, vol. 6 (Princeton, NJ: Princeton University Press, 1987), doc. 45, pp. 434–43.

[6]The title refers to the physicist Paul Epstein, who had recently extended Sommerfeld's condition in a relatively minor way. The reference to Epstein in the context of the Sommerfeld rule has not survived; it is now known as Bohr-Sommerfeld quantization, or as the WKB (Wenzel-Kramers-Brillouin) quantization in reference to an approximation method for the Schrödinger equation that leads to the same condition, as will be discussed.

[7]See, for example, J. U. Nöckel and A. D. Stone, "Ray and Wave Chaos in Asymmetric Resonant Optical Cavities," in *Nature* 385 (1997), p. 45.

[8]E. Schrödinger, "Quantisierung als Eigenwertproblem [Quantization as a Problem of Eigenvalues (part II)]," in *Annalen der Physik* 79 (1926), pp. 489ff.

[9]J. B. Keller, "Corrected Bohr-Sommerfeld Quantum Conditions for Nonseparable Systems," in *Annals of Physics* 4 (1958), pp. 180–88; an initial report of this title was published by Keller as a research report in 1953. See also J. B. Keller and S. I. Rubinow, "Asymptotic Solution of Eigenvalue Problems," in *Annals of Physics* 9 (1960), pp. 24–75.

[10]Interestingly, a letter from Einstein to Paul Ehrenfest suggests that Einstein believed that any real mechanical system was not ergodic and that his version of Bohr-Sommerfeld quantization might apply to all real mechanical systems: "It seems to me that real mechanics is such that the existence of the integrals [of motion] . . . are already secured by virtue of its general foundations. But how to implement this?" From doc. 350, June 1917, *Collected Papers of Albert Einstein*, vol. 8, p. 339.

[11]Keller, "Corrected Bohr-Sommerfeld Quantum Conditions"; Keller and Rubinow, "Asymptotic Solution."

[12]Einstein to Ehrenfest, June 1917, *Collected Papers*.

[13]Schrödinger, "Quantization as a Problem of Eigenvalues."

[14]The author intends this section to serve as only a partial historical account because of his inability at this point to claim a complete study of the relevant historical literature.

[15]L. de Broglie, *Recherches sur la théorie des quanta* [Studies on the Theory of Quanta]. Ph.D. thesis (1924); reprinted in *Annales de la Fondation Louis de Broglie* 17 (1992).

[16]M. L. Brillouin, "Remarques sur la mécanique ondulatoire [Remarks on Wave Mechanics]," in *Journale de Physique et le Radium* 7 (1926), pp. 353ff.

[17]Keller, "Corrected Bohr-Sommerfeld Quantum Conditions"; Keller and Rubinow, "Asymptotic Solution."

[18]F. Reiche, *The Quantum Theory*, 2nd ed., trans. H. S. Hatfield and H. L. Brose (New York: Dutton, 1930), p. 93: "Somewhat later, A. Einstein put forward a quantum hypothesis which has the advantage of being independent of the choice of coordinates. But a closer discussion of these abstract investigations would lead us too far here."

[19]Keller's 1982 paper "Semiclassical Mechanics," in *SIAM Revue* 27 (1985), pp. 485ff., does make the limitations of the method clear; in private communication, he related that these limitations became clear to him during the 1960s through his interactions with the mathematician Jürgen Moser at the Courant Institute.

[20]I. C. Percival, "Regular and Irregular Spectra," in *Journal of Physics, B* 6 (1973), L 229.

[21]M. C. Gutzwiller, "Periodic Orbits and Classical Quantization Conditions," in *Journal of Mathematical Physics* 12 (1971), pp. 343ff. See also his *Chaos in Classical and Quantum Mechanics* (New York: Springer Verlag, 1990).

[22]M. C. Gutzwiller, ed., "Resource Letter ICQM-1: The Interplay of Classical and Quantum Mechanics," *American Journal of Physics* 66 (1998), pp. 304–24.

Notes to Chapter 19

[1]Albert Einstein, "Principle of Research," originally published in 1918; reprinted in *Ideas and Opinions* (New York: Dell Publishing, 1954), pp. 219–22.

[2]Albert Einstein, "Autobiographical Notes," in *Albert Einstein: Philosopher-Scientist*, ed. and trans. P. A. Schipp (Evanston, IL: Library of Living Philosophers, 1949).

[3]Einstein, *Ideas and Opinions*, pp. 306–307.

[4]Albert Einstein, "Letter to Theodor Kaluza," *Collected Papers of Albert Einstein, Vol. 9* (Princeton, NJ: Princeton University Press, 2004), p. 21.

[5]A. Einstein and W. Pauli, "On the Non-existence of Regular Stationary Solutions of Relativistic Field Equations," *Annals of Mathematics* 44 (1943), pp. 131–37.

[6]Albert Einstein, "Letter to L. de Broglie," Einstein Archives, 1954; quoted in A. Pais, "Einstein and the Quantum Theory," *Reviews of Modern Physics* 51 (1979), p. 910.

[7]Einstein and Pauli, pp. 131–37.

[8]Quoted in K. C. Cole, *Los Angeles Times*, November 16, 1999.

[9]Albert Einstein.

CONTRIBUTORS

Leon Botstein has been president of Bard College since 1975. He is also the Leon Levy Professor in the Arts and Humanities at Bard. He received his B.A. degree with special honors in history from the University of Chicago, and his M.A. and Ph.D. degrees in European history from Harvard, as well as several honorary degrees. He was a National Arts Club Gold Medal recipient in 1995, and in 1996 he was awarded the Centennial Medal of the Harvard Graduate School of the Arts and Sciences. In 2003 he received the Award for Distinguished Service to the Arts from the American Academy of Arts and Letters. Botstein formerly served as president of Franconia College, lecturer in history at Boston University, and special assistant to the president of the New York City Board of Education. He is past chairman of the Harper's Magazine Foundation and of the New York Council for the Humanities, a member of the National Advisory Committee for the Yale–New Haven Teachers Institute, and a member of the board of the Central European University and of numerous other boards and professional associations. He has been a visiting professor at the Manhattan School of Music in New York and at the Hochschule für Angewandte Kunst in Vienna.

Lorraine Daston is a Director at the Max Planck Institute for the History of Science, Berlin, and has taught at several universities in the United States and Europe, including Chicago, Göttingen, Princeton, Harvard, and the Humboldt-Universität zu Berlin. She has written on a wide range of topics in the history of science, including the history of probability theory, wonders, evidence, objectivity, and the moral authority of nature. She is interested in how categories of scientific rationality and practice—facts, proof, natural law—emerge and develop in historical context. Recent publications include:

Wonders and the Order of Nature, 1150–1750 (with Katharine Park, 1998); *Biographies of Scientific Objects* (2000); *Wunder, Beweise und Tatsachen: Zur Geschichte der Rationalität* (2001); *Eine kurze Geschichte der wissenschaftlichen Aufmerksamkeit* (2001); *The Moral Authority of Nature* (with Fernando Vidal, 2004); *Things That Talk: Object Lessons from Art and Science* (2004). Together with Peter Galison, she is finishing a book on *Images of Objectivity*.

E. L. Doctorow was born in New York in 1931. He was raised in the Bronx by second-generation parents of Russian Jewish descent. After graduating in 1952, he began his career as a reader at Columbia Pictures, moved on to become an editor for New American Library in the early 1960s, and worked as chief editor at Dial Press from 1964 to 1969. Although he had written books for years, it was not until the publication of *The Book of Daniel* in 1971 that he was acclaimed. *Ragtime*, his next book, was a superb commercial and critical success. His critically acclaimed novels blend history and social criticism. On May 23, 2004, he delivered a commencement address critical of President George W. Bush at Hofstra University. Currently, he holds the Glucksman Chair in American Letters at New York University. Major publications: *The Book of Daniel* (1971); *Ragtime* (1975); *Drinks Before Dinner* (1979); *Loon Lake* (1980); *City of God* (2000); *Reporting the Universe* (2003); *Sweet Land Stories* (2004); *The March* (2005).

Yehuda Elkana was born 1934 and is President and Rector of the Central European University in Budapest-Warsaw. He received his Ph.D. at Brandeis University in 1968. He taught at Harvard University, in the Department of History and Philosophy of Science at the Hebrew University of Jerusalem, and at the ETH in Zurich. He was Director of the Van Leer Jerusalem Institute, 1968–93, and of the Cohn Institute for the History and Philosophy of Sciences and Ideas, Tel Aviv University, 1981–91. He is Permanent Fellow of the Wissenschaftskolleg zu Berlin. His fields of specialization include history, philosophy, and sociology of Science. Selected publications: *The Discovery of the Conservation of Energy* (1974, 1975); *Albert Einstein, Historical and Cultural Perspectives* (editor, with Gerald Holton, 1982); *Anthropologie der Erkenntnis: Die Entwicklung des Wissens als episches Theater einer listigen Vernunft* (1986); "Programmatic Attempt at an Anthropology of Knowledge," in E. Mendelsohn and Y. Elkana, eds., *Sciences and Cultures: Anthropological and Historical Studies of the Sciences* (1981); "Rethinking—not Unthinking—the Enlightenment," in Wilhelm Krull, ed., *Debates on Issues of Our Common Future* (1989); "Some Thoughts on the Future of Research," in R. Cassling and G. Fragniere, eds., *Social Sciences and Political Change: Promoting Innovative Research in Post-Socialist Countries* (2003).

Yaron Ezrahi is a professor of political science at the Hebrew University of Jerusalem. He has published extensively on the relations among science, culture, and politics. His publications include: *The Descent of Icarus: Science and the Transformation of Contemporary Democracy* (1990); *Rubber Bullets, Power and Conscience in Modern Israel* (1998); "Science and the Political Imagination in Contemporary Democracies," in S. Jasanoff, ed., *States of Knowledge: The Co-Production of Science and the Social Order* (2004). Professor Ezrahi's forthcoming book is on the crisis in the contemporary democratic political imagination.

Michael L. Friedman was born in 1947 in Brookline, Massachusetts, received his B.A. from Queens College in 1969, and his Ph.D. from Princeton University in 1973. He is currently Frederick P. Rehmus Family Professor of Humanities at Stanford University. His publications include: *Foundations of Space-Time Theories: Relativistic Physics and Philosophy of Science* (1983); *Kant and the Exact Sciences* (1992); *Reconsidering Logical Positivism* (1999); *A Parting of the Ways: Carnap, Cassirer, and Heidegger* (2000; German 2004); *Dynamics of Reason* (2001; Italian, 2004). He has served as President of the Central Division of the American Philosophical Association and of the Philosophy of Science Association. He was awarded the Matchette Prize of the American Philosophical Association and the Lakatos Award in Philosophy of Science. He was elected Fellow of the American Academy of Arts and Sciences in 1997 and Membre titulaire de l'Institut international de philosophie in 2000.

Jürg Fröhlich has been Full Professor of theoretical physics at ETH Zurich since 1982. He specializes in mathematical physics. Born in 1946, Fröhlich studied physics at ETH Zurich from 1965 to 1969; he was awarded his Ph.D. with a thesis on the infrared problem in quantum field theory. In 1973–74 he was Research Fellow at Harvard University and became Assistant Professor at Princeton University immediately afterward. In January 1978 he accepted the position of professeur permanent at the Institute of the Hautes Etudes Scientifiques in Burques-sur-Yvette, where he taught until the summer of 1982. Jürg Fröhlich's research is focused on quantum field theory, quantum theory of liberal systems, and static physics, as well as various mathematical methods of theoretical physics. His works have been awarded the national Latsis-Preis (1984) and the Dannie-Haineman-Preis (1991). Since 1993 he has been a member of the Academia Europea.

Peter L. Galison is the Joseph Pellegrino University Professor (Department of the History of Science and Department of Physics) at Harvard University. In 1997 he was named a John D. and Catherine T. MacArthur Foundation Fellow; in 1999 he was a winner of the Max Planck Prize awarded by the Max

Planck Gesellschaft and Humboldt Stiftung. Galison is interested in the intersection of philosophical and historical questions such as these: What, at a given time, convinces people that an experiment is correct? How do scientific subcultures form interlanguages of theory and things at their borders? More broadly, Galison's main work explores the complex interaction among the three principal subcultures of 20th-century physics—experimentation, instrumentation, and theory. His books include: *How Experiments End* (1987); *Image and Logic* (1997); *Einstein's Clocks, Poincaré's Maps* (2003); and (with Lorraine Daston) *Objectivity* (2007). In addition, he has launched several projects examining the powerful cross-currents between physics and other fields; these include a series of coedited volumes on the relations among science, art, and architecture. He coproduced a documentary film on the politics of science, *Ultimate Weapon: The H-bomb Dilemma,* and is now working on a second, with Robb Moss, *Secrecy,* about the architecture of the classification and secrecy establishment.

David Gross was born in 1941 in Washington, D.C., and is a string theorist. Along with Frank Wilczek and David Politzer, he was awarded the 2004 Nobel Prize in Physics for his discovery of asymptotic freedom. He received his Ph.D. in physics from the University of California–Berkeley in 1966 and was a Junior Fellow at Harvard University and a Professor at Princeton University until 1997. He is the recipient of a MacArthur Foundation Fellowship in 1987, the Dirac Medal in 1988, and currently is the director and holder of the Frederick W. Gluck Chair in Theoretical Physics at the Kavli Institute for Theoretical Physics of the University of California–Santa Barbara. In 1973, Gross, working with his first graduate student, Frank Wilczek, at Princeton University, discovered asymptotic freedom, which holds that the closer quarks are to each other, the less the strong interaction (or color charge) between them, and in extreme proximity, the nuclear force between them is so weak that they behave almost as free particles. Asymptotic freedom, independently discovered by David Politzer, was important for the development of quantum chromodynamics. Gross, with Jeff Harvey, Emil Martinec, and Ryan Rohm, also discovered the heterotic string.

Hanoch Gutfreund is a graduate of the Hebrew University, where he obtained his M.Sc. (1962) and Ph.D. (1966) in Physics. Since 1985, he has been the incumbent of the André Aisenstadt Chair in Theoretical Physics, and was formerly active in various academic and administrative leadership roles, such as Head of the Racah Physics Institute, Head of the Institute for Advanced Studies, Rector and President of the Hebrew University. He was among the initiators and founders of the Interdisciplinary Center for Neural Computation and remains a member and lecturer at this Center. He is chair of an Israeli Academy of Science's Committee on University-Industry relations. In the last few years, he has coordinated the Hebrew University's worldwide activities related to Albert Einstein.

Linda Dalrymple Henderson earned her Ph.D. at Yale University and is the David Bruton Jr. Centennial Professor in Art History and Distinguished Teaching Professor at the University of Texas at Austin. She is the author of *The Fourth Dimension and Non-Euclidean Geometry in Modern Art* (1983, 2008) and *Duchamp in Context: Science and Technology in the Large Glass and Related Works* (1998). With literature scholar Bruce Clarke, she co-edited the anthology *From Energy to Information: Representation in Science and Technology, Art, and Literature* (2002). She guest-edited the winter 2004 special issue of *Science in Context*, which includes her two-part introduction, "I. Writing Modern Art and Science. An Overview" and "II. Cubism, Futurism, and Ether Physics in the Early Twentieth Century." Among her other science-oriented essays are contributions to two anthologies: "Four-Dimensional Space or Space-Time: The Emergence of the Cubism-Relativity Myth in New York in the 1940s," in Michele Emmer, ed., *The Visual Mind II* (2005); and "Modernism and Science," in V. Liska and A. Eysteinsson, eds., *Modernism*, in the International Comparative Literature Association series *Comparative History of Literatures in European Languages* (2007).

Dudley Herschbach was born in California (1932), attended Stanford University (B.S. 1954 in mathematics, M.S. 1955 chemistry) and Harvard (M.A. 1956 physics, Ph.D. 1958 chemical physics, 1957–59 junior fellow), and was a faculty member at University of California–Berkeley (1959–63) and Harvard (since 1963). He has taught graduate courses in quantum mechanics, chemical kinetics, molecular spectroscopy, and collision theory, as well as undergraduate courses in physical chemistry and general chemistry, and recently a seminar for first-year students on thermodynamics and kinetics of enzyme processes. He is engaged in efforts to improve science education and the public understanding of science. His current research is devoted to molecular beam experiments, particularly producing molecules so slow that wave properties become prominent; high-pressure experiments examining abiotic generation of methane and hydrocarbons under conditions like those in the Earth's mantle; theoretical models for enzyme-DNA interactions; and dimensional scaling treatment of strongly correlated many-body systems. In 1986 he was awarded the Nobel Prize in Chemistry.

Gerald Holton is Mallinckrodt Research Professor of Physics and Research Professor of History of Science at Harvard University. He is a Fellow of the American Physical Society and of the American Academy of Arts and Sciences, and a member of the American Philosophical Society and of several European organizations including the Deutsche Akademie der Naturforscher Leopold-ina and the Academie internationale d'histoire des sciences. He served as President of the History of Science Society of the U.S.A. and on several U.S. National Commissions, including that on UNESCO. Among his book publications are: *Thematic Origins of Scientific Thought: Kepler to Einstein* (1973, 1988); *Science and Anti-Science* (1993); *Einstein, History, and Other Passions* (2002); and *Victory and Vexations in Science* (2005). He was the Founding Editor of the quarterly journal *Daedalus*, and on the Editorial Committee of the *Collected Papers of Albert Einstein*. His honors include the Sarton Medal, the Herbert Spencer Lecturer at Oxford University, and the Jefferson Lecturer by the National Endowment of the Humanities.

Caroline A. Jones is Professor of art history and Director of the History, Theory, and Criticism discipline group in the School of Architecture and Urban Planning at the Massachusetts Institute of Technology, Cambridge. She has been a Professeur invité, l'Institute national d'histoire de l'art (Paris), as well as a fellow of the Wissenschaftskolleg zu Berlin, The Max Planck Institute for the History of Sciences (Berlin), the Boston University Humanities Foundation, and the Institute for Advanced Study (Princeton). She has made two documentary films and has curated many exhibitions; her publications include *Machine in the Studio: Constructing the Postwar American Artist* (1996, 1998; Winner of the 1999 Charles Eldredge Prize for "outstanding scholarship in American art," National Museum of American Art, Smithsonian Institution.); *Eyesight Alone: Clement Greenberg's Modernism and the Bureaucratization of the Senses* (2005); and the edited volume *Sensorium: Embodied Experience, Technology, and Contemporary Art* (2006).

Susan Neiman is Director of the Einstein Forum. Born in Atlanta, Georgia, Neiman studied philosophy at Harvard and the Freie Universität Berlin, and taught philosophy at Yale (1989–96) and Tel Aviv University (1996–2000). Her areas of specialization include moral and political philosophy, and the history of modern philosophy. She is the author of *Slow Fire: Jewish Notes from Berlin* (1992), *The Unity of Reason: Rereading Kant* (1994), and *Evil in Modern Thought* (2002; translated into eight languages). Her books won several American and international awards. She is a member of the Berlin-Brandenburg Academy of Sciences. Her forthcoming book is *Moral Clarity* (2008).

Lisa Randall is a professor of theoretical physics at Harvard University, where she studies elementary particle physics and cosmology. Her research is about the fundamental nature of particles and forces, and how matter's basic elements relate to the physical properties of the world that we see. She earned her Ph.D. from Harvard University and held professorships at the Massachusetts Institute of Technology and Princeton University before returning to Harvard in 2001. She is a member of the American Academy of Arts and Sciences and is a past winner of an Alfred P. Sloan Foundation Research Fellowship, a National Science Foundation Young Investigator Award, and the Westinghouse Science Talent Search (now sponsored by Intel). In 2003, she received the Premio Caterina Tomassoni e Felice Chisesi Award, presented at the University of Rome, La Sapienza. In 2005, she published *Warped Passages: Unraveling the Mysteries of the Universe's Hidden Dimensions*.

Jürgen Renn was born in 1956 in Moers, Germany. He received degrees in physics from Freie Universität Berlin (1983) and in mathematics from Technische Universität Berlin (1987). He was collaborator and co-editor of *Collected Papers of Albert Einstein* (1986–92). He has held the following positions: Assistant Professor, Boston University (1989); Associate Professor, Boston University in philosophy and history of science, physics (1990–93); Simon Silvermann Visiting Professor at Tel-Aviv University in history of science (1993); Visiting Professor at ETH Zürich in philosophy (1993–94); Adjunct Professor, Boston University (since 1994); Honorary Professor, Humboldt-Universität zu Berlin in history of science (since 1995). Since 1994, he has been director at the Max Planck Institute for the History of Science. He specializes in the history of early modern mechanics, the history of relativity theory, and the interaction between cognitive and contextual factors in the history of science.

Matthew Ritchie's work investigates parallels between the mythography of science and art through large-scale installations. His projects have been exhibited at numerous museums and galleries around the world, including the Museum of Modern Art (New York), Massachusetts Institute of Technology (Cambridge), San Francisco Museum of Modern Art, the Contemporary Arts Museum (Houston), the Musée d'Art Moderne (Paris), Kunstmuseum Wolfsburg, the Palais de Tokyo (Paris), Portikus (Frankfurt am Main), the Massachusetts Museum of Contemporary Art, the 21st Century Museum of Contemporary Art (Kanazawa), the Dallas Museum of Art and the Guggenheim Museum (New York).

Silvan S. Schweber was born in Strasbourg in 1928. He earned his Ph.D. at Princeton University 1952, held a Postdoctoral Fellowship at Cornell University (1952–54), was a Senior Research Physicist at the Carnegie Institute of Technology (1954), and at Brandeis University was Associate Professor (1955–59) and Full Professor of Physics (1961 on) and the Richard Koret Professorship of the History of Ideas (1982 on). Since 1980, he has been an Associate of the department of the history of science at Harvard University.

He is a fellow of the American Academy of Arts and Sciences, the American Association for the Advancement of Science, and the American Physical Society, and a member of the History of Science Society. His publications include several books and numerous articles: *Mesons and Fields*, Vol. 1 (with H. A. Bethe and F. de Hoffman, 1955); *Introduction to Relativistic Quantum Field Theories* (1961); *QED and the Men Who Made it* (1994); *In the Shadow of the Bomb*: *Bethe, Oppenheimer and the Moral Responsibility of the Scientist* (2000).

A. Douglas Stone is a theoretical physicist specializing in condensed matter and optical physics. He is currently the Carl A. Morse Professor of Applied Physics and Physics at Yale University, where he has served as Chairman of Applied Physics and Director of the Division of Physical Sciences. His work on quantum fluctuation phenomena in micron-scale conductors has led to a new subdiscipline of solid-state physics known as mesoscopic physics. He has also worked extensively on laser physics and optical devices, patenting three inventions in this field. He is a General Member and former Trustee of the Aspen Center for Physics and a Fellow of the American Physical Society, and has received the McMillan Award of the University of Illinois for "outstanding contributions to condensed matter physics." He obtained an Honors B.A. in Physics and Philosophy from Balliol College in Oxford (1978) and a Ph.D. in Physics from the Massachusetts Institute of Technology (1983). He has recently lectured widely on Einstein's contributions to quantum theory and is writing a popular book on this subject.

INDEX

Note to readers: *Italic* page numbers indicate illustrations or tables. *Passim* indicates numerous mentions over the given range of pages.

Boltzmann, Ludwig, 186, 204, 232, 241, 251, 338n7

Boltzmann's constant, 258, 260

Born, Max, 6, 9, 38, 95

Bose-Einstein condensate, 5, 8, 270

Bourdon, David, 124

Bragdon, Claude, 124

Braque, Georges, 103

Breton, André, 117, 326n52

Bridgman, Percy W., 95

Briggs, Lyman, 77–78

Briggs Committee, 77–78

Brillouin, Leon, 284

Brinton, Crane, 85

Brownian motion, 171, 233–235, 241, 261, 267, 338n10

Brush, Stephen, 39

brushstrokes, 140–142, 146, 330n13

Bucherer, Albert, 302

Burckhardt, Jacob, 14

Bush, Vannevar, 78, 81, 83

Busoni, Ferruccio, 173–174

Calaprice, Alice, 47

Calder, Alexander, 124

capitalism, 69

Carter, Brandon, 39

Cassidy, David, 10, 110

Centenary Symposium (Jerusalem, 1979), 311n1 (ch.1)

Cézanne, Paul, 137, 329n6

Chadwick, James, 321n75

chain reaction, 74, 318n10

chamber music, 162

Chamberlain, Austin, 85

chaotic motion. *See* motion, regular and chaotic

"charismatic authority," 14

chronologies and timelines
AE's childhood and education, *218*, *221*, *224*

ECHO timeline, *255*
quantum theory, development of, *263*
relativity theory, development of, *242*

Churchill, Winston, 81–82, 85–86

citizenship, United States, 64

civil disobedience, 320n66

clock coordination work, 189, 195–198

Cockcroft, John, 303

Codex Aureus, 329n3

common sense as a cultural system, 53

common-sense realism and contemporary politics, 48–58

communal knowledge, 52–53

communal reality and journalism, 54–56

communism, 67, 69

community of scientists, 16–17, 176–177

completeness as a guiding category, 6

comprehensibility of the universe, 36–37, 41

concepts, first- and second-order, 44–46

conformal field theory, 267

constructive theories of world government, 96

Constructivism. *See* Russian Constructivism

contemporary politics and common-sense realism, 48–58

continuity equation, 281

continuum as a guiding category, 6

continuum theory of matter, 338n7

contraction. *See* distortion-contraction of form in modern art

Co-Operation Press Service for International Understanding, 85

Copernicus, 240, 249

Cosmic Religion, 13

Cosmologia, 248

cosmological constant, 145, 304–307, 330n18

Coudenhove-Kalergi, Count, 7

Haller, Friedrich, 229
Halsman, Philippe, 121, 326n64
Hamilton, William, 272
Hamilton-Jacobi equation, 281
Hammer, Martin, 128
Handel, 166–167
Hanford reactor, 80
Hanslick, Eduard, 169
harmony, 174
Harrington, Anne, 11
Harris, Sidney, *122*, 122
Hartley, John, 54–55
Hawking radiation, 267
Hebrew University, 32–34
Hecht, Selig, 92
Heifetz, Jascha, 161
Heisenberg, Werner, 3, 263
Hellpach, Willy, 28, 29–30
Helmholtz, Hermann von, 206–216
Helmholtz-Lie theorem, 208–210, 213–214
Henderson, Linda, 130, 152
Henry, Holly, 128
Heraclitus, 331n24
Herbert, Lynn, 129
Hertz, Gustav, 299–300
Hilbert, David, 243–244, *243*
Hilbert space, 132, 151, 207
Hindenburg, Paul von, 12
Hiroshima and Nagasaki, aftermath, 84–92
Hobbes, Thomas, 52–53
Höch, Hannah, 106–107, *108*, 110
Hoffmann, Banesh, 38
Hogness, Thorfin, 92
Holler, Carsten, *Plate 8*, 155
Holton, Gerald, 39
Hooft, Gerard 't, 267
House Un-American Activities Committee, 65, 93
Hubble, Edwin, 304

hubris-humility dichotomy, 36–47 *passim*
Hume, David, 158–159
humility-hubris dichotomy, 36–47 *passim*
humor, 181
Hupfeld, Hermann, 118
Hutchins, Robert M., 84
hydrogen bomb, 89–92

iconography in modern art, 122–126, 127
ideal gas law, 283, 338n6
identity. *See* Jewish identity; scientist identity
images of AE in modern art, *Plate 2*, *107*, 110, *122*, 127, 323n23, 323n24
Impressionism, 135–143, 329n6
indeterminism. *See* determinism/indeterminism
Infeld, Leopold, 95, 253
influence, nature of, 3–4, 14
information represented in art, 153–154, 157–158
Institute for Advanced Study (Princeton University), 5, 64, 79
integrable and non-integrable dynamical systems, 278
International Atomic Energy Authority, 93
internationalism, 8
Internet, 255
invariance as objectivity, 26
Israel, AE's proposed presidency, 62
Italian Baroque, 167

Jakobson, Roman, 42
James, Henry, 179
James, Kathleen, 109, 128
Jeans, James, 119, *120*, 128
Jefferson, Thomas, 49–50

Pauli's view, 294
United Nations, 92
United Nations, 1947 address, 63
universalism, 69–70
University of Bern, 229
Urey, Harold, 92

van Doesburg, Theo, 112–115, *113*, 325n38
van Fraassen, Bas C., 56
vanishing-point perspective, 131
vanity, 18
Varo, Remedios, 327n66
Veblen, Oswald, 75, 79
Veltmann, Martinus, 267
Verallgemeinerung (unity), 10–11
violin, 161–162
von Laue, Max, 4, 6, 74
von Neumann algebras, 265

Wallace, Henry, 65
Walton, Ernest, 303
Warhol, Andy, *Plate 2*, 120–121
Watson, Edwin M., 77–78
Weber, Heinrich, 223–225
Weber, Max, 14
Weinberg, Steven, 40, 238
Weisskopf, Victor, 92
Weizmann, Chaim, 30, 66
Weizsäcker, Carl F. von, 78
Wells, H. G., 74

Wenzel-Kramers-Brillouin (WKB) method, 280–282, 284, 339n6
Weyl, Hermann, 26, 79, 268, 305
Wheeler, John, 125–126
Wien's law for black body radiation, 261
Wigner, Eugene, 74–78, 83
Wilczek, Frank, 267
Wilkinson Microwave Anisotropy Probe (WMAP), 145, 330n19, 330n20, 331n22
will and testament, 34
Wilson, Robert, 123–124, 143–144
Winteler, Jost, 220, 221
Witten, Edward, 296
Wittgenstein, Ludwig, 170–171, 181–182
WKB method. *See* Wenzel-Kramers-Brillouin method
WMAP. *See* Wilkinson Microwave Anisotropy Probe
world court, 96, 317n4
world government, 85–87, 96, 320n51
World War I, 68–69

Zahn, Charles Thomas, 302
Zangger, Heinrich, 188–189
Zeilinger, Anton, 159
Zionism, 12–13, 29–34
Zola, Emile, 329n6
zonal time, 197–198
Zurich notebook, *244, 256*

Milton Keynes UK
Ingram Content Group UK Ltd.
UKHW051319150924
448319UK00013B/57